Teaching Cases in Tourism, Hospitality and Events

Dedicated to

My lovely kids Rakshit and Rayaan

Saurabh Kumar Dixit

My father Late Piramanayagam and my mother Late Kaliammal, my wife Sumathi and my daughter Amridha, for always loving and supporting me

Senthilkumaran Piramanayagam

Teaching Cases in Tourism, Hospitality and Events

Edited by

Saurabh Kumar Dixit

Professor and Founding Head of the Department of Tourism and Hotel Management
North-Eastern Hill University, Shillong (Meghalaya) India

Senthilkumaran Piramanayagam

Professor
Welcomgroup Graduate School of Hotel Administration
Manipal Academy of Higher Education
Manipal, India

CABI

CABI is a trading name of CAB International

CABI
Nosworthy Way
Wallingford
Oxfordshire OX10 8DE
UK

Tel: +44 (0)1491 832111
E-mail: info@cabi.org
Website: www.cabi.org

CABI
200 Portland Street
Boston
MA 02114
USA

Tel: +1 (617)682-9015
E-mail: cabi-nao@cabi.org

A catalogue record for this book is available from the British Library, London, UK.

ISBN-13: 9781800621008 (hardback)
9781800621015 (ePDF)
9781800621022 (ePub)

DOI: 10.1079/9781800621022.0000

Commissioning Editor: Claire Parfitt
Editorial Assistant: Emma McCann
Production Editor: James Bishop

Typeset by SPi, Pondicherry, India
Printed and bound in the USA by Integrated Books International, Dulles, Virginia

Contents

Contributors

Julia N. Albrecht is a senior lecturer in tourism at the Department of Tourism, University of Otago, New Zealand. Her research interests include tourism and destination management, and visitor management, both with a focus on nature-based tourism. Julia's work has been published in leading journals such as the *Journal of Sustainable Tourism*, *Annals of Tourism Research*, *Tourism Management*, *Journal of Cleaner Production* and *International Journal of Tourism Research*. Julia is a co-editor of the *Journal of Outdoor Recreation and Tourism* as well as a member of the editorial board of the *Journal of Hospitality and Tourism Research*. Julia is the sole editor of the books *Visitor Management in Tourism Destinations* (2017, CABI) and *Managing Visitor Experiences in Nature-based Tourism* (2021, CABI). Email: julia.albrecht@otago.ac.nz

Andres Artal-Tur is a full professor in the Department of Economics, Accounting and Finance at the Technical University of Cartagena (Spain) and a research associate in the Institute of International Economics, University of Valencia (Spain). His main research interest is in tourism, regional and international economics. Publications in the area of tourism include the analysis of tourist's behaviour (stay duration, satisfaction and loyalty to destinations, expenditure patterns), with applications in the field of tourism marketing and destination planning. Other topics of interest are those of culture and tourism, and tourism sustainability. In the field of regional and international economics he has focused on the analysis of trade and migration flows, and location theories (firms, people). He has published his research findings extensively in international impact journals, and has edited books produced by well-known publishers such as Springer, Routledge, Elsevier, Emerald and CAB International. Email: Andres.Artal@upct.es

Delly M. Chatibura is a senior lecturer in the Department of Tourism and Hospitality Management at the University of Botswana. She has contributed broadly to scholarly work in hospitality and tourism journals and has presented some of her academic research in Botswana, Hungary, Malaysia, South Africa and Zambia, to list a few. However, her main interests are in culinary tourism and hospitality management. She also serves as a manuscript reviewer for journals such as the *Journal of Foodservice Business Research* and serves on the editorial board of the *Journal of Hospitality and Tourism*. Among other duties, Delly serves as an external reviewer for tourism and hospitality programmes for two institutions in Botswana. Email: mahachid@ub.ac.bw

Willem Coetzee is a senior lecturer in the Department of Tourism, University of Otago. Willem completed a PhD in sustainable tourism in 2004 and joined the department in

2015. His research focus is on sustainable tourism and events and festivals. In this context, he has explored specific issues among stakeholders in and around areas in Southern Africa, the impacts of nature-based tourism on surrounding communities, and the impacts of tourism in water-stressed destinations. His recent publications revolve around perceptions and consumer behaviours of attendees at sporting events and festivals. Willem has been involved in a range of tourism supply-and-demand projects in small towns and the formulation of tourism feasibility studies for communities. Email: willem.coetzee@otago.ac.nz

Clara Cubillas-Para is a PhD student at the Business Management Department of Universidad Politécnica de Cartagena (Spain). Her research interests include consumer behaviour, knowledge management and travel and tourism. She has participated in a number of international conferences, and published some papers and book chapters. Email: claracubilasparact@gmail.com

Aarti Dangwal is a PhD holder in management. She is working as an assistant professor with the University School of Business, Department of Commerce, Chandigarh University, Gharuan, Punjab. She is actively involved in research, and academic co-curricular and educational development activities. Her research areas are tourism, finance and business. Email: aartidangwal13@gmail.com

Sushant M. Desale has completed his International MBA at the Technical University of Cartagena (Spain), and is currently aiming to enrol in the PhD studies programme of the Faculty of Business Science Department under the direction of Professor Andres Artal-Tur. His main research interest is in data management and analysis in the area of tourism and healthcare. Other topics of interest are those of culture, travel and the automobiles industry. He has extensively worked on data analysis on national and international clinical research projects with an impact on people's lives. Email: sushantmadhukar.desale@edu.upct.es

Saurabh Kumar Dixit is Professor and founding Head of the Department of Tourism and Hotel Management, North-Eastern Hill University, Shillong (Meghalaya), India. His research interests include consumer behaviour, gastronomic tourism, and experiential management and marketing in hospitality and tourism contexts. He has 15 books to his credit, including *The Routledge Handbook of Consumer Behaviour in Hospitality and Tourism* (2017), *The Routledge Handbook of Gastronomic Tourism* (2019), *The Routledge Handbook of Tourism Experience Management and Marketing* (2020) and *Tourism in Asian Cities* (2020), *Tourism in India: Marketing Perspectives* (2021), *Consumer Behavior in Hospitality and Tourism* (2021) *The Emerald Handbook of Luxury Management for Hospitality and Tourism* (2022) and *The Routledge Handbook of Wine Tourism* (2023). Presently he is editing *The Routledge Handbook of Collaborative Consumption in Tourism*. He has contributed many research papers in reputed peer-reviewed journals and books. He is an editorial board member, active reviewer and guest editor of many international journals such as the *International Journal of Contemporary Hospitality Management*, *British Food Journal*, *Tourismos*, *Tourism Recreational Research*, *Anatolia* and *Journal of Global Scholars of Marketing Science: Bridging Asia and the World (JGSMS)*. His recent profile can be seen at https://www.routledge.com/authors/i15903-saurabh-dixit (accessed 7 August 2022). Email: saurabh5sk@yahoo.com

Kalliopi Fouseki is Professor in Sustainable Heritage Management and the MSc Sustainable Heritage course director at the UCL Institute for Sustainable Heritage. She holds a BA in archaeology and art history, an MA in cultural heritage studies and a PhD in heritage management. She's worked in museums including the New Acropolis Museum, the Archaeological Museum of Ancient Olympia in Greece and the Science Museum, London and also at the University of York, the Open University in Greece, UK and Cyprus. Her research interests are: (i) participatory approaches to heritage management; (ii) heritage values; (iii) heritage and sustainable development; (iv) heritage and conflict; and (v) heritage and energy. Email: kalliopi.fouseki@ucl.ac.uk

Joo-Ee Gan is the Associate Dean (Education) at the School of Hospitality and Service Management, Sunway University. Her research is socio-legal in nature. Focusing on employment in the hotel sector and sustainable practices in the hotel and tourism industry, her studies evaluate the impact of legal measures. Her research on governance examines tourism as a vehicle for protecting the environment and advancing the socio-economic well-being of local communities, particularly the indigenous people. Prior to joining academia, Joo-Ee was a solicitor specializing in corporate matters. She was also a legal editor of the *Hong Kong Law Reports & Digest*. She is an accredited mediator of the Malaysian Bar Council and is on the Panel of the Malaysian Mediation Centre. Email: jegan@sunway.edu.my

Su Gibson currently serves as the Assistant Director for the University of South Carolina Beaufort's (USCB) Center for Event Management and Hospitality Training. She is also an instructor for the Hospitality Management Department at USCB. She has more than 20 years of industry experience in the hospitality industry, including travelling nationally and internationally as an inspector for the *Forbes Travel Guide*. Her research interests include service quality evaluation and management, and the emotions, setting, language and actions related to feelings of being welcome or unwelcome in a service environment. Email: gibsonsl@uscb.edu

Matt Gnagey is an associate professor of economics in the Goddard School of Business and Economics at Weber State University (WSU), Utah, USA. His research uses economics, ArcGIS mapping and statistics to study environmental and urban issues involving land use and property values, and his work has been published in the *Journal of Environmental Management*, *Review of Economics of the Household*, *Games*, *Annals of Regional Science*, the *Bulletin of Indonesian Economic Studies* and the *Journal of Economic Behavior and Organization*. He has also received research grants from the WSU's Center for Corporate Social Responsibility and the Utah Division of Water Resources. Email: mattgnagey@weber.edu

Kathryn Hayat is Associate Professor in Tourism Education and Deputy Head of the Department of Hospitality and Tourism at University College Birmingham, UK. Her experience includes over 10 years of lecturing tourism and aviation management, with a specialization in digital marketing and social media. Before her career in academia, Kathryn worked in digital marketing and social media roles at both tour operators and airline wholesalers, pioneering the use of Facebook for both business-to-consumer (B2C) and business-to-business (B2B) purposes. Email: k.hayat@ucb.ac.uk

Jamie Hoffman is an associate professor at California State University, Sacramento. She has worked in the field as a recreational therapist for many years in diverse settings including non-profit organizations, adaptive sports, municipal parks and recreation, veterans affairs, campus recreation, in-home healthcare, acute rehabilitation and higher education. Her research includes: (i) working with diverse populations and has had an international focus looking at disability and how that influences recreation and sports participation; (ii) temple stays; and (iii) adaptive sports participation and quality of life. Email: jamie.hoffman@csus.edu

María Pilar Jiménez-Medina is an associate professor of tourism at Universidad Politécnica de Cartagena (Spain), which she joined in 2000 after working in positions of responsibility in the Spanish tourism sector. She obtained degrees in tourism and anthropology, a master's in tourism management and PhD in business administration. Her research interests cover the study of the sociocultural dimension of tourism sustainability – the residents' perception of the impacts of tourism on their communities and the relationship between the resident-tourist binomial, the analysis of tourist and resident behaviour in the context of tourism marketing and destination planning, and the consequences of the coronavirus disease 2019 (COVID-19) crisis on tourism. She has led different research projects, which has resulted in both tourism products and various publications in journals, books and conferences. Email: mpilar.jimenez@upct.es

Rohan Jugran is an assistant professor of revenue management in the Welcomgroup Graduate School of Hotel Administration at Manipal Academy of Higher Education, Manipal, India. He is actively involved in teaching and research related to revenue management, revenue forecasting and front-office operations in hotels. Email: rohan.90r@gmail.com

R.M. Kabila is a doctoral candidate at the Department of Management Studies, National Institute of Technology, Tiruchirapalli, India. She studied for a master's degree in business administration with specialization in marketing at the Thiagarajar School of Management, India and has 2 years of work experience. Her research interests include bottom of the pyramid and consumer behaviour. Email: kabilavisalakshi@gmail.com

Ilias Krystallis is Assistant Professor in Enterprise Management, at the School of Sustainable Construction, UCL. He has worked in civil engineering within the built environment sector for more than 10 years, including securing two Policy Fellowships funded by Research England, the Department for Business, Energy and the Industrial Strategy (BEIS) and the Capabilities in Academic-Policy Engagement (CAPE) project. Sponsors of his research have included: (i) the Association for Project Management (APM); (ii) Department of Health and Social Care (DHSC); (iii) the Institution of Civil Engineers (ICE); (iv) the International Construction Project Management Association; (v) Research England; (vi) the Royal Institution of Chartered Surveyors (RICS); and (vii) private firms. Email: i.krystallis@ucl.ac.uk

Mohit Kukreti is a PhD holder in human resource development (HRD) with more than 23 years of international experience in higher education administration, academic quality assurance and higher education institution (HEI) strategic planning. He has worked as the Program Director of the IBA Program with the Directorate General of 6 CAS colleges at the Ministry of Higher Education, Oman and is currently associated with the University of Technology and Applied Sciences (College of Applied Sciences, Ibri). He has produced several policy documents, funded projects and refereed journal articles. His research areas are HRD, HEI policies and academic quality, and tourism, etc. Email: mohit.ibr@cas.edu.om

Vasantha Lakshmi is an assistant professor in the Department of International Business in the Institute for Technology and Management (ITM), Mumbai, India. She is a fellow (Human Resource and Organisational Development (HR&OD)) from the Academy of Human Resource Development. She holds a master's in law as well as a master's in human resources and industrial relations from Osmania University, Hyderabad. She has 16 years of experience teaching courses in human resources management, corporate governance, and international business in the business schools (B-schools). She has hands-on experience in international placements and corporate relations in the B-schools. She has also published research papers in national and international journals. Email: vasanthap@itm.edu

Kimberley Camelia S. Langstieh is a research scholar in the Department of Tourism and Hotel Management at North-Eastern Hill University, Shillong, India. She completed her bachelor of arts from Mount Carmel College, Bangalore, India and majored in history, journalism and travel and tourism, and her master's in travel and tourism management from North-Eastern Hill University, Shillong. Apart from work experience in the tourism industry, she has also been a guest lecturer at North-Eastern Hill University. Her area of research is indigenous tourism, religious tourism and sustainable tourism. Email: klangstieh@gmail.com

Rachel Hyunkyung Lee is a PhD candidate at the University of Houston's Conrad N. Hilton College of Global Hospitality Leadership. She has a bachelor's degree in business administration from Chung-Ang University in Korea, a master's degree in business administration with a focus on marketing from Korea University, and a master's degree from the Conrad N. Hilton College of Global Hospitality Leadership at the University of Houston.

Her research focuses on sustainability, food service and management, and consumer psychology. Email: hlee59@cougarnet.uh.edu

Tiffany S. Legendre is an associate professor at the Conrad N. Hilton College of Global Hospitality Leadership at the University of Houston. Her research focuses on branding and sustainability, and her teaching concentrates on food and beverage management, sustainability and research methods. Email: jslegend@central.uh.edu

Sarat Kumar Lenka is currently working as a professor in tourism at the Indian Institute of Tourism and Travel Management (IITTM), Bhubaneswar. He has more than 33 years of teaching and 14 years of administrative experience. He has undertaken and completed ten government-sponsored research projects as the director/chief investigator. He has published 30 research papers in peer-reviewed, University Grants Commission's (UGC) Consortium for Academic and Research Ethics (CARE)-listed and Scopus-indexed journals of national and international repute and also authored two books. His areas of interest include rural tourism, ecotourism, tourism marketing, community-based tourism and tourism planning. Email: profsaratlenka@gmail.com

Kamakshi Maheshwari is working as Assistant Professor in the Indian Institute of Tourism and Travel Management (IITTM), Gwalior. She has a PhD on the research topic 'Development and management of heritage tourism in the Gwalior region' from Jiwaji University. She is the Governing Council Member of INTACH (Indian National Trust for Art and Cultural Heritage) since 2009 and the president of WICCI (Women's Indian Chamber of Commerce & Industry) Rural Tourism Council (Madhya Pradesh). She is coordinating many prestigious projects of the Government of India including the Swachhta Action Plan, Ek Bharat Shreshta Bharat, Incredible India Tourist Facilitator and many more. In 2021 she was given an Outstanding Young Person of India award by JCI (Junior Chamber International). She is a LIMCA book award holder in 2008, 2010 and 2012. Email: kamakshi.maheshwari@iittm.ac.in

Drew Martin is Professor of Tourism and Hospitality Marketing at the University of South Carolina. Formerly, he was a professor of marketing at the University of Hawaii at Hilo. His research examines advertising, services marketing, consumer/visitor behaviour, and comparative cultural studies. He has served as Senior Associate Editor for the buyer behavior section of the *Journal of Business Research* (2012–2015), Associate Editor for the *International Journal of Culture, Tourism and Hospitality Research* (2007–2013), and as a member of the boards for the Hawaii Small Business Development Center (2016–2017) and the South Carolina Restaurant and Lodging Association (2017–2020). Email: martin11@mailbox.sc.edu

Miguel Mayol-Tur is a PhD student at the Technical University of Cartagena (Spain). He has participated in a number of international conferences in tourism, economics and management. He has published various papers in academic journals. His main research lines are in information and communications technology (ICT) and hospitality management, and free software applications for small and medium enterprises (SMEs). Email: mitcoes@gmail.com

Amitabh Mishra has over 15 years of academic and administrative expertise in international higher education. He is an assistant professor in the Department of Business Administration at the Sultanate of Oman's Government University of Technology and Applied Sciences (College of Applied Sciences, Ibri). He has vast experience in tourism, hospitality and business administration, having published over 20 research papers/articles in a variety of peer-reviewed and Scopus-indexed publications. He has also authored a couple of books. Tourism, hospitality and business management are among his areas of interest. Email: amitabh.ibr@cas.edu.om

Snigdharani Mishra is currently handling a dual role, as a professor, and head of department in the HR department and Registrar at the Institute for Technology and Management (ITM), Mumbai, India. She has been an academic for more than 25 years. She has published articles and research papers in various professional and refereed journals such as *Vikalp*, *Indian Journal for Training and Development* (ISTD) and *Merc Data Consulting* (MDC) journal of management published by L&T to name a few. Under her guidance five students have successfully completed a PhD and three are in the middle of their journey. She has been associated with a number of workshops, training and research programmes in various organizations and educational institutions. Snigdha has also a couple of consulting assignments in organizations such as Tata Consulting Services (TCS), National Payments Corporation of India (NPCI) and Radha Krishna Hospitality Services (RKHS). Her research interest lies in leadership, work–life balance, personality, employee wellness, stress management and Prevention of Sexual Harassment (POSH). She has been associated with Stockholding Corporation of India Limited, Mumbai for addressing their new joiners during their orientation. Her passion is in mentoring and counselling students. Email: snigdharani@itm.edu

Cass Morgan is the Associate Dean of the Jerry and Vickie Moyes College of Education at Weber State University, Utah, USA. She also serves as a faculty member in the Outdoor and Community Recreation Education programme where she teaches courses in adventure travel and sustainable tourism, event planning, outdoor education, and recreation administration and management. Morgan's research interests include sustainable tourism, programme evaluation methods, and youth development. Her work has been published in *Leisure Sciences*, *Journal of Outdoor Recreation, Education and Leadership*, *Journal of Parks and Recreation Administration*, and *Journal of Wilderness and Environmental Medicine*. Email: cassmorgan@weber.edu

Nametsegang Motshegwa is a hospitality management lecturer in the Faculty of Hospitality and Sustainable Tourism at Botho University, Botswana. She holds an MCom degree in tourism and hospitality management (Midlands State University) and a BBA degree in tourism and hospitality management (University of Botswana). As a researcher, Ms Motshegwa's interests are food safety, foodways, events tourism, sports tourism, knowledge management. Nametsegang actively participates in research, has presented at numerous international conferences and published several journal articles. Email: nametsegang.motshegwa@bothouniversity.ac.bw

Preeti Narendra is an assistant professor and the Area Coordinator of Human Resource at the Institute for Technology and Management (ITM), Mumbai, India. She is a passionate teacher, researcher, mentor and wellness expert spearheading the wellness centre on campus with various initiatives. She has more than 14 years of experience teaching in the business management college handling various subjects such as general management, human resource management, organizational behaviour, corporate social responsibility, employee counselling, industrial relations, labour law, learning and development, business ethics, talent management, change management, organizational development and entrepreneurship. She has presented more than 20 research papers at national and international conferences and published papers in national and international journals. Also she has handled various training and consulting programmes with Tata Consulting Services (TCS), Radha Krishna Hospitality Services (RKHS), Quality Circle Forum of India (QCFI), Dewan Housing Finance Corporation Limited (DHFL) and Bohler India. Email: preetin@itm.edu

José Miguel Navarro-Azorín is a lecturer in applied economics at the Universidad Politécnica de Cartagena (Spain). He holds a master's degree in industrial economics and markets, and a PhD in economics. His research interests lie in the broad areas of regional and urban economics, and tourism economics. He has published book chapters and several papers in scientific journals such as *Social Indicators Research*, *Applied Economics*, *Tijdschrift voor Economische en Sociale Geografie*, *Environment and Planning B: Planning and Design*, *European Planning Studie* and *Maritime Economics and Logistics*. Email: jmiguel.navarro@upct.es

Naresh P. Nayak is currently working for his PhD with the Department of Tourism, University of Otago, New Zealand. He has been associated with the Welcomgroup Graduate

School of Hotel Administration, Manipal Academy of Higher Education, Manipal, India. Naresh received a University of Otago Doctoral Scholarship in 2020. He has published his work in the *International Journal of Logistics System and Management* and the *International Journal of Services and Operational Management*. His primary research interests are tourism and hospitality education, sustainable tourism and consumer behaviour. Email: nareshp.nayak@postgrad.otago.ac.nz

Evarisa M. Nengnong is a research scholar at the Department of Tourism and Hotel Management, North-Eastern Hill University, Shillong, India. She completed a master's in tourism and travel management and BSc in Hotel Management and Administration. Apart from 5 years of work experience in the hotel industry, media, events, sales and marketing, she has also been a guest lecturer at North-Eastern Hill University and ICFAI University, Shillong. Her research focuses on tourism marketing, hospitality and tea tourism. Email: evarisa.nengnong@gmail.com

Bình Nghiêm-Phú is an assistant professor at the School of Economics and Management, University of Hyōgo, Japan. The majority of his research aims at understanding of consumers' perceptions and evaluations of the characteristics and images of products, services, organizations and places. He adopts the approaches of applied psychology theories to the implementation of marketing and management activities. Email: binhnghiem@gmail.com

Maria Isabel Osorio-Caballero is an assistant professor at the School of Economics, Universidad Nacional Autónoma de México-UNAM (México). Her research interests include international economics, economic growth and inequality, remittances and regional economic integration, with a special focus on Latin America and Asia. She has published a number of articles in impact journals and books including *The Pacific Alliance* (2017) and *Economia Mexicana*, usually acting as a research contributor at international conferences. She is also a consultant on industry projects, being a member of the National System of Research in México. Email: isabel.osorio@economia.unam.mx

Senthilkumaran Piramanayagam is a professor in hospitality and tourism at the Welcomgroup Graduate School of Hotel Administration, Manipal Academy of Higher Education, Manipal, India. He has authored many articles, book chapters and case studies on consumer behaviour in tourism and hospitality. He has also contributed as a statistician and data analyst in various collaborative research projects in business management, hospitality and health science. Email: senthil.kumaranp@manipal.edu

Andrea Salustri is a researcher at Sapienza University of Rome, in the Department of Juridical and Economic Studies (DSGE), where he teaches public finance. He holds a degree in economics and a PhD in economic theory and institutions from the University of Rome 'Tor Vergata'. He is currently interested in sustainable development, local tourism, territorial development, and social solidarity economy. Email: andrea.salustri@uniroma1.it

Shwetasaibal Samanta Sahoo is currently working as an assistant professor in the Faculty of Management Studies at Sri Sri University, Cuttack, Odisha, India. He has more than 16 years of teaching experience in different colleges and universities in travel, tourism and management subjects. His areas of interest include tourism planning, sustainable tourism, rural tourism, human resources and hospitality management. Email: catchsaibal@gmail.com

Noelia Sánchez-Casado is an associate professor of marketing at Universidad Politécnica de Cartagena (Spain). She holds a degree and PhD in business administration, and a master's degree in tourism management. She has taken part in international marketing conferences, including the Association for Consumer Research Conference and the European Marketing Academy Conference, and published articles and chapters in international journals and books. Her research is focused on the use of social media as a marketing tool, relationship and tourism marketing. Email: noelia.sanchez@upct.es

Kader Şanlıöz-Özgen holds a PhD in tourism management from Dokuz Eylül University, Türkiye. She acts as an assistant professor at the Hotel Management Department in

Özyeğin University, Türkiye, and has extensive professional experience in the field. Her research interests include hospitality experience, revenue management and service excellence in hotels. Email: kader.sanlioz@ozyegin.edu.tr

M. Selvalakshmi has over 20 years of experience in teaching marketing and strategy to management students at the postgraduate level. Her area of research interest includes services management and education. She has published cases in books produced by Ivey Publishing and The Case Centre. She has held administrative positions in the past and currently serves as the Principal of Thiagarajar School of Management, a reputed business school in Madurai, India. Email: selvalakshmi@tsm.ac.in

Colin Seymour is Senior Postgraduate Teaching Assistant for the MSc Sustainable Heritage course at University College London (UCL), involving face-to-face and online teaching, dissertation supervision and giving feedback to students. He is a qualitative researcher, trained and experienced in semi-structured interviews, in addition to: (i) NVivo coding; (ii) video-based research and education; (iii) focus groups; (iv) online research; (v) ethnography; and (vi) writing research proposals. With 20 years' work experience in the education sector, he is fascinated by cultural heritage tourism and engaging students in learning. His research interests are heritage/arts management, business models, tangible and intangible assets, knowledge societies/economies, heritage tourism and genealogical/legacy tourism. Email: c.seymour@ucl.ac.uk

Abhijeet Shirsat is an assistant professor in hospitality and tourism in the Department of Recreation, Parks, and Tourism Administration at California State University, Sacramento Campus. He has more than 15 years of experience in the hospitality industry. He is passionate about learning and education in its different forms, and his teaching philosophy grew out of his wide range of experience in the industry. His research interests relate to human behavioural topics in hospitality, tourism, events and social media. He intends to build upon the literature that identifies the problems and challenges of the hospitality and tourism industry. Email: shirsat@csus.edu

Chandradeep Singh is currently working as a project associate in Incredible India Tourist Facilitator Certification (IITFC) Heritage Advance Course in the Indian Institute of Tourism and Travel Management (IITTM), Gwalior. He is a governing council member of the Indian National Trust for Art and Cultural Heritage (INTACH). He has worked as an assistant professor in tourism and management for more than 3 years at Aditya College and the Vikrant Group of Institutions. Working actively in academic tourism he also has a number of research publications and research conference presentations at international and national levels. He is on the board of reviewers in tourism journals such as the *Journal of Education, Society and Behavioural Science*. Email: chandradeep240@gmail.com

V.J. Sivakumar earned his PhD from Anna University, Chennai, India. His doctoral study was focused on the area of customer relationship management in the Indian retail banking industry. He is a professor in the Department of Management Studies at the National Institute of Technology, Tiruchirappalli, India. His articles have been published in numerous refereed international and national journals. He has also presented papers in many international and national conferences. His research interests include customer relationship management, mobile banking and online retailing. Email: vjs@nitt.edu

Hằng Thị Bích Trần is a senior lecturer at the Faculty of Hospitality and Tourism, Thuongmai University, Vietnam. She is Director of the Scientific Research Management Office at this university. Her research and teaching activities are mainly about managerial issues in hospitality and tourism companies in Vietnam. Email: tranbichhang@tmu.edu.vn

Ranjana Tiwari is a researcher, case writer and academic practitioner specializing in tourism and hospitality management. She is currently working as Assistant Professor in JAIN (Deemed-to-be University), India. She completed her PhD in 2016 from Mohanlal Sukhadia University, India which investigated the links between tourist destination attributes, perception of service quality, and loyalty intentions of tourists. Her current research and publishing interests are in the areas of emerging trends in tourism and hospitality marketing, destination promotion, destination brand development, gamification in tourism, wellness, gender issues in tourism, the circular economy, and tourism social entrepreneurship. Email: ranjanatiw88@gmail.com

Alan Vijay has a PhD in management. He is working as an assistant professor with the Department of Hotel Management at Jamal Mohamed College, Tiruchirappalli, India. His primary research interests are international tourism and hospitality and destination marketing. Email: alanvijay@yahoo.com

Preface

The increased number of international tourism service providers, hotels, branded restaurants and event management organizations around the world have created a high demand for hospitality, tourism and event management graduates with strong technical and managerial competencies. While the technical skills part of the education is incorporated through hands-on experience by laboratory and internship training, managerial knowledge is mostly incorporated through the lecture method of teaching. A volume of content is delivered in a short time to many individuals using the lecture method. The lecture method of teaching is good and efficient at transferring information; however, it has its limitations. The most important limitation is it does not encourage learners to think critically, link and apply the theories and concepts into an organizational context, which is essential for managing the hospitality, tourism and event management organizations.

On the other hand, case-based learning affords learners the ability to understand, think and apply the concepts and theories in an organizational context. Case studies pave the foundation for understanding realistic situations where it helps the learner develop critical thinking, interest and enthusiasm to solve problems. Case studies provide much-needed practical exposure, which is difficult to deliver within the four walls of the classroom. Case studies are very effective in incorporating several tenets of the finer aspects of management. They also help the learners test their analytical skill and abilities to solve complex management issues. Case studies are ideal for active learning and avoiding passive memorizing learning. Case studies are also an ideal evaluation instrument in virtual and blended learning.

In many tourism, event management and hospitality textbooks available on the market, the case studies are presented as slender vignettes, appearing to be straightforward narratives, and mostly insisting that the students end up with finding answers for handling a problematic situation. With a wide variety of case studies pertaining to tourism, hospitality, and event management, this book will fill the gap in the availability of extensive case studies in these areas of management. The variety of cases presented in the book also encourages critical thinking and enables great management competencies among hospitality management graduates. The book aims to present case studies that reflect the current practices and challenges in the tourism, hospitality and event management industry in the international context from both functional and operational perspectives. The teaching cases presented in this book not only reflect the common perennial issues, but even challenges faced by tourism, hospitality and events industries due to the coronavirus disease 2019 (COVID-19) pandemic.

In this book, Case study 1 by Andrea Salustri elaborates on a quantitative method to solve complexity in planning daily excursions supported by different stakeholders. The author analysed a set of rural destinations and developed a daily excursion plan based on rural destinations' attractiveness and accessibility. Daily excursions and local tourism could be key factors in increasing people's well-being and supporting the post-pandemic recovery of Italy's rural tourism territories.

Case study 2 Delly Mahachi Chatibura and Nametsegang Motshegwa narrate the adoption of virtual tourism by a national tourism organization in Botswana (Botswana Tourism Organisation (BTO)) for destination marketing purposes. BTO recently introduced a web campaign known as the Botswana Live Virtual Tours platform. The case study aims to help learners understand the meaning and advantages and disadvantages of three virtual tourism formats, as well as make sound strategic decisions on adopting virtual tourism as a means for destination promotion.

In Case study 3 by Abhijeet R. Shirsat and Jamie Hoffman, the authors have identified reasons for a phenomenon called 'The Great Resignation' which is expected to have wider and long-term implications for the hospitality industry, especially after the pandemic. The authors are able to provide strategic solutions from the lens of Maslow's hierarchy of needs. The case will be an eye-opener for young hospitality professionals, both as an employee and an employer.

Bình Nghiêm-Phú and Hằng Thị Bích Trần illustrate a method to examine the macro- and micro-environmental factors influencing concept hotels in Case study 4. The authors aim to explain the challenges and implications of the business environment in adopting service robots as a unique selling proposition (USP) of concept hotels, using *Henn na Hotels* in Japan, known for their robot-based services, as an example.

Gamification is another buzzword in tourism and hospitality marketing. Gamification has been adopted as a strategy to create an engaging and motivating experience for users. In Case study 5, Ranjana Tiwari elaborates on the Tourism Pursuit app, developed using the concept of gamification to enhance the tourist experience and engagement by tourism authorities with the support of students from local institutions.

In Case study 6, Sushant M. Desale, Andres Artul-Tur and Isabel Osorio-Caballero explain analysing lodging experiences from user-generated content available at Tripadvisor. They have elaborated the sentiment analysis method by content analysis of about 70,000 hotel visitors' reviews in Barcelona. Besides demonstrating the application of ordered probit regression analysis, the teacher can use this case to train the students in preparing the data set, running the statistical model, and inference of analysis.

Joo-Ee Gan examines a successful tourism governance framework adopted by the Royal Belum, a well-managed ecotourism destination in Malaysia, in Case study 7. It helps the tourism teacher to explain the need, significance and complexity of the tourism governance framework that guides the stakeholders and ensures environmental sustainability to the students.

In Case study 8, Colin Seymour, Kalliopi Fouseki and Ilias Krystallis explain the business model canvas by adopting the Victoria and Albert Museum (V&A), founded in 1852 as a heritage organization. It also helps the learners to understand how the political, economic, social, technological, legal and environmental (PESTLE) and sustainability factors influence an organization's business model.

In Case study 9, Shwetasaibal Samanta Sahoo and Sarat Kumar Lenka focus on real-time events in the state of Odissa in India, which is highly prone to natural disasters. The case is expected to help learners understand, identify and implement a disaster management strategy to reduce the hazards of natural disasters due to cyclones.

The Tourism Area Life Cycle (TALC) model propounded by Richard W. Butler in the 1980s is most widely used by tourism scholars. The TALC model retains its relevance even today after four decades. In Case study 10, Senthilkumaran Piramanayagam, Saurabh

Kumar Dixit and Alan Vijay adopt a hypothetical context to explain the application of the TALC model. The case study helps teachers of tourism management explain to learners the life stages of a destination and its implication in tourism planning and marketing.

In Case study 11, Naresh P. Nayak, Julia N. Albrecht and Willem Coetzee demonstrate how the Ishikawa technique or fishbone diagram invented by Ishikawa in 1943 could be adopted to reduce the water consumption in hotels using real-time examples. Also they illustrate how the outcomes of the Ishikawa approach can be assimilated into standard operating procedures (SOPs) in the accommodation business.

Paradise Splendour (Case study 12) is a hypothetical case study written by Snigdharani Mishra, Vasantha Lakshmi and Preeti Narendra. They describe the sensitive aspect of the work environment, employee perception, attitudes and behaviour, which have significant implications on the customer experience at a hotel. The case study helps learners understand the various facets of human behaviour such as perception, individual decision making, inclusion, and the importance of healthy relationships in the workplace.

There is a need to promote a sustainable tourism model as the hotel industry accounts for around 2% of the 5% global carbon dioxide (CO_2) emitted by the tourism sector (UNWTO, 2022). In Case study 13, Kimberley Camelia Langstieh and Evarisa Mary Nengnong highlight the sustainable practices of Café Cherrapunjee, a colonial guest house located in Meghalaya, India and its impact on the community and environment.

In Case study 14, Rohan Jugran intends to familiarize learners with forecasting, pricing and displacement concepts from the perspective of revenue management. Rohan creates a hypothetical family-owned hotel unit named the Twinhill Hotel to explain these concepts. The case study emphasizes the need for knowledge of the market, competitors and the application of quantitative methods to reach a data-driven decision.

In Case study 15, Hanik Kader Şanlıöz-Özgen discusses the challenges and the complexity of decisions involved during pre-opening and opening of a new hotel, and its impact on marketing, promotion and customer experience, using a hypothetical scenario. Also the case study explains the significance of the collaborative effort of hotel-operation team members to offer unforgettable experiences to guests.

In Case study 16, Senthilkumaran Piramanayagam and Saurabh Kumar Dixit discuss the issues faced by a family-owned hospitality business organization called the Raymond Hotel Company (RHC). The problems faced by the RHC are common among family-owned business firms and they affect their survival and profitability. The case study will help learners become familiar with the issues related to the need for professionalism and the challenges of nepotism in managing a family-owned hospitality business.

Corporate social responsibility (CSR) has been high on the agenda of the hospitality and tourism industry. In Case study 17, Rachel Hyunkyung Lee and Tiffany S. Legendre explain three archetypes of the CSR model using various CSR initiatives of Hilton Worldwide, one of the world's largest and fastest growing hospitality companies.

In Case study 18, Mohit Kukreti, Amitabh Mishra and Aaarti Dangwal demonstrate how research can contribute to solving the national-level issues of skilled workforce development in tourism and hospitality in the context of Oman. Through simple research, they explain the cultural and societal issues affecting the current and future tourism workforce.

Artificial intelligence (AI), machine learning and robotics – these terms are no longer deemed futuristic or restricted to the realms of science fiction. The hospitality and tourism businesses have embraced and adopted AI technologies. Upon completion of Case study 19, developed by Kathryn Hayat, learners can understand the relevance of AI in the hospitality and tourism industry and know how it can be a significant aspect of business decision making and strategic implementation.

The cause of a service failure in a service-based organization is not straightforward because delivery systems are complex, and customers are participants in the process. In Case study 20, Drew Martin and Su Gibson explain the role of a service blueprint in solving service failures in quick-service restaurants. This case study helps learners understand the application of service blueprinting and systems thinking to improve the quality of services.

In Case study 21, Noelia Sánchez-Casado and colleagues demonstrate how a simple customer satisfaction survey and subsequent analysis can help service providers to segment the market. This case study introduces the learner to the statistical methods required and their application in developing marketing strategies.

In Case study 22, Miguel Mayol-Tur and Andres Artal-Tur introduce a desk-top secondary-data-based research approach to solve a problem related to the decision on the right mix of management information system (MIS) software for hospitality and tourism companies by combining both proprietary and open-source software. It is essential for practising managers and future managers to decide on the right mix of MIS that best suits their business.

Event tourism is positioned as a powerful way to drive local, regional and national economies. In Case study 23, Cass Morgan and Matt Gnagey demonstrate the process of a quantitative method to analyse the economic impact of small events, which is essential for data-driven decisions. The quantitative method of calculating economic impact is illustrated using a non-profit youth ski foundation as an example.

Operations research (OR) is a science that supports rational decision making. In tourism and hospitality, OR helps with demand forecasting, market segmentation, pricing and revenue management, productivity analysis, crowd management, network and value chain analysis, event scheduling and tourism circuit development. In Case study 24, Senthilkumaran Piramanayagam aims to familiarize the tourism students with an OR tool known as Traveling Salesman Problem (TSP) in creating tourist circuits.

Apart from being one of the most essential components of tourism, the handicraft and handloom sector also contribute to employment generation to a great extent. In Case study 25, Kamakshi Maheshwari and Chandradeep Singh illustrate the effort of Kalakriti, an organization that promotes and protects the artisans through its business model in India.

In Case study 26, Senthilkumaran Piramanayagam and Saurabh Kumar Dixit narrate the significance of employee equity and applications of Adams' equity theory of motivation through a hypothetical hospitality organization called Sage Bliss. This case also helps learners to distinguish between symptoms, surface causes and underlying causes for problems in any organization.

In Case study 27, M. Selvalakshmi, R.M. Kabila and V.J. Sivakumar present the post-pandemic challenges faced by a hypothetical budget hotel located in the heart of Madurai city in the state of Tamil Nadu, India. They demonstrate how applying the concept of the 'flower of service' can be used to solve a managerial dilemma.

All the case studies in this book also have teaching notes that reflect theoretical perspectives, detailed learning activities, and solutions for problems that are excellent, mediocre and even incorrect ones. We believe that this book, *Teaching Cases in Tourism, Hospitality and Events*, will prove a valuable resource to tourism, hospitality and event management educators and learners and ensure the best teaching and learning practices are followed.

Saurabh Kumar Dixit and Senthilkumaran Piramanayagam, January 2023

Reference

UNWTO (World Tourism Organization) (2022) Hotel Energy Solutions (HES). Available at: https://www.unwto.org/hotel-energy-solution (accessed 15 December 2021).

Foreword

Over the past two decades more than ever before, tourism, hospitality and events has transformed itself into a global, diverse, innovative and highly competitive service industry. Being at the grass-roots level within every economy, historically this industry has a profound influence on every country. As a unique service industry tourism, hospitality and events has its roots grounded firmly on customer-perceived value and experience. Therefore, superior customer experience has become an essential ingredient to gain customer loyalty and positive recommendations from customers by word of mouth. Scholars have long argued that customers are not passive recipients of value but rather active co-creators of value and the experience. Experiential purchases are highly intangible in nature and therefore they can exist only in the memories of the customer. For example, The Ritz-Carlton no longer positions itself as a hotel company, rather, it now considers itself an experience and memory creator. Customer experience management, through the integration of people, resources and knowledge, has become one of the key preoccupations of all managers and firms. There is a growing recognition that employees, customers, society and the environment are critical in creating a unique value integration proposition in the tourism, hospitality and events industry.

Technological advancements have brought about many changes in the market and in consumer behaviour, however, it is important to recognize that tourism, hospitality and events is fundamentally a people-focused industry. People's emotional involvement in offering services has a profound influence on the experience of the customers. While various forms of technology may assist in connecting customers with the firm, it is indeed the human element within each firm that is capable of creating a trustful relationship between the firm and the customer. Consequently, tourism, hospitality and event firms must focus their energies on creating value both through human interaction and technological contributions. Furthermore, this interconnected industry exists within the larger ecosystem that includes employees, customers, society, environment and nature. This complex and interconnected industry must be understood as an evolving ecosystem where each element contributes to a symbiotic coexistence.

In the midst of this new focus the world abruptly changed, because of the global pandemic that traumatized the industry, impacting both small and large firms alike. Given its grass-roots-level involvement in the economy, scholars and managers have the view that our industry will provide new direction and context for other sectors to learn from. Many firms within the tourism, hospitality and events industry, in fact, made use of this adversity to innovate and to transform themselves into a new species of resilient enterprises.

This book provides an excellent collection of cases that addresses many critical issues in the tourism, hospitality and events industry. The authors of this edited book have brought together many distinguished scholars to provide a clear overview of multiple factors that contribute to our understanding and to the successful management of the tourism, hospitality and events industry globally.

Jay Kandampully PhD
Professor of Service Management and Hospitality
Editor-in-Chief: *Journal of Service Management*
The Ohio State University
USA

Acknowledgements

An endeavour such as *Teaching Cases in Tourism, Hospitality and Events* cannot reach its logical end without the support, assistance and enthusiastic cooperation of many. We would like to extend our special thanks to all authors who contributed case studies, and assimilated the review comments and suggested modifications, within the stipulated timeline.

We extend our special thanks to Dr Jay Kandampully, Professor of Service Management and Hospitality, The Ohio State University, USA, for his inspiration and for writing the Foreword to this case study book.

We sincerely owe a huge debt of gratitude to all educational administrators, heads of institutions and faculty colleagues from our affiliating universities. We want to acknowledge the support of different institutions and publishers for approving the use of figures and other materials owned by them in this book.

We would like to express our sincere appreciation to Alison Smith, Managing Editor, Emma McCann, Editorial Assistant and other editorial team members from CABI for their support and timely guidance from beginning to end of the project for bringing this book to this present form.

Last but not least, we would also like to acknowledge the constant inspiration, encouragement and motivation offered by our parents, better halves, kids and other family members for allowing us to concentrate diligently on this book project. Without their support, it would have been impossible to reach the book's current form.

Saurabh Kumar Dixit and Senthilkumaran Piramanayagam, January 2023

1 Discovering Southern Latium: a Daily Excursion from Rome

Andrea Salustri*
Sapienza University of Rome, Italy

Abstract
Planning complex daily excursions is a challenging task, as value is co-designed and co-produced by a variety of stakeholders. Moreover, attention must be paid to territorial development and accessibility when considering the tourist attractiveness of rural and natural areas. Based on a horizontal screening of a sample of rural destinations, students are invited to elaborate a set of policy recommendations to facilitate the implementation of a tourism-related bottom-up process of value creation. After the sharp fall in international tourism triggered by the coronavirus 2019 disease (COVID-19) outbreak, daily excursions and local tourism could be key factors in increasing people's well-being and supporting the recovery of Italy's rural territories.

Subjects: sustainable rural development, local cultural tourism, local tourism offer, nested markets

Learning Objectives

At the end of the case study, the learner will be able to:

1. Identify minimum requirements of tourism attractiveness of rural destinations through desk research.
2. Assess the accessibility of rural destinations through desk research.
3. Balance tourism attractiveness and accessibility of a set of rural destinations to identify the best localities for daily excursions.

Case

Ms Coraddi, the Assessor for Tourism of the Latium Region, launched a public consultation on the design of a tourism route among the historical villages of Ciociaria, a cultural region located south of Rome (the capital of the region). The rationale is that, since the coronavirus 2019 disease (COVID-19) outbreak caused a sharp fall in international tourism, the resilience of tourism-related businesses at the regional level has been mostly related to the design and promotion of local excursions, both towards the urban centres and the countryside. As COVID-19 had harsher effects in densely populated areas than in remote places, local demand for tourism services shifted from urban areas to the countryside, for reasons concerning health safety, but also in search of compensation for the sedentary life imposed by lockdowns. The increased demand for

*andrea.salustri@uniroma1.it

DOI: 10.1079/9781800621022.0001

tourism services in rural and natural areas of the Latium Region opened an unprecedented opportunity for focusing attention on rural tourism and its crucial role in generating new jobs and diffused benefits in the localities, when designed and implemented in close cooperation with public administrations and local communities.

So far, the tourism attractiveness of the regional peripheries has been often neglected because of their lower attractiveness for international tourism and the lower returns they may offer with respect to tourism initiatives developed within Rome. In fact, there is no doubt that Rome offers a diversified and qualified tourism offer based on fruition of the rich local cultural heritage, however, also rural and natural territories are often endowed with unexploited cultural and natural resources, as well excellent culinary traditions. It is Ms Coraddi's responsibility to raise awareness among local public administrators of small municipalities, located in regional peripheries, of local tourists' emerging needs. Also, it is Ms Coraddi's responsibility to encourage local entrepreneurs operating in peripheral contexts to develop a local tourism offer that may be showcased on the regional tourism portal and promoted among regional residents through initiatives involving associations and other consumer networks.

As the achievement of these goals may require time, Mr Zingarotti, President of the Latium Region, asks Ms Coraddi to develop a pilot project involving the rural and natural areas surrounding Ciociaria, irrespective of whether these areas are formally inside or outside its administrative and cultural borders. If it is successful, the project may be replicated in other regional territories.

As a first step, Ms Coraddi forms an interdisciplinary team of experts involving a geographer, a sociologist and a political scientist. The task of the team is as follows:

1. Development of a qualitative and interdisciplinary analysis of Ciociaria and its surroundings based on field research.
2. Identification of a sample of 'representative' small and medium-sized municipalities of homogeneous areas on the borders (or surrounding) Ciociaria's valley.
3. Collection of data concerning the local cultural and natural heritage, as well as the culinary traditions of the municipalities included in the sample.
4. Collection of data concerning the local accessibility and demographic development of the municipalities included in the sample.
5. Analysis of the local supply of tourism services, with a focus on agrotourism enterprises.

Ms Coraddi's expert team writes a report, and the results of the analysis are published on the Web. (This is the focus for the students in this case study.) Then, Ms Coraddi launches a contest of ideas to collect the opinions of Latium residents.

The activity of the expert team

Dr Saviori, currently serving as Professor of Human Geography at Tor Vergata University of Rome, has been asked to organize a 1-week tour of Ciociaria's borders and surroundings and identify homogeneous areas of destination (i.e. clusters of municipalities with similar characteristics). Dr Sanela, currently serving as a Professor of Sociology at the University of Southern Latium, has been asked to identify a strategy to develop a sustainable local tourism offer in small municipalities located in the peripheral areas of the region and to complete the analysis with a collection of administrative and economic data. Dr Barcon, serving as Professor of Political Sciences at Sapienza University, has been asked to illustrate the issues concerning the development of the Italian 'Inner areas' (i.e. of those peripheral areas lacking adequate welfare services – public transport, health and education). Finally, Dr Apolli, expert in tourism marketing at Tor Vergata University of Rome, has been asked to identify a methodology to elaborate a horizontal screening of the localities identified and assess their potential as local tourism destinations.

Description of the homogeneous areas of destination

After a short journey in Ciociaria and desk research to complement observational data with statistical analysis, Dr Saviori has identified six homogeneous areas that may serve as destinations for daily excursions from downtown Rome (the most populated area in the region). They surround Ciociaria's valley (i.e. in fact an industrial pole of the region) and they are characterized by homogeneous elements regarding their morphology, cultural heritage and socio-economic development.

After a brief introduction on the geography of the Latium Region and its relationship with Rome's territory, Dr Saviori introduces the results of his research.

Latium

The region hosts a plurality of territories that are heterogeneous by nature, history, population and economic development. The region is dominated by its centre, the city of Rome, Italy's capital and most populous city, but also a world tourist destination. In recent history Rome has always attracted inhabitants and resources from the rest of the region, polarizing economic activities and generating depopulation trends in peripheral areas. However, the territory of the region outside Rome is rich in natural and historical artistic resources that may become part of a polycentric process of local development triggering more balanced regional development.

Ciociaria

Ciociaria is a wide valley that extends south-east of Rome, beyond the Colli Albani and as far as Lazio's border with Abruzzo and Molise. Broadly speaking, Ciociaria includes the Liri and Sacco valleys, the Simbruini, Ernic, Meta and Mainarde mountains that border it to the east and, on the opposite side, the inner part of the Lepini, Ausoni and Aurunci mountains.

Roman Castles

The so-called 'Roman Castles' (Castelli Romani) are a volcanic complex located a few kilometres south-east of Rome. The first people to permanently inhabit Roman Castles were the Latins in about 2000 BC. Many patricians of Ancient Rome chose the Colli Albani to build their residence (Caligula kept his two 'floating villas' anchored in Lake Nemi). During the Middle Ages, the Church of Rome gradually became the owner of all the properties in the area. From the mid-16th century to the end of the 18th century, Roman Castles enjoyed a long period of peace and stability, during which luxurious country residences and elegant parks and gardens were built. The improvement of transport infrastructure initiated in the second half of the 19th century made Roman Castles an ideal place for the famous Roman out-of-town trips and summer holidays. Then, during the Second World War almost all the towns in the Roman Castles were destroyed. Today Roman Castles are one of the most appreciated tourist places in Lazio, both for the well-known traditional food and wine and for the pleasant landscapes.

Aniene valley

The territory of the Aniene valley presents a great variety of natural environments. The cultivated fields, the so-called minor architecture, the history of relations and exchanges between the populations, make this area a typical cultural landscape. The predominance of agricultural or pastoral activities in the past depended on the changing fortunes of the river margins: if these were neglected, the plain would flood and pastoral life would take over, otherwise, when the river returned to its bed, agricultural activities would re-emerge. Nowadays, the valley is characterized by the cultivation of fruit and vegetables, vineyards and, above all, olive trees. In addition to olive production, religious and cultural tourism is growing, due to the nearby presence of Subiaco with its Benedictine monasteries, as well as nature and sports tourism, due to the nearby presence of the Lucretili and Simbruini parks.

Simbruini mountains

The Simbruini mountains cover about 30,000 ha, between the Aniene valley, the Sacco valley, the Abruzzo border and the Ernic mountains. It is an area of Apennine mountains characterized by peaks of up to 2000 m, extensive forests, large plains and spring waters. The area is populated by small villages rich in historical and artistic evidence, sometimes thousands of years old.

Ernic mountains

The Ernic mountains are located between the Simbruini Mountains Regional Park and the Abruzzo, Latium and Molise National Park. Almost the entire area is covered with forests of considerable ecological importance, on the slopes of which are villages such as Veroli, Sora, Alatri, Collepardo, Guarcino and Vico nel Lazio, with a significant history, well testified by the numerous cultural assets located inside the historical centres (but not limited to these).

Abruzzo, Latium and Molise National Park

The territory of the Abruzzo, Latium and Molise National Park consists mainly of a group of mountains ranging from 900 m to 2200 m above sea level. The park is home to numerous villages and small towns, which have largely preserved their identity, given the stratification of historical events and traditional ways of life. Common features of the historic centres, often dominated by the ruins of an ancient castle, are the irregular urban layout, with the focal point being the square with the church, towards which converges the network of narrow streets parallel to the slope, often linked by steep stairways.

Ausoni mountains

The Ausoni mountains are part of Latium Antiappennines and, together with the Aurunci mountains and the Liri valley, separate from the Tyrrhenian coast. Its proximity to the sea is the main feature that determines the landscape and all the naturalistic aspects of the territory. Part of the forest that once covered the Ausoni mountains has been preserved and is home to some of the most beautiful cork oaks in Italy. The Ausoni mountains are home to numerous villages of high historical and cultural value, and important natural resources such as caves and karst plateaus.

Lepini mountains

The Lepini mountains are part of Latium Antiappennines. They cover about 80,000 ha among Rome, Frosinone and Latina, just behind the resurgences of Ninfa. Numerous woods and a great variety of flora and fauna make them a beautiful natural area. Of more general interest to tourists are places such as the village of Sermoneta, Cori and Segni, with its ancient polygonal walls surrounding the acropolis.

Quantitative analysis

Dr Sanella, after a short meeting with Dr Saviori, has compiled the following database containing administrative, territorial, cultural and socio-economic data for each municipality included in the sample supplied by Dr Saviori (Tables 1.1 and 1.2).

A brief description of the 'Inner areas'

Ms Coraddi asks Dr Barcon to explain to the team the meaning of the 'Inner areas' and how this concept is related to local tourism. Dr Barcon explains that the Italian Inner areas are territories that, due to a noticeable distance from the regional centralities, lack immediate access to essential welfare services (education, health and public transport). As a result of secular processes of anthropization that characterizes the Latium Region, the regional Inner areas are often endowed with important environmental and cultural resources. In some cases, the extreme variety and abundance of territorial capital has

Table 1.1. Demography, territory and accessibility. Author's own table.

Municipality	Population (number)	Altitude (m asl)[a]	Territory (km^2)	Depopulation (% 10 years)[b]	Distance from Rome by car (time)[c]	Distance from Rome by car (km)[d]	Public transport[e]
Alatri	27,651	502	96.96	-0.05811	M	M	na
Amaseno	4,252	112	77.73	-0.03386	M	L	na
Anagni	21,071	424	112.81	-0.02304	S	M	na
Capranica Prenestina	315	915	20.36	-0.19643	S	S	na
Castro dei Volsci	4,606	385	58.45	-0.08101	M	L	na
Cervara di Roma	440	1,053	31.75	-0.09465	M	M	na
Collepardo	912	586	24.68	-0.05198	M	L	na
Fiuggi	10,342	747	32.98	0.06421	M	M	na
Fumone	2,043	783	14.84	-0.07640	M	M	na
Jenne	341	834	31.45	-0.18029	M	M	na
Nemi	1,862	521	7.33	-0.07132	S	S	na
Palestrina	21,995	450	47.02	0.03098	S	S	na
Pastena	1,333	318	42.16	-0.13722	M	L	na
Picinisco	1,122	725	62.15	-0.10669	L	L	na
Posta Fibreno	1,088	430	9.8	-0.10526	M	L	na
Rocca Massima	1,101	735	18.17	-0.00272	M	M	na
Rocca Priora	11,861	768	28.26	-0.00101	S	S	na
Sambuci	861	434	8.3	-0.10125	S	S	na
Segni	9,139	668	60.86	-0.02694	M	M	na
Subiaco	8,699	408	63.23	-0.07369	S	M	na
Veroli	19,743	615	119.64	-0.04894	M	M	na

[a]m asl, metres above sea level.
[b]% 10 years, percentage change over 10 years.
[c]L, long (> 120 min); M, medium (60–120 min); S, short (< 60 min).
[d]L, long (> 100 km); M, medium (50–100 km); S, short (< 50 km).
[e]na, not available.

Table 1.2. Local amenities and tourism development.[a] Author's own table.

Municipality	Agrofood resources and typical food products	Festivals	Other resources of tourist interest	Accommodation and restaurants (number)	Agrotourism enterprises	Supra-local networks[b]
Alatri	N	N	Y	10	Y	N
Amaseno	Y	Y	N	4	N	Y
Anagni	Y	Y	Y	26	Y	N
Capranica Prenestina	Y	Y	Y	0	Y	Y
Castro dei Volsci	Y	N	Y	5	Y	Y
Cervara di Roma	Y	Y	Y	2	Y	Y
Collepardo	Y	Y	Y	3	Y	Y
Fiuggi	N	Y	Y	171	Y	Y
Fumone	Y	Y	Y	3	Y	Y
Jenne	Y	Y	Y	1	N	Y
Nemi	Y	Y	Y	10	Y	Y
Palestrina	Y	Y	Y	25	Y	Y
Pastena	Y	Y	Y	4	Y	Y
Picinisco	Y	Y	Y	5	Y	Y
Posta Fibreno	Y	Y	Y	7	Y	Y
Rocca Massima	N	Y	Y	3	Y	Y
Rocca Priora	Y	Y	N	6	N	Y
Sambuci	N	Y	Y	2	Y	Y
Segni	Y	Y	Y	11	Y	Y
Subiaco	Y	Y	Y	20	Y	Y
Veroli	Y	Y	Y	16	Y	Y

[a] Y, yes (the municipality has local amenities, e.g. local agrofood resources, typical food products and festivals); N, no (no apparent local amenities found through desk research).
[b] Supra-local networks of tourism interest include, among others, cultural routes and itineraries and natural protected areas that overlap several municipalities.

fueled a socio-economic dynamism that has inverted the ongoing peripheralization processes. However, where territorial and social distances have prevailed, Inner areas are suffering an economic decline, often associated with the dilapidation of the natural and cultural heritage. This in turn has led to depopulation and a 'fierce resistance' of indigenous inhabitants to exogenous innovations (Musotti, 2020). All Inner areas which successfully resisted and inverted the process of peripheralization (Kühn, 2015) are ideal places for developing local cultural or industrial agri-food systems, as they are rich in cultural and environmental resources, and food and wine tourism can benefit from rich local culinary traditions. However, the obstacles that such extreme variety generates in the organization of an attractive and sustainable tourist offer should not be overlooked.

What is a multidimensional analysis?

Ms Coraddi asks Dr Apolli to explain to the team what is a multidimensional analysis and how it may help to identify the most attractive destinations for daily excursions in Ciociaria. Dr Apolli explains that multidimensional analysis is a decision support tool that can be used in all cases where a single-criterion approach (such as cost–benefit analysis) is insufficient. Multidimensional analysis allows decision makers to include a full range of cultural, social, environmental, technical, economic and financial criteria in their assessment. To compare the contribution of each alternative to meeting the objectives, a selection of criteria is identified. Each criterion must be measurable, in the sense that it must be possible to assess, at least qualitatively, how well each alternative is expected to perform in relation to the criterion. A standard feature of multidimensional analysis is a performance matrix, in which each row describes an option, and each column describes the performance of the options against each criterion. The expected consequences of each alternative are assigned a numerical score on a preference scale. After having inverted negative polarities, more preferred options score higher on the scale, and less preferred options score lower. Finally, numerical weights are assigned to define, for each criterion, the relative valuations of a shift between the top and bottom of the chosen scale.

Discussion Questions

1. Identify the tourist attractiveness of each municipality and represent your assessment through a composite index.
2. Identify obstacles to tourism development that each municipality may face due to depopulation phenomena and lack of local accessibility.
3. Compare tourism attractiveness and territorial development of each municipality and select the best alternative(s) for a daily excursion outside Rome towards Southern Latium.
4. Select the best alternative identified in the multidimensional analysis, then through online research identify emotional elements that may contribute to qualify its tourism attractiveness and organize the trip.
5. In your opinion, which regional policies may incentivize regional dwellers to plan daily excursions in Ciociaria?

Teaching Notes

Summary

The case intends to familiarize learners about the plurality of issues that should be considered when organizing daily excursions in the regional countryside. The analysis is based on the use of a multidimensional analysis to perform an initial assessment of the tourism potential of a sample of small and medium-sized municipalities located in the regional territory. The main objective is to incentivize undergraduate students and practitioners of tourism to identify, through desk research, ideal rural and natural locations for planning a daily excursion or a short stay in the countryside, contributing to the culturally sustainable development of the locality, while recovering from the sedentary life imposed by the COVID-19 pandemic. An initial screening of the localities may lead to more in-depth research on one or two of them, and to the design of a diversified tourism activity targeted to satisfy individual tastes.

Teaching method and equipment

To help students understand the complexity of planning a complex activity such as a daily excursion into a peripheral region, data on various factors need to be considered. Such factors include: (i) administrative, socio-economic and cultural statistics at the municipal level concerning depopulation trends; (ii) local accessibility; (iii) the relevance of the tourism offer; (iv) the availability of natural and cultural amenities; and (v) the existence of specific culinary traditions and food and wine products. The calculation of the scores obtained by each municipality on two composite indicators, measuring the level of territorial development and of local tourism attractiveness, using a spreadsheet is an initial step towards the use of simple statistical methodologies in analysing tourism demand and offer at the local level. Students are expected to learn from the comparison of the achieved results within a collective debate.

The tutor is invited to discuss the (limited) scope of the analysis, its potential role in fostering daily excursions and local tourism and in triggering the culturally sustainable development of the selected destinations. The case is designed for a class ranging from a few students to a group of no more than 20, where each student develops the full statistical analysis and reports a string of results and a short discussion. The instructor of the case study is expected to take the role of Ms Coraddi.

Teaching instructions

The case can be taught in 4 hours (that may be a split over more than one teaching session) using the teaching plan illustrated below. The suggested plan assumes that learners are familiar with the use and interpretation of elementary descriptive statistics and with the use of a spreadsheet. If the students are not familiar with these tools, the tutor could begin by asking learners' experience on the use of statistical analyses in other aspects of daily life. The discussion will help the tutor to ensure a common understanding of what the methodology proposed is about. After an initial phase of the case study, the tutor may organize another session to stimulate a debate among students on their preferred destinations. The instructor may incentivize the students to collect data for one additional destination in the region of their interest. Students may also contribute by identifying other variables of interest that may be added in future classes.

Teaching instruction plan

Discussion point	Time required (min)	Teaching strategies
Introduction	10	The instructor introduces the teaching case by asking students about recent excursions or short stays in the countryside, focusing on the elements of attractiveness of the destination and its accessibility.
Briefing on concepts related to local tourism awareness and accessibility	10	The instructor briefly summarizes concepts related to local tourism awareness, focusing on the added value that may result from their joint and culturally sustainable development. Specifically, the instructor invites students to appreciate the local identity as a multidimensional construct involving a mix of diversified elements (cultural, gastronomic, demographic, natural ...). Then, the instructor invites students to think about those factors that may devalue the local tourism experience (reduced accessibility of the destination, lack of basic services, lack of an adequate tourism offer ...)
Summary of the case	10	The instructor illustrates the background information and the scope of the analysis. The instructor also introduces the use of the spreadsheet to calculate the aggregated scores assumed from the composite indexes for each unit of observation.
Identify the tourist attractiveness of each municipality	45	The instructor asks students to identify the tourist attractiveness of each municipality using a spreadsheet to perform a multidimensional analysis. Then, the instructor asks students to send her/him (by email)[a] the results achieved and a short note (a maximum of ten lines of comment and the identification of the five best alternatives). Then, the instructor computes absolute frequencies and communicates to students the alternatives that gained the most votes for being attractive to tourists. Finally, the instructor asks students to justify the rationale behind their answers.

Continued

Teaching instruction plan Continued.

Discussion point	Time required (min)	Teaching strategies
Identify obstacles to tourism development that each municipality may face	45	The instructor asks students to identify the level of tourism development of each municipality using a spreadsheet to perform a multidimensional analysis. Then, the instructor asks students to send her/him (by email)[a] the results achieved and a short note (a maximum of ten lines of comment and the identification of the five worst alternatives). Then, the instructor computes absolute frequencies and communicates to students the alternatives that gained the most votes for having obstacles to tourism development. Finally, the instructor asks students to justify the rationale behind their answers.
Compare tourism attractiveness and territorial development of each municipality	20	The instructor asks students to represent in a graph the two aggregate scores computed for each municipality (tourism attractiveness and territorial development). Students are invited to send her/him (by email)[a] the graph and a short comment (maximum ten lines) illustrating the relationship between territorial development and tourism attractiveness (i.e. Are they positively or negatively related? Are there groups of municipalities sharing similar characteristics?).
Identify the best alternative and through online research organize the trip	45	The instructor asks students to compute for each municipality an overall performance score by weighting the scores related to tourism attractiveness and territorial development. Then, the instructor asks students to send her/him (by email)[a] the results achieved and a short note (maximum of ten lines of comment and the identification of the five best alternatives). Finally, the instructor computes absolute frequencies and communicates to students the alternatives that gained the most votes as the best municipalities to be included in a daily excursion from Rome.
Propose regional policies that may incentivize regional dwellers to plan daily excursions in Ciociaria	45	The instructor asks students to elaborate one or two policy recommendations that may incentivize Rome's residents to plan daily excursions towards Ciociaria. Students are invited to send her/him (by email)[a] their proposal and a short note (maximum of ten lines of comment). The instructor collects students' answers and tries to summarize results.
Concluding remarks and discussion	10	The instructor summarizes the main results achieved in the case study exercise and invites students to express their opinions on the overall organization of the case study and what they have learned.

[a]If the class is large, the instructor may have difficulties in collecting all the emails in a short time. Consequently, the instructor is invited to think about more efficient ways of collecting information from students. As an example, students may use an online spreadsheet and text editor to write their answers and notes in a shared document.

Duration

The expected duration of the case study exercise is about 4 hours. The effective time needed may vary according to the number of students in the class.

Analysis

In this section for each analytical step presented in the teaching instruction plan more detailed instructions are provided.

1. Identify the tourist attractiveness of each municipality

The first task in this analysis is that of performing a horizontal screening of the municipalities surveyed to identify their tourism attractiveness. The latter can be measured by transforming the variables illustrated in Table 1.2 through a composite index capturing: (i) local food and wine products and events; (ii) the availability of cultural and natural amenities; (iii) the development of the local tourism offer (with a focus on the presence of agrotourism enterprises); and (iv) the participation of supra-local networks of tourism interest. Students may represent their preferred mix of culture, food and wine, and nature by organizing the variables collected in domains and assigning weights to the indexes built at each scale.

As an example, the variables collected may be aggregated as suggested in the scheme shown in Table 1.3. In the example tourism attractiveness (TOA) depends on the availability of elementary indicators which are: (i) local amenities (LAM) (local culinary traditions and food and wine products, local natural and cultural resources); (ii) supra-local amenities (SAM) (supra-local cultural routes and itineraries and natural protected areas); and (iii) the development of the local tourism offer (LTO) (measured as the number of places offering accommodation and restaurants, the existence of local festivals and gastronomic events, the presence of agrotourism enterprises). Each elementary indicator represents a sub-domain of tourism attractiveness and is measured by variables (shown in column 2 of Table 1.3).

The same scheme could be used to:

- specifiy the **polarity** (if it is a 'good', the polarity is positive, if it is a 'bad' the polarity is negative);
- define a system of **weights** whereby, for example, each elementary indicator is equally weighted and the weights total 1, and the variables that make up the elementary indicator are allocated a proportion of this according to their importance (so two variables of equal importance may be weighted 0.5 as together the weights total 1). Regarding the system of weights, if it is not explicitly added it means that elementary scores are aggregated at each step into a composite score by computing an unweighted average; and
- a **scoring rule** for each variable – in the current case, use of 'normalization criteria' are not necessary, as the scores already assume values between 0 and 1.

In the example (Table 1.4) all variables have a positive polarity (they measure desirable aspects of the locality). A system of weights has been specified, both at variable and indicator level. Specifically, while AFW (agrofood resources and typical local products) and OTH (other resources of tourism interest) have received an equal weight (0.5) in measuring LAM, SLN (supra-local networks) is the only variable associated with elementary indicator SAM so it has the weight 1. ACR (accommodation and restaurants) and AGR (agrotourism enterprises) have received a higher weight (0.4 each) than FES (festivals) (0.2) in measuring LTO.

Table 1.3. Tourism attractiveness composite index (TOA): a hypothetical structure.[a] Author's own table.

Elementary indicator	Variable
Local amenities (LAM)	Agrofood resources and typical local products (AFW) Other resources of tourism interest (OTH)
Supra-local amenities (SAM)	Supra-local networks (concerning natural and cultural resources) (SLN)
Development of the local tourism offer (LTO)	Accommodation and restaurants (ACR) Festivals (FES) Agrotourism enterprises (AGR)

[a]The composite index of TOA is made up of elementary indicators (column 1), each of which are measured by the variables listed in column 2.

Table 1.4. Tourism attractiveness index (TOA): polarity, weights, scoring rules. Author's own table.

Elementary indicator[a]	Weight of the indicator	Variable[b]	Polarity	Weight of the variable	Scoring rule[c]
LAM	0.4	AFW	+	0.5	Y = 1, N = 0
		OTH	+	0.5	Y = 1, N = 0
SAM	0.2	SLN	+	1	Y = 1, N = 0
LTO	0.4	ACR	+	0.4	> 10 = 1; 5–9 = 0.5; 1–4 = 0.25; 0 = 0
		FES	+	0.2	Y = 1, N = 0
		AGR	+	0.4	Y = 1, N = 0

[a] LAM, local amenities; SAM, supra-local amenities; LTO, development of the local tourism offer.
[b] AFW, agrofood resources and typical local products; OTH, other resources of tourism interest; SLN, supra-local networks (concerning natural and cultural resources); ACR, accommodation and restaurants; FES, festivals; AGR, agrotourism enterprises.
[c] Y, yes; N, no.

Finally, the scoring rules adopted are very simple: in the case of binary variables, 'Yes' has been associated to a score of 1, 'No' to a score of 0 (zero). There is only one continuous variable, and it has been converted into a discrete one, using a scoring rule that 'rewards' municipalities with at least ten tourism-related activities with the highest score of 1 ('> 10 = 1' in Table 1.4, column 'Scoring rule'). The reason is that for a daily excursion having ten tourism-related activities in the municipality may guarantee an acceptable level of development of the local tourist offer.

Consequently, for each municipality, the following indexes are computed (where * means 'times', i.e. ×):

$$LAM = 0.5^{*}AFW + 0.5^{*}OTH$$

$$SAM = 1^{*}SLN$$

$$LTO = 0.4^{*}ACR + 0.2^{*}FES + 0.4^{*}AGR$$

$$TOA = 0.4^{*}LAM + 0.2^{*}SAM + 0.4^{*}LTO$$

For each unit of observation, the scores computed are reported in Table 1.5. It is worth noting how, due to the high level of standardization (i.e. in this exercise only the *existence* of food and wine traditions, cultural and natural amenities, etc. have been assessed, but not their relevance), the methodology used may overlook relevant alternatives. This privileges the 'satisfactory one', that is, those offering a mix of nature, food and wine, and culture, independently from the importance of the assets identified (i.e. the mix is fully recognized but not the relevance of a single amenity. For example a municipality with a mix of nature, food and wine, and culture receives a high score. On the other hand, a municipality that hosts a very important cultural resource but that does not have food and wine traditions or natural resources receives a lower score. Consequently the index rewards 'satisfactory' alternatives characterized by the optimal mix but not necessarily the most attractive resources). Given these premises, the best alternatives for a daily excursion seem to be Nemi, Palestrina, Segni, Subiaco and Veroli.

2. Identify obstacles to tourism development that each municipality may face

The second task in this analysis is that of performing a horizontal screening of the municipalities surveyed to identify potential obstacles to tourist development, and more specifically, to the design of a daily excursion in the locality. Specifically, the analysis focuses on the fact that peripheral areas are often scarcely populated areas that may be subject to depopulation phenomena. Low and decreasing 'demographic mass' may be associated with lack of elementary services (absence of a bar, an ATM/cash machine, a good Internet connection, etc.). Furthermore, peripheral areas often suffer from scarce local accessibility, consequently even

Table 1.5. Tourism attractiveness composite index (TOA): scores.[a] Author's own table.

	LAM	SAM	LTO	**TOA**[b]
Alatri	0.50	0.00	0.80	**0.52**
Amaseno	0.50	1.00	0.30	**0.52**
Anagni	1.00	0.00	1.00	**0.80**
Capranica Prenestina	1.00	1.00	0.60	**0.84**
Castro dei Volsci	1.00	1.00	0.60	**0.84**
Cervara di Roma	1.00	1.00	0.70	**0.88**
Collepardo	1.00	1.00	0.70	**0.88**
Fiuggi	0.50	1.00	1.00	**0.80**
Fumone	1.00	1.00	0.70	**0.88**
Jenne	1.00	1.00	0.30	**0.72**
Nemi	1.00	1.00	1.00	**1.00**
Palestrina	1.00	1.00	1.00	**1.00**
Pastena	1.00	1.00	0.70	**0.88**
Picinisco	1.00	1.00	0.80	**0.92**
Posta Fibreno	1.00	1.00	0.80	**0.92**
Rocca Massima	0.50	1.00	0.70	**0.68**
Rocca Priora	0.50	1.00	0.40	**0.56**
Sambuci	0.50	1.00	0.70	**0.68**
Segni	1.00	1.00	1.00	**1.00**
Subiaco	1.00	1.00	1.00	**1.00**
Veroli	1.00	1.00	1.00	**1.00**

[a]LAM, local amenities; SAM, supra-local amenities; LTO, development of the local tourism offer.
[b]As an example, for Alatri the TOA is computed as:
$0.4 \times LAM + 0.2 \times SAM + 0.4 \times LTO = 0.4 \times 0.5 + 0.4 \times 0 + 0.4 \times 0.8 = 0.2 + 0 + 0.32 = 0.52$.

if relatively close to Rome, they cannot be reached using the regional public transport system due to lack of services in the 'last mile'. Also, using private transport to cover the last mile (e.g. car or motorbike) may take time.

A composite index of territorial development (TED) can be measured using a subset of the variables illustrated in Table 1.1 through a composite index capturing demographic size (and depopulation trends), territorial size and accessibility. Students may develop an overall assessment by organizing the variables collected in domains and assigning weights to the indexes built at each scale.

As an example, the variables collected may be aggregated as suggested in the scheme shown in Table 1.6. It is worth noting how altitude has been used as a proxy of territorial underdevelopment, as at least part of the area may be inhabited.

As explained previously in '1. Identify the tourist attractiveness of each municipality', the scheme could be used to specify: (i) the polarity (if it is a 'good', the polarity is positive, if it is a 'bad' the polarity is negative); (ii) a system of weights; and (iii) a scoring rule for each variable included in the analysis and normalization criteria (in this case it is not necessary as the scores already assume values between 0 and 1).

In the example (Table 1.7) some variables have a positive polarity, while others have a negative one. A system of weights has been specified, both at variable and indicator level. Specifically, while POP (number of residents) has a positive polarity, DEP (demographic trend in the last 10 years) has a negative one. Also, while SKM (area in square kilometres) has a positive polarity, ALT (altitude) has a negative one. Finally, while PTR (public transport) has a positive polarity, DIS (distance from Rome by car (km)) and TIM (distance from Rome by car (time)) have a negative one. In all cases with positive polarity the highest score has been assigned to the highest values observed, while the opposite has been done in case of negative polarity. Finally, the scoring rules

Table 1.6. Territorial development composite index (TED): a hypothetical structure. Author's own table.

Elementary indicator	Variable
Demographic size (DEM)	Number of residents (POP)
	Demographic trend in the last 10 years (DEP)
Territorial size (TES)	Area in square kilometres (SKM)
	Altitude (m asl)[a] (ALT)
Accessibility (ACC)	Distance from Rome by car (km) (DIS)
	Distance from Rome by car (time) (TIM)
	Public transport (PTR)

[a]m asl, metres above sea level.

Table 1.7. Territorial development index (TED): polarity, weights, scoring rules. Author's own table.

Elementary indicator[a]	Weight of the indicator	Variable[b]	Polarity	Weight of the variable	Scoring rule[c]
DEM	0.2	POP	+	0.4	< 1,000 = 0; 1,001–5,000 = 0.4; 5,000–10,000 = 0.8; > 10,000 = 1
		DEP	–	0.6	> 0 = 1; –0.00– –0.10 = 0.5; < –0.10 = 0
TES	0.2	SKM	+	0.5	< 10 = 0; 11–50 = 0.4; 51–100 = 0.8; > 100 = 1
		ALT	–	0.5	< 300 = 1; 301–600 = 0.75; 601–900 = 0.5; 901–1,200 = 0.25; > 1,200 = 0
ACC	0.6	DIS	–	0.3	'L'= 0, 'M' = 0.5, 'S' = 1
		TIM	–	0.3	'L'= 0, 'M' = 0.5, 'S' = 1
		PTR	+	0.4	Y = 1, N = 0

[a]DEM, demographic size; TES, territorial size; ACC, accessibility.
[b]POP, number of residents; DEP, demographic trend in the last 10 years; SKM, area in square kilometres; ALT, altitude (m asl); DIS, distance from Rome by car (km); TIM, distance from Rome by car (time); PTR, public transport.
[c]L, long; M, medium; S, short; Y, yes; N, no.

adopted are very simple and lead to discrete score variables ranging from zero (0) to one (consequently, normalization rules are not needed).

For each municipality, the following indexes have been computed (where * means 'times', i.e. ×):

$$DEM = 0.4^{*}POP + 0.6^{*}DEP$$

$$TES = 0.5^{*}SKM + 0.5^{*}ALT$$

$$ACC = 0.3^{*}DIS + 0.3^{*}TIM + 0.4^{*}PTR$$

$$TED = 0.2^{*}DEM + 0.2^{*}TES + 0.6^{*}ACC$$

For each unit of observation, the scores computed are reported in Table 1.8. Considering the level of territorial development, the best alternatives for a daily excursion seem to be Palestrina, Rocca Priora, Anagni, Subiaco and Nemi. It is worth noting how none of the municipalities surveyed can be reached using only public transport, due to the difficulties in covering the 'last mile' (it seems, however, that local public transport is now available for Anagni, Fiuggi and Castro dei Volsci).

3. Compare tourism attractiveness and territorial development of each municipality

The scores computed for the two aggregate indexes of tourism attractiveness (TOA) and territorial development (TED) can be contrasted using a scatterplot diagram (Fig. 1.1), then an aggregate index can be computed by weighting the scores of the two indicators.

A small club of five municipalities (Palestrina, Subiaco, Nemi, Veroli and Segni) seem to possess both requirements for being an ideal destination for a daily excursion

Table 1.8. Territorial development composite index (TED): scores.[a] Author's own table.

	DEM	TES	ACC	TED[b]
Alatri	0.70	0.78	0.30	**0.48**
Amaseno	0.46	0.90	0.15	**0.36**
Anagni	0.70	0.88	0.45	**0.59**
Capranica Prenestina	0.00	0.33	0.60	**0.43**
Castro dei Volsci	0.46	0.78	0.15	**0.34**
Cervara di Roma	0.30	0.33	0.30	**0.31**
Collepardo	0.30	0.58	0.15	**0.27**
Fiuggi	0.70	0.45	0.30	**0.41**
Fumone	0.46	0.45	0.30	**0.36**
Jenne	0.00	0.45	0.30	**0.27**
Nemi	0.46	0.38	0.60	**0.53**
Palestrina	0.70	0.58	0.60	**0.62**
Pastena	0.16	0.58	0.15	**0.24**
Picinisco	0.16	0.65	0.00	**0.16**
Posta Fibreno	0.16	0.13	0.15	**0.15**
Rocca Massima	0.46	0.45	0.30	**0.36**
Rocca Priora	0.70	0.45	0.60	**0.59**
Sambuci	0.00	0.38	0.60	**0.44**
Segni	0.62	0.65	0.30	**0.43**
Subiaco	0.62	0.78	0.45	**0.55**
Veroli	0.70	0.75	0.30	**0.47**

[a]DEM, demographic size; TES, territorial size; ACC, accessibility.
[b]As an example, for Alatri the TOA is computed as:
0.2 × DEM + 0.2 × TES + 0.6 × ACC = 0.2 × 0.70 + 0.2 × 0.78 + 0.6 × 0.30 = 0.48.

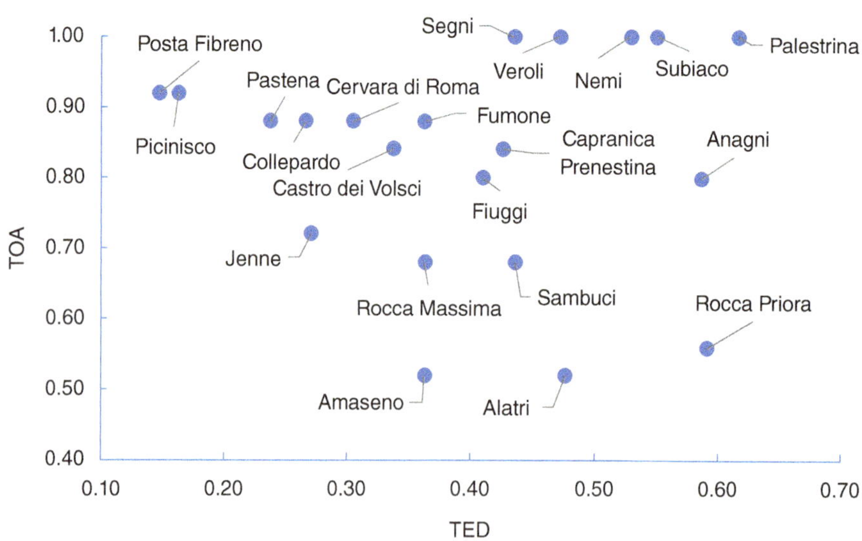

Fig 1.1. Territorial development (TED) versus tourism attractiveness (TOA). Author's own graph.

around Ciociaria. Then, for most of the municipalities, touristic attractiveness and territorial development seem to be negatively correlated. Consequently, tourists interested in a daily excursion in Posta Fibreno or Picinisco, should pay attention to issues concerning the organization of the trip (such as transport, quality of roads) and the local availability of services of general interest (such as Internet connection, banks and when travelling with fragile people, hospitals). Furthermore, at least in the case of Posta Fibreno, due to the reduced territorial size, tourists may include in the excursion a visit to other places in the neighbourhood.

4. Identify the best alternative and through online research organize the trip

The two composite indexes of touristic attractiveness and territorial development can be used to compute an aggregated index representing the individual preferred locations for a daily trip in Southern Latium. People who do not want to spend too much time in travelling around the region will assign a higher weight to TED, while people caring more about the quality of the local tourism offer will assign a higher weight to TOA. As an example, an overall index of daily excursions practicability (DEM) may be the following:

$$DEM = 0.6^{*}TED + 0.4^{*}TOA$$

By weighting territorial development more than tourism attractiveness, the municipalities located in the northern part of Ciociaria are privileged, if compared with those located in the south. This choice is consistent with the fact that it may take more than 2 hours from Rome to reach the localities at the borders of the region. Consequently, when planning a daily excursion, this 'privilege' is justified by the fact that spending 4 hours travelling by car may discourage many tourists from planning a daily excursion to a remote locality. However, Southern Latium is a densely populated area, consequently territories located in the southern part of Ciociaria may be ideal touristic places for local dwellers.

Based on these premises, the best places for planning a daily excursion from Rome towards Ciociaria seem to be Palestrina, Subiaco, Nemi, Veroli and Segni (see Table 1.9). Numerous municipalities may deserve local tourists' attention, however, as far as they will be affected by lack of local accessibility and depopulation trends, many tourists (especially families, the elderly, people with disabilities, etc.) will be disincentivized from planning a daily excursion.

Table 1.9. An aggregated index of daily excursions praticability (DEX). Author's own table.

	TED[a]	TOA[a]	DEX
Palestrina	0.62	1.00	**0.76**
Subiaco	0.55	1.00	**0.72**
Nemi	0.53	1.00	**0.71**
Veroli	0.47	1.00	**0.68**
Anagni	0.59	0.80	**0.67**
Segni	0.43	1.00	**0.66**
Capranica Prenestina	0.43	0.84	**0.59**
Rocca Priora	0.59	0.56	**0.58**
Fumone	0.36	0.88	**0.57**
Fiuggi	0.41	0.80	**0.57**
Castro dei Volsci	0.34	0.84	**0.54**
Cervara di Roma	0.31	0.88	**0.54**
Sambuci	0.44	0.68	**0.53**
Collepardo	0.27	0.88	**0.51**
Pastena	0.24	0.88	**0.49**
Alatri	0.48	0.52	**0.49**
Rocca Massima	0.36	0.68	**0.49**
Picinisco	0.16	0.92	**0.47**
Posta Fibreno	0.15	0.92	**0.46**
Jenne	0.27	0.72	**0.45**
Amaseno	0.36	0.52	**0.43**

[a]TED, territorial development; TOA, tourism attractiveness.

However, places distant from Rome but characterized by high touristic attractiveness should not be overlooked, as they may offer an attractive touristic proposal to Ciociaria's residents. The organization of a daily trip in the selected localities through online research is left as an exercise for the students.

5. Propose regional policies that may incentivize regional dwellers to plan daily excursions in Ciociaria

Based on the results achieved, at least two policies may incentivize local dwellers to plan daily excursions in Ciociaria. These are: (i) improve the local accessibility of small municipalities located at the borders of Ciociaria and its surroundings by increasing the local public transport system; and (ii) promote a diversified touristic offer mixing culture, food and wine, and nature, instead of 'monochromatic' packages that valorize only one aspect, often on a wider geographic scale.

Concerning the first issue, Ciociaria is well connected to Rome by both the highway and the railway, consequently improvements in covering the 'last mile' may consistently reduce the average time of travel (and the cost and the risk of delays due to traffic congestion). This may induce tourists to make a more consistent use of public transport instead of travelling by car, reducing overtourism, as often happens during gastronomic festivals.

Concerning the second issue, a complex excursion focusing on a mix of culture, food and wine, and nature may contribute to valorize the local scale, and to appreciate secondary elements that 'selective tourists' may overlook. Furthermore, this wider perspective may simplify the organization of the travel, as almost every municipality surveyed hosts a diversified mix of resources, so the only travel needed is from the place of residence to the tourist destination and back. This may facilitate tourists' interaction with the local community, increasing local benefits and making daily excursions more sustainable.

References

Kühn, M. (2015) Peripheralization: theoretical concepts explaining socio-spatial inequalities. *European Planning Studies* 23(2), 367–378.

Musotti, F. (2020) Produzione agro-alimentare locale e aree interne. In: Storti, D., Provenzano, V., Arzeni, A., Ascani, M. and Rota, F.S. (eds) *Sostenibilità e Innovazione delle Filiere Agricole Nelle aree Interne. Scenari, Politiche, Strategie.* FrancoAngeli, Milan, Italy, pp. 27–40.

Further Reading

Barca, F., Casavola, P. and Lucatelli, S. (2014) *A Strategy for Inner Areas in Italy: Definition, Objectives, Tools and Governance*, 1st edn. Materiali Uval, Rome. Available at: https://www.agenziacoesione.gov.it/wp-content/uploads/2020/07/MUVAL_31_Aree_interne_ENG.pdf (accessed 15 May 2022).

Celata, F. (2007) Geographic marginality, transport accessibility and tourism development. In: Celant, A. (ed.) *Global Tourism and Regional Competitiveness*. Patron, Bologna, Italy, pp. 37–46.

Cresta, A. (2021) L'emergenza Covid-19 e il riposizionamento del turismo nelle aree interne: prime riflessioni sull'Irpinia. *Documenti geografici* 2, 29–51.

Department for Communities and Local Government (2009) *Multi-criteria Analysis: a Manual*. Communities and Local Government Publications, Wetherby, UK. Available at: http://www.communities.gov.uk/documents/corporate/pdf/1132618.pdf (accessed 10 October 2021).

2 Developing Virtual Tourism in the Wake of COVID-19: a Critical Function of Tourism Destination Management Organizations

Delly M. Chatibura[1]* and Nametsegang Motshegwa[2]
[1]*University of Botswana, Gaborone, Botswana;*
[2]*Botho University, Gaborone, Botswana*

Abstract

The emergence of the coronavirus disease 2019 (COVID-19) has brought unprecedented impacts on the global tourism industry. Starting in 2020, the virus spread from Wuhan, China, where it was first detected, to more than 216 countries around the world (WHO, 2021). To date, there has been more than 240 million confirmed cases and more than 4 million deaths arising from the virus (WHO, 2021). To reduce the spread of the disease, most countries initiated nationwide or community lockdowns, border closures, suspension of flights, or closure of public spaces and restrictions of public gatherings. These restrictions have had negative ramifications for the tourism industry, which stands as an important economic sector in most countries. In pre-pandemic 2019, international global tourist arrivals clocked 1.5 billion and the industry contributed US$3.5 trillion to the total direct gross domestic product (TDG-DP) (UNWTO, 2022). However, by 2021, international tourist arrivals had dropped to 415 million and the TDGDP had also dropped to US$1.9 trillion (UNWTO, 2022). Because of the fragility of the tourism industry arising from the pandemic, there is a need for tourism organizations to effectively manage their destinations. Due to the pandemic, the World Tourism Organization (UNWTO) has prompted governments to rethink and restart tourism in a more resilient manner. To achieve this goal, among other activities, from its innovation, investment and digital transformation agenda, the UNWTO advocates for the digitalization of the tourism sector and encourages destinations to harness innovation and digital advances (UNWTO, nd). Among the innovative and digital guidelines that UNWTO suggests for restarting tourism, is the promotion of new technological products and experiences that include the implementation of augmented and virtual reality, to improve visitor experiences before, during and after a visit (UNWTO, 2020). Given this background, this case study narrates the adoption of virtual tourism by a national tourism organization in Botswana, for destination marketing purposes.

Subjects: destination marketing, tourism development, tour operations management

Learning Objectives

At the end of the case study, a learner will be able to:

1. Explain the concept of virtual tourism.
2. Appreciate the relationship between virtual tourism and destination marketing.
3. Discuss the advantages and disadvantages of three virtual tourism formats.
4. Make sound strategic decisions regarding the implementation of virtual tourism.

*mahachid@ub.ac.bw

(eds. S.K. Dixit and S. Piramanayagam)
DOI: 10.1079/9781800621022.0002

Case

Tourism in Botswana is heavily reliant on international tourists. The country attracts international tourists, from mainly the USA, Germany and the UK. It also attracts regional tourists, from South Africa, Zimbabwe and Zambia, to name a few countries. However, from 2020, tourist flows (at international and regional level) to the country, were disrupted due to the coronavirus disease 2019 (COVID-19). The country was under a state of emergency from April 2020 until September 2021 where intermittent border restrictions were introduced (United Nations Botswana, 2020). COVID-19 has brought operational and financial challenges for the tourism industry. Predictions of a 32.2% contraction of the trade, hotel and restaurant sector, for instance are expected (United Nations Botswana, 2020).

The role of Botswana's national tourism organization, known as Botswana Tourism Organisation (BTO), has never been more prominent at such a time. BTO is a parastatal corporate entity that was established in 2009 through the Botswana Tourism Organisation Act (Botswana Tourism Organisation, 2021a). Previously known as the Botswana Tourism Board (BTB), the organization commenced its operations on 30 April 2010 (Botswana Tourism Organisation, 2021a). BTO's primary mandate is to market tourist products in the country. Due to the importance of tourism for Botswana and the inevitable challenges brought by COVID-19, the tourism industry, and hence BTO, is required to be more receptive to the use of technology, in the form of virtual tourism applications, among others, as a safe alternative to direct human interactions. The use of virtual tourism technologies in the tourism sector for marketing tourism destinations has the potential to provide sensory information to prospective tourists who are seeking destination information and a safe alternative to actual destination visitation. Virtual tours (or VTs), in particular, are very effective in advertising tourism destinations and can increase one's interest for visiting real places of interest in future (El-Said and Aziz, 2022).

The term virtual tour (VT) originates from 1994, when a guided tour of the 1550 Dudley Castle in England was developed as a three-dimensional (3D) construction (Messmer, 2016). Lacking a generally accepted definition, the term can be defined as a simulation of an existing location with the help of special videos or still images (El-Said and Aziz, 2022). The simulation is accompanied by multimedia elements such as sound, music, narration or descriptive text. A VT package is therefore a combination of high-resolution photos or images and multimedia elements. VT videos are created using spin photography, a technique that uses a group of images, called a spinset (1World Sync, 2022), which are stitched together, by the camera or using software, such as Pixtra Panostitcher.

The intention of VTs is to authentically recreate the real site experience. They are developed for interaction on the World Wide Web, using a visual output device, such as a computer, as with video tours or still photo tours. The visual representation of the destination conveys a positive image that increases the number of visitors and assists with decision making (El-Said and Aziz, 2022). Today, virtual tourism is used to simulate protected areas such as cave systems and hotel rooms and beaches (Shaikh *et al.*, 2018). Apart from advertising, VTs can be used for educational and cultural functions. The list of cultural or heritage attractions that are accessible using VTs on the Internet is growing. With advancements in technology and a drive towards sustainable tourism, more tourism agencies are advocating for innovative virtual tourism.

BTO recently introduced a web campaign known as the Botswana Live Virtual Tours platform. The platform was introduced to profile, market and promote various destinations during the COVID-19 pandemic. The platform was officially launched by the President of Botswana, His Excellency President Mokgweetsi Eric Masisi on 2 September 2021 at Goo-Moremi, a popular tourism destination in the country (Botswana Tourism Organisation, 2021c). The product is intended to livestream key tourism destinations, which include Eastern Botswana, the Central Kalahari Game Reserve, the Makgadikgadi and

Nxai Pans, Chobe-Savuti-Linyanti, the Okavango Delta, Kgalagadi, Gaborone and surroundings and the Tsodilo Hills and Gcwihaba Caves using Facebook Live, YouTube or Botswana Live (a VTs website).

A group of eight students from the University of Botswana reviewed the campaign website, in one of their assignments on VTs and destination marketing, in the Tourism and Hospitality Sales and Marketing module. They focused on the website's profile of the Okavango Delta which is captured on the website as shown in Box 2.1.

The students also noticed that the website has a video (with audio) of the delta (which lasts 1.08 min). The video has subsections on Maun (lasting 26 seconds), Moremi Game Reserve (21 seconds) and Khwai (13 seconds). However, these videos which open up in YouTube are very short (lasting less than 30 seconds for each area in the delta) and they lack detail. Although there is an option to view the video using 3D Vista Player, this link is non-functional.

However, through thorough research, the students were able to gather additional information on the Okavango Delta, which they perceived should be included in the area's description on the website. For instance, they noted that the Okavango Delta is located in the north-western part of the country. It owes its existence to the Okavango River which flows from Angola into the Kalahari Basin (Botswana Tourism, 2022). The delta spans an area of 15,000 km^2 in the dry season and 22,000 km^2 in the wet season (Botswana Tourism, 2022), earning a place as the largest inland delta in the world (Bradley, 2009). The delta is vast and it is difficult to access in some places. Because of these reasons, the area was never subjected to significant development (IUCN and UN Environment World Conservation Monitoring Centre, 2015). However, the delta teems with over a 100 species of mammals (including the 'big five' game animals – leopard, African buffalo, lion, rhinoceros and the African bush elephant), and some of the world's most endangered mammals like the cheetah, the wild dog and the white rhinoceros (IUCN and UN Environment World Conservation Monitoring Centre, 2015). The delta is also home to about 1300 species of flowering plants, 89 species of fish and 64 species of reptiles (Botswana Tourism, 2022). Birdlife is abundant and over 400 species have been identified, including globally threated species of vulture, the southern ground hornbill, and the wattled crane (IUCN and UN Environment World Conservation Monitoring Centre, 2015). The natural habitat is diverse and includes clear rivers and lagoons, swamps, seasonally flooded grasslands, forest, deciduous woodland and island communities (IUCN and UN Environment World Conservation Monitoring Centre, 2015). Each of these habitats supports a unique collection of distinct species of reptiles, birds, mammals and reptiles. Part of the Okavango Delta is protected within the Moremi Game Reserve, which is managed by the Department of Wildlife and National Parks. The rest of the delta is operated as Wildlife Management Areas (WMAs) and Controlled Hunting Areas (CHAs), which are open for hunting safaris. Although tourism in the delta is perceived to be exclusively reserved for high spenders, close to 50,000 tourists visit the delta every year, mainly for the wildlife (Mbaiwa, 2005).

Box 2.1. Okavango Delta. From Botswana Tourism Organisation, 2021b.

Declared the UNESCO 1000th World Heritage Site, the iconic Okavango Delta is the largest inland delta in the world. This unique link between large bodies of water and the sand-covered or savannah landscape of the Kalahari has produced a wide range of flora and fauna which makes the delta one of the largest nature paradises in the world. About one-third of the area of the delta forms part of Moremi Game Reserve, the oldest game reserve in Botswana, the best game-viewing area and the first of its kind to be established by a community in Africa in March 1963.

Activity highlights:

- Ride in a dugout canoe (mokoro)
- Game drives
- Boat cruises
- Sundowners in the wild
- Relaxation in camp
- Dinner under the stars

Despite the limited tourism activities, the delta is home to some prestigious lodges and camping sites, such as the andBeyond Nxabega Okavango Tented Camp, an outstanding camping site in the delta, well known for its international standards.

After the review, the students developed three reports they intended to present to the Chairman of the BTO, on the development of virtual tourism activities for the destination, to augment the short-form videos currently offered. The first report is based on a proposal for the development of 360° tours of the Okavango Delta. Such 360° or panoramic tours give the users access to the actual environment, from any direction. They are created using a single or multiple 360° cameras or omnidirectional cameras, as a combination of camera shots of the real place or as one continuous shot (Program Ace, 2022). Though they lack real-time interaction with the environment and thus fall short of the definition of virtual reality (VR) which requires sensory immersion (Goldman and Falcone, 2016), 360° videos offer some 'real' or lifelike experience of the location, which can be viewed using personal computers, head-mounted displays or mobile devices.

The second team of students proposed the use of 3D VTs to promote the Okavango Delta. 3D VTs are designed in a digital environment (Yang *et al.*, 2021). They are an animated simulation or models of the actual destination.

The third group of students proposed an option for VR tours, which require real-time interaction of the viewers with the destination. VR is an advanced technology which is only achievable with special headsets (Cameron *et al.*, 2020). The user navigates the virtual environment (i.e. moves around and explores) and interacts with it, resulting in real-time simulation of one or more of the user's senses (Guttentag, 2010). VR gives users the sensation of being physically present at the environment.

VTs can be published using YouTube, Facebook or other platforms, and can be directly linked to a facility's website or an attraction site for bookings.

Having outlined the three projects, the students further suggested that they were willing to work with a special task team appointed by the BTO to spearhead the review of the VTs platform. Having no specifically dedicated personnel to run any of the projects at the organization, the BTO is faced with a tall order. The organization also needs to deliberate on the three ideas, internally, before recruiting any special task teams to develop the projects. In addition, they can only select one project for now, due to limited funding.

Discussion Questions

1. In your opinion, why should the Chairman of BTO deliberate on the three ideas internally with other members, before selecting special task teams?

2. Of the three virtual tourism formats, which one do you think best applies to the Okavango Delta, as a prime tourism destination in Africa, and BTO's mandate, as a national tourism organization?

3. If you were the Chairman, requested to select three people to comprise the special task team, whom would you select? Give reasons to justify your answer.

4. What role do you think the students would play if selected into the task teams?

5. Which of the three virtual tourism formats would you suggest for a tourism destination of your choice in your country and why?

Teaching Notes

The main objective of the case study is to explain the concept of virtual tourism and to equip learners with knowledge on three key formats of virtual tourism used for destination marketing. The 360° tours, 3D and virtual reality (VR) tours are formats that are increasingly used in tourism marketing. For example, the Singapore Tourism Board, official tourism marketing organization in Singapore, showcases 360° virtual experiences of the Marina Bay Sands, the Raffles Hotel, the Esplanade Concert Hall, the Resorts World Sentosa, Merlion Park, Sentosa Island and many more key tourism attractions, on the

Visit Singapore website (Singapore Tourism Board, 2022). From a management perspective, the case study also expects students to estimate the human resources and skills required to set up task teams or work groups for developing such projects. The facilitator will guide the students in identifying critical skills sets required to work in teams at tourism organizations.

The students will also apply critical thinking skills to analyse the best and appropriate approach to use when addressing a challenging management situation. Using pros and cons analysis, students will be able to suggest the best available option, among varied options, that could appropriately be used to address an underlying situation.

Target audience

The case study is intended for students at bachelor's degree level and those taking master's programmes in tourism or hospitality management. The case study is recommended for the following subjects or topics: (i) tourism and hospitality marketing; (ii) tourism destination management; and (iii) strategic management. Basic knowledge of tour package design and computer applications is also required.

Teaching methods and equipment

The case study is a group-based activity. The activity can be conducted by the students as a group discussion, a role play, as a class presentation using PowerPoint slides or as a debate during a lesson period, depending on the students' specialization and level of study. A group discussion, debate and role play would be more suitable for bachelor's students, while the PowerPoint presentations can be expected from master's students, as they have better oral, research and writing skills. Their analysis can be supported by scientific evidence when justifying their responses.

With group discussions, students can work in three groups (for each VT format) to develop responses to each of the questions. For role play, they can then select one member to serve as the Chairman of BTO. As for the debate, students will work in two groups but will also select a chairman (the lead debater) to lead the discussion for their decisions.

Teaching instructions

The proposed time for the case is 120 min for bachelor's students. However, for graduate students, the case can be taught in 3 hours, especially if PowerPoint slides are to be used for presentations. The proposed time assumes that students have familiarized themselves with VTs online that are based on 360°, 3D or VR formats. The familiarization activity allows students to be on a common ground of understanding with the facilitator. The case study is therefore delivered in two phases, with the first phase taking place outside the normal lesson time.

Teaching instruction plan

Teaching point	Time required	Teaching strategy/activity
Pre-lesson preparation	1 week before the lesson	The facilitator requests students to review existing literature on the advantages and disadvantages of 360°, 3D and VR tours. The facilitator also requests students to participate in a sample virtual tour (VT) online.
Introduction	10 min	The facilitator begins the lesson by asking students when they last travelled, and whether the last visit was an international or domestic visit. The facilitator then requests students to suggest how COVID-19 has impacted tourism experiences across the world.

Continued

Teaching instruction plan Continued.

Teaching point	Time required	Teaching strategy/activity
Case study introduction	10 min	The facilitator introduces the topic on VTs and randomly asks students if they have ever taken a VT and how they would feel if they were to take a VT of a prestigious tourism destination in Africa. The students are then allowed to read through the case.
Case summary	10 min	The facilitator requests two students (volunteers) to summarize the case by highlighting the background information, the current scenario with BTO's Botswana Live Virtual Tours platform, the protagonists (the group of students) and the dilemma or decisions that need to be taken.
Discussion of question 1	10 min	A discussion, led by the facilitator, should take place.
Discussion of question 2	30 min	After the discussion on the first question, students are requested to form three groups, one for each VT format. The groups will participate in either (i) role plays, group discussions and presentations; or (ii) a debate.
Discussion of question 3	15 min	A discussion, led by the facilitator, should take place.
Discussion of question 4	20 min	After the discussion on questions 3 and 4, students are requested to form three groups, one for each VT format and discuss the question in 5 min. After that, groups will be requested to provide their findings in 15 min.
Discussion of question 5	10 min	The facilitator allows the students to brainstorm on a tourism destination that can be showcased using any one of the virtual tourism formats. The facilitator can then request a volunteer from the class (or more than one depending on the class size), to describe how the format could be used for the selected destination.
Lesson conclusion	5 min	The facilitator concludes the lesson by summarizing the three VT formats and the factors that could be considered to develop them.

Duration

120 min

Analysis

1. In your opinion, why should the Chairman of BTO deliberate on the three ideas internally with other members, before selecting special task teams?

The employees within BTO should be aware of the discussion. This is an essential component of internal marketing. The Chairman should be able to sell the idea successfully, to win internal buy-in and support. The employees are potential customers of the VTs and as such they can promote the product, once launched through word of mouth. This process also allows internal employees to have a common understanding of the project with the Chairman and the students, of the benefits and features of such projects. In the process, new ideas and suggestions may arise.

2. Of the three virtual tourism formats, which one do you think best applies to the Okavango Delta, as a prime tourism destination in Africa, and to BTO's mandate, as a national tourism organization?

To respond to this question, students need to understand two things. First, the Okavango Delta is a pristine attraction that is largely undeveloped because of the conservation efforts of the Government of Botswana. Most of the areas in the delta are inaccessible or only accessible using small light aircraft or mekoro (dugout canoes; mokoro singular; mekoro plural). As such, some regions of the delta remain undeveloped. Secondly, while BTO is the official tourism organization in Botswana, the management of areas in the Okavango falls under different government entities such as the Department of Wildlife and National Parks (for the Moremi Game Reserve which falls under the Okavango Delta) and community-based tourism organizations, trusts and private concessions. Access to take pictures of the real locations using 360° cameras, for developing 360° videos, needs to be sought from all or some of these custodians. It therefore means BTO needs to address issues of accessibility which could be bureaucratic. This could delay the process or reduce the geographical coverage of areas captured in the videos. Despite these disadvantages, the videos will provide detailed and real information on game drives, hunting sessions, mekoro rides or boat cruises across the delta, and unlike 3D videos they offer a sense of immersion, when viewing the videos. Viewers will see some of the nearly extinct species of birdlife and mammal life (including the 'big five' game, on a typical game drive). While taking a mokoro ride, they will be able to see the clear Okavango waters. The viewers will have the opportunity to view the videos from any direction, meaning they can control what they see. However, the development of 360° VTs in inaccessible areas will be very limited. So students need to think about accessibility and remoteness of some of the sites, which would then probably leave them to opt for a 3D or VR tours.

With the 3D tours, viewers can remotely view a virtual model of specific areas or facilities in the Okavango Delta. The viewers can navigate the delta by clicking or tapping their screen and they are able to move their screen from side to side or zoom it in or out. With 3D models, BTO can select certain areas of the Okavango Delta to showcase. For example, models of some of the most outstanding lodges and camping sites, such as the andBeyond Nxabega Okavango Tented Camp, can be developed. The viewers can also have a look at 3D virtual models of the boats used for cruises on the delta. Because most of the delta is pristine and unexplored, the 3D video tours would be ideal. 3D tours would also be ideal due to the sensitive nature and fragility of the ecosystem in the delta.

The students can also base their discussion on the degree of interaction required. If the viewer is expected to perform several activities, like loading a gun in a hunting session, then a 3D or VR tour would suffice as they provide the depth in detail that is required. However, where the viewer is an observer and there is no or limited interaction between them and the video, the 360° video would be advisable to use.

The exclusive and high-value cost of tourism in the Okavango Delta could assist the argument for the adoption of VR tours. VR users will experience the destination as if they were physically present. This technology is expensive and can be offered on a small scale to viewers who can afford it.

3. If you were the Chairman, requested to select three people to comprise the special task team, whom would you select? Give reasons to justify your answer.

To create VTs, the Chairman could select the following people for the team:

- Someone with photography skills, who is able to use their imagination to capture the best photos/images to be included in a tour. This person has good knowledge of which photography or video tools to use for creating the best VTs.
- A person with computer-based skills. This person should use the latest technology to develop 360° and VR videos. The candidate should be a digital content user or developer and must have video editing tools. This person must have a good understanding of video editing software and the right mix between multiple media elements (e.g. audio and video mix).
- Someone who is innovative and creative should be included in the team. This is important to increase the authenticity and enhance the content of the videos online. Innovation requires a special mindset and someone who is opportunity focused.
- Someone with good research skills should be incorporated in the team. This person should continuously review literature for exciting content to include in the videos. They need an understanding of who the most likely users of the VTs will be, what their characteristics are and what their interests would be. Also this person should be available to

monitor activities on online social sites including Facebook, Google, etc.

- A representative from BTO should be included in the team. This person will be responsible for acting as a liaison between the team and the local and business communities and the government. This person will also be responsible for marketing the VTs through BTO or through the BTO website.

4. What role do you think the students would play if selected into the task teams?

The students are the idea generators. The idea generator is an important member of the team, as they own the idea. They are the original owners of the ideas associated with the 360°, 3D and VR tours for the Okavango Delta. They have the option of copyrighting their ideas, especially if they are to be commissioned by a third party (BTO). They can also receive monetary rewards from BTO, for their participation in idea generation.

5. Which of the three virtual tourism formats would you suggest for a tourism destination of your choice in your country and why?

The students should identify a tourism destination in their country that can meet the criteria suggested for the 360° tour, the 3D tour or the VR tour in Question 2.

References

Botswana Tourism (2022) Okavango Delta. Available at: https://www.botswanatourism.co.bw/explore/okavango-delta (accessed 15 March 2022).

Botswana Tourism Organisation (2021a) About BTO. Available at: https://www.botswanatourism.co.bw/about-bto (accessed 5 September 2021).

Botswana Tourism Organisation (2021b) Botswana Live. Available at: https://botswanalive.co.bw/ (accessed 12 August 2022).

Botswana Tourism Organisation (2021c) Botswana Live. Press Release. Available at: https://m.facebook.com/BotswanaGovernment/photos/a.336021353147196/4319031164846175/?type=3&source=48 (accessed 5 September 2021).

Bradley, J.H. (2009) Gliding in a Mokoro through the Okavango Delta, Botswana. Available at: http://www.capetowntocairo.com/travelogue/gliding-in-a-mokoro-through-the-okavango-delta.htm (accessed 15 March 2022)

Cameron, J., Gould, G. and Ma, A. (2020) *360 Essentials: a Beginner's Guide to Immersive Storytelling*. Ryerson University Library. Available at: https://pressbooks.library.ryerson.ca/360essentials/ (accessed 15 March 2022).

El-Said, O. and Aziz, H. (2022) Virtual tours a means to an end: an analysis of virtual tours' role in tourism recovery post COVID-19. *Journal of Travel Research* 61(3), 528–548. https://doi.org/10.1177/0047287521997567

Goldman, J. and Falcone, J. (2016) Virtual reality doesn't mean what you think it means. Available at: https://www.cnet.com/tech/mobile/virtual-reality-terminology-vr-vs-ar-vs-360-video/ (accessed 16 March 2022).

Guttentag, D.A. (2010) Virtual reality: applications and implications for tourism, *Tourism Management* 31(5), 637–651. https://doi.org/10.1016/j.tourman.2009.07.003

IUCN (International Union for Conservation of Nature) and UN (United Nations) Environment World Conservation Monitoring Centre (2015) Okavango Delta. Available at: http://www.yichuans.me/datasheet/output/site/okavango-delta/ (accessed 15 March 2022).

Mbaiwa, J.E. (2005) Enclave tourism and its socio-economic impacts in the Okavango Delta, *Botswana. Tourism Management* 26(2), 157–172. https://doi.org/10.1016/j.tourman.2003.11.005

Messmer, H. (2016) The beginnings of digital visualization of historical architecture in the academic field. In: Hoppe, S. and Breitling, S. (eds) *Virtual Palaces, Part II. Lost Palaces and their Afterlife*. Palatium, Munich, Germany, pp. 21–54.

Program Ace (2022) What is a Virtual Tour in 2022? Types, Features, Development. Available at: https://program-ace.com/blog/what-is-virtual-tour/ (accessed 15 March 2022).

Shaikh, S., Bokde, K., Ingale, A. and Tekwani, B. (2018) Virtual tourism. *International Research Journal of Engineering and Technology* 5(4), 2044–2046.

Singapore Tourism Board (2022) Experience Singapore now. Travel Later. Available at: https://www.visitsingapore.com/virtual-experiences/ (accessed 13 March 2022).

United Nations Botswana (2020) Socio-economic impact analysis of covid-19 in Botswana. In: *UN Botswana Socio-Economic Impact Analysis, Analysis Brief No.1 06052020*. UN Botswana, Gaborone.

UNWTO (World Tourism Organization) (nd) Innovation, education and investments. Available at: https://www.unwto.org/digital-transformation (accessed 15 March 2022).

UNWTO (World Tourism Organization) (2020) Global Guidelines to Restart Tourism. Available at: https://webunwto.s3.eu-west-1.amazonaws.com/s3fs-public/2020-05/UNWTO-Global-Guidelines-to-Restart-Tourism.pdf (accessed 26 January 2022).

UNWTO (World Tourism Organization) (2022) UNWTO World Tourism Barometer. January 2022. Available at: https://webunwto.s3.eu-west-1.amazonaws.com/s3fs-public/2022-01/220118-Barometersmall.pdf?_PBIQdr4u_qM0w56.I0NpfGPzylGu6Md (accessed 15 March 2022).

1World Sync (2022) What is the difference between 360-degree and 3D photography? Available at: https://1worldsync.com/resource-center/blog/what-is-the-difference-between-360-and-3d-photography/) (accessed 12 August 2022).

WHO (World Health Organization) (2021) WHO Coronavirus (COVID-19) Dashboard: Global situation. Available at: https://covid19.who.int/ (accessed 13 March 2022).

Yang, T., Lai, I.K.W., Fan, Z.B. and Mo, Q.M. (2021) The impact of a 360° virtual tour on the reduction of psychological stress by COVID-19. *Technology in Society* 64: 101514. https://doi.org/10.1016/j.techsoc.2020.101514

3 Why Workers Are Not Returning to the Hospitality and Tourism Industry: a Turning Point

Abhijeet R. Shirsat* **and Jamie Hoffman**
California State University-Sacramento, Sacramento, California, USA

Abstract
An unexpected challenge has taken over the hospitality industry where the current workforce is not returning to jobs since the coronavirus disease 2019 (COVID-19) pandemic started. There are a variety of reasons and motivations why people are not working including poor wages, working conditions, lack of benefits, schedule flexibility, work-from-home (WFH) opportunities, childcare facilities, mental health facilities, transportation, and the anxiety of catching the virus. In this case study, we will identify reasons and provide strategic solutions from the lens of Maslow's hierarchy of needs. Recommendations and strategies for practitioners to motivate and attract workforces will be provided in this case study.

Subjects: human resources, operations management, contemporary issues and trends

Learning Objectives

At the end of the case study, the learner will be able to:

1. Identify the reasons workers are not returning to their jobs since the pandemic started.
2. Explain and provide reasons for employers to meet the needs, wants and demands through the lens of Maslow's hierarchy of needs.
3. Explain the future implications of the changes brought about by the pandemic for the workforce.

Case

As an international phenomenon, workforces have been impacted by coronavirus disease 2019 (COVID-19). Very few times in the history of the world have seen employees overpower the employers, and hence we call this phenomenon 'a turning point'. After December 2019 at the beginning of the pandemic, many hospitality and tourism employees were laid off or furloughed. Employer expectations during and post-pandemic were that employees would return to 'normal'. However, we are seeing significant deficits

*shirsat@csus.edu

DOI: 10.1079/9781800621022.0003

in the workforce as tourism has increased after the initial phase of the pandemic. One of the major reasons for this phenomenon is that workers from the hospitality and tourism industry found other avenues for employment. Based on this, management has to consider how it can provide services for employees and accommodate some of their demands in order to attract and retain employees. A survey from the US job market report of 12,909 US workers revealed that at least 50% of former hospitality workers won't return to their work (Joblist, 2021). Similar phenomena have been experienced across the globe.

In the UK, one in five hospitality workers is not returning to work citing the COVID-19 pandemic and Brexit as the culprit. However, national statistics provide further clarification of the situation in the industry and reveal that the situation was dire before the pandemic, based on the industry culture: (i) long working hours; and (ii) low pay. Similarly, employers in Australia are desperate to attract international workers. They are willing to pay for the airfare, quarantine accommodation and visa fees for immigrants who want to work in Australia's hospitality and tourism industry. European countries including Germany, Spain, France and Austria are all facing a dramatic loss of workforce causing major concern for the overall industry. Workers in India have not returned to work because of the anxiety of the economic slowdown and uncertainties due to the pandemic.

According to Joblist (2021), workers in the hospitality and tourism industry are quitting their jobs because of low pay and they want to build a new career in a different sector for stability. It is evident that workers in the hospitality and tourism industry are re-evaluating their choices since the drop in COVID-19 cases and alternative work options have become available. Workers are finding new skills that will assist them to build careers in an industry where they have opportunities for growth and autonomy. The phenomenon of quitting and finding new careers was recently coined 'The Great Resignation' by Dr Anthony Klotz, an organizational psychologist at Texas A&M University (NPR, 2021a).

The psychology behind 'The Great Resignation'

Humans have a need for autonomy and the pandemic provided people with an opportunity to re-evaluate their past and current career choices. Working from home, flexibility in schedules and extra time to spend with family are some of the perks that the masses enjoyed during the global lockdown. Many are resigning and finding new employment where they can utilize their skills and cherish a better life. Dr Anthony Klotz (NPR, 2021a) gives four reasons why people are quitting their jobs: (i) a backlog of resignations from the past; (ii) burnout; (iii) pandemic epiphany; and (iv) the need for work-from-home (WFH) opportunities. This has forced employers to rethink strategies to hire and retain employees including hiring bonuses, increased salaries, and benefits being offered like never before. However, many companies are still unsure of emerging issues post-pandemic, which will be discussed in this case study.

The hospitality and tourism industry has historically faced high employee turnover, however, the pandemic significantly increased this rate. More than 50% of employees planned to quit their jobs in the hospitality and tourism industry before the end of 2021 according to the US Job Market Report (Joblist, 2021). Workers were citing the following reasons driving them away from the industry: (i) low job satisfaction; (ii) difficult customers; (iii) lack of schedule flexibility; (iv) physical demands; (v) poor wages; (vi) working conditions; (vii) lack of benefits; (viii) WFH opportunities; (ix) lack of childcare facilities; (x) lack of or poor mental health facilities; (xi) difficulties with transport to and from work; and (xii) the anxiety of catching the virus. To understand the needs, expectations, motivations and demands of the workers who are choosing not to return to the industry, we will discuss them in two main sections: first, long-standing issues; and secondly, issues that have emerged since the start of the pandemic. In Fig. 3.1 the authors outline the issues for retaining employees.

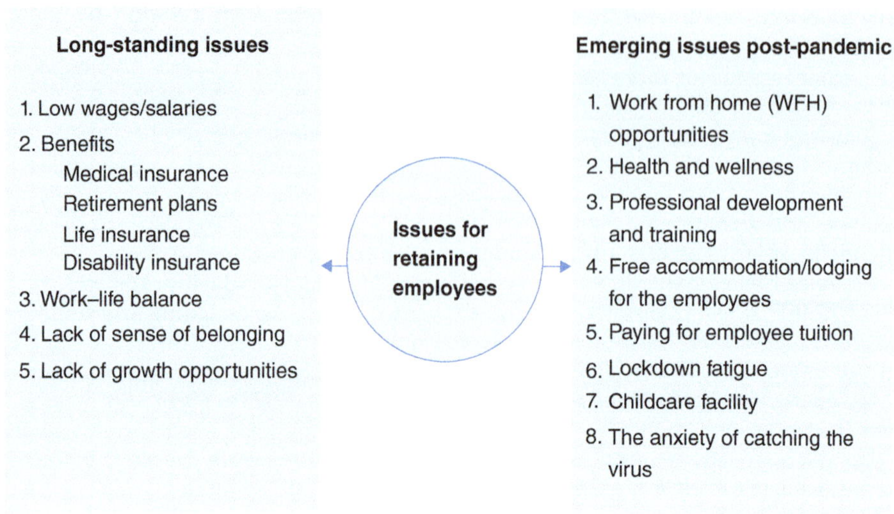

Fig 3.1. Issues for retaining employees. Author's own figure.

Long-standing issues

1. Low wages/salaries

Low salaries are a long-standing issue in the industry. Employers are facing unprecedented challenges due to the industry's salary standards. To overcome this known issue the industry is offering higher wages and benefits compared with before the pandemic in order to entice workers. Wages for workers in restaurants, hotels, theme parks and other entertainment venues rose by 2.3% during the pandemic according to the US Department of Labor Department (NPR, 2021b). However, the data from the US Bureau of Labor Statistics (BLS) illustrate in their report that a record 1.7 million jobs are open in the industry and nearly a million people are quitting (Molla, 2021).

Workers who were laid off or furloughed found alternative employment outside the industry and are questioning returning to the industry due to a new work–life balance along with equal or better wages and benefits. Hence it will be important to see how the industry standards change and if it helps in bringing the employees back to work. Many restaurants are now providing hiring bonuses, higher than minimum wages, and benefits to incentivize opportunities. Hotels are also trying to attract employees with similar steps; however, these benefits and salary hikes are due to the current lack of workforce in the industry. Experts believe that the industry needs to reinvent the culture and standards for a sustainable better future.

2. Benefits

The industry is known for few to no employee benefits and little job security. Retirement plans, medical insurance, sick leave, paid time off, disability insurance and stock options are rarely provided in the workplace. Workers have evaluated their situations and found alternative employment providing higher pay, fewer demands, job flexibility and better benefits.

3. Work–life balance

Employees working in the industry have frequently complained about work–life balance. Long working hours, holiday work, no paid leave, late-night working requirements and inconsistent work schedules have led to a lack of job satisfaction. Several studies in the hospitality and tourism industry report different strategies to acquire work–life balance, however, this has never been a priority for employers. While it is a topic of debate for many, hospitality employees are currently pushing the barriers to force employers to reconsider the status quo of work–life balance if they want the workers to return.

4. Lack of sense of belonging

Social belonging is a basic human need, and it is engraved in our DNA. And yet, more than 40% of the employees feel isolated at their workplace and are less committed and engaged at work. This is true in the hospitality and tourism industry. During COVID-19 these relationships were further strained due to the lack of one-to-one interactions resulting from the lockdown measures by governments globally.

5. Lack of growth opportunities

Growth is important for employees as it provides motivation to work. Stagnancy at work is one of the most influential reasons for burnout. During COVID-19 many promotions and growth opportunities were eliminated or not provided. This created burnout in the employees who diligently worked to achieve consideration for promotions, continuing industry retention challenges.

Emerging issues post-pandemic

1. Work-from-home (WFH) opportunities

The labour markets are extraordinarily tight, and workers have scheduling options. Talented and multi-skilled workers receive multiple job offers and decide on employment based on their needs and requirements. In addition, the pandemic has revealed that people can perform equally, if not better, from home and this has become more lucrative for people of colour and highly educated women with kids (Bloom, 2021). A study of 35,000 Americans has revealed that 32% of the workers never want to work in an office again. In the hospitality and tourism businesses, employees related to human resources, marketing and other back-office jobs are exercising this option. This is a significant management concern, managing workers from home versus work in-office teams.

2. Health and wellness

According to Wellable, a marketing firm, the pandemic has created an increase in employee bereavement and isolation. Loss of income and fear are triggering mental health conditions or exacerbating existing ones. Wellable Labs (2021) identified a few current trends: employers are investing most in mental health (88%), telemedicine (87%), stress management/resilience (81%), mindfulness and meditation (69%), and COVID-19 risk intake/wellness passport (63%) programmes. These significant concerns for employee health and wellness may help guide employers in allocating specific line items in future budgets to address employee needs.

3. Professional development and training

While many employers have transitioned to virtual work environments, the need for professional development and training has not disappeared. The need for information has expanded and allowed for a shift in the presentation of information allowing researchers to reach a broader audience virtually. Virtual online formats have created numerous technical and organizational challenges, but also offer opportunities for reaching wider audiences, increasing accessibility, reducing the carbon footprint of meeting travel, and improving diversity and equity. As access has increased to a broader audience through virtual delivery, the direct financial impact on cities with large convention centres and chambers of commerce that rely on these large groups to visit and utilize their industries has been crippling. While this is directly impacting the hospitality and tourism industry, it is time for a reinvention of services and opportunity for growth in development that otherwise would never have occurred. This creates an opportunity for employers to offer training and development to motivate staff and increase retention utilizing virtual formats.

4. Free accommodation/lodging for the employees

During the pandemic, many workers lost their jobs, eventually resulting in losing their homes or residences. This created a new challenge for the hospitality and tourism organizations that were already scrambling to find employees. Omni Hotels and Resorts and similar small and large hospitality organizations provided free accommodation over the summer of 2021 to employees, and continue to offer free nights to some employees, in an effort to recruit and increase retention.

5. Paying for employee tuition

To create a complete package of offerings for potential employees, the industry began to offer reimbursement for employee education (tuition) to entice them to take jobs during and after the pandemic. This kind of support is unprecedented in the industry.

6. Lockdown fatigue

Several countries implemented strict lockdowns during the pandemic and some countries are still implementing lockdowns for residents who are unvaccinated. Humans are social by nature and restrictions like the lockdown created fatigue, resulting in individuals having to deal with a rollercoaster of emotions. Not being able to go out, meet people, socialize, and being isolated from family and friends created a sense of loneliness. This combined with not having the liberty to work on their own terms is one of the reasons that people quit their jobs and/or are not returning to work.

7. Childcare facility

The lockdown and other societal changes saw an immediate surge in birth rates. This has now created an immediate need for childcare for new parents who need to return to work but now need childcare. These new parents are looking at their employers to provide flexibility and possible solutions for their childcare needs in the workplace so that they can be close to their babies during their initial growing years while they are working. This is a headache for the already challenged hospitality and tourism management who are struggling to recruit and retain employees.

8. The anxiety of catching the virus

As anticipated, the pandemic created a sense of fear in citizens as COVID-19 has taken more than 6,429,611 lives worldwide and counting (Johns Hopkins University, 2022). The virus is highly contagious and constantly mutating, creating challenges for the scientific community. The vaccines have proved to be effective but are still not readily available in many countries. Only the most developed nations have seen a considerable number of people getting the vaccine while others remain unvaccinated. People have anxieties about catching the virus and the threat of dying. Anxiety and fear are motivating factors for employees to stay at home, or find other employment or entrepreneurial opportunities which allow them to work from home or isolate themselves.

Industry changes during and after the pandemic

Current employers and employees in the industry are reporting a significant shift in the attitudes of the guests who are visiting, staying or considering to stay/eat. Guests are expecting higher standards of service, hygiene and infrastructure. While it is anticipated that the industry will establish a 'new normal', it may be an uphill battle with a new culture of enhanced hygiene and sanitation requirements. The current demanding customer will require the industry to reinvent the customer service skills of the employees who will represent the organizations and enhance the guest experience. One example is the complete removal of touch of menus in restaurants; restaurants are now using QR codes, as well as apps for ordering food and paying for services. Therefore, technology may play a significant role in the paradigm shift of the industry. However, it will have implications for the continued need for labour and a balance between the use of technology and employees will be necessary to maintain the human connection and personalized services available. Employee training (including soft skills, politeness, and patience with the guests – which may assist businesses) may be needed to stand out in the market. These skills cannot be implemented via technology and only humans can help in providing such personalized service and experiences.

Maslow's hierarchy of needs

Based on the analysis of the current phenomenon of workers not returning to work and changing their jobs, Maslow's hierarchy of needs will be used as a framework to understand the current needs, wants and demands of employees. Maslow (1943) stated that people are motivated by their needs and some needs take priority over others. Maslow presented a hierarchy which will be used to explain the current phenomenon that is unfolding, in terms of employee shortages. Some of these situations in the phenomenon could be overlapping in terms of need.

1. Physiological needs

According to Maslow's theory food, clothing and shelter are basic human needs and in order to secure these needs, most people seek employment. Recent inflation throughout the globe and wage stagnation has caused significant challenges for society. In the industry, wages have not risen in decades as compared with other sectors. Decent wages/salaries provide resources to meet our primary needs and if not met it is obvious the workers will look for alternative employment.

In addition, work–life balance is a primary need of humans, including sex, sleep and relaxation. Long working hours cause tremendous issues for workers trying to meet the aforementioned primary needs. Hence it is vital for employers to re-evaluate the wages and work–life balance to provide opportunities to satisfy the basic physiological needs of their employees.

2. Safety needs

People want to experience a safe and secure life after their psychological needs are met. They expect benefits such as medical care, life insurance, retirement plans, mental health and wellness, and protection from any type of viruses or workplace illnesses. During the pandemic, it has become more evident that safety needs are vital in employee satisfaction and autonomy. Employers who are providing such benefits are preferred over others. In addition, during the pandemic, employees are seeking employers who are putting their health and safety first, including mask and vaccine mandates.

3. Love and belongingness need

One of the leading factors associated with dissatisfaction in the workplace includes not having a sense of belonging. Once the psychological and security needs are met, people start to think about love and belongingness. Job satisfaction also includes a sense of teamwork, comradery and connection. During the lockdown, employees working remotely experienced anxieties and lockdown fatigue causing burn out and many quit their jobs as they were concerned with collegiality, belongingness and job satisfaction in the workplace.

4. Esteem needs

Once the top three needs are satisfied, people start thinking about the fourth need of esteem; people like to feel valued, including a sense of achievement and a desire for respect. Hospitality and tourism employees have long complained about stagnancy. The current boom in the job market is alluring for employees looking for growth, professional training and development, and this creates competition in the market.

5. Self-actualization needs

The highest level of need includes self-actualization, where individuals identify what ultimately provides fulfilment. During the pandemic, WFH opportunities and schedule flexibility options reintroduced people to their inner desires. For example, individuals travelled more, pursued new hobbies, or spent more time with their families learning to cherish and indulge in these activities. As the numbers of COVID-19 cases are declining, employers are expecting employees to return to the office creating havoc for those who are in the self-actualization phase where they believe that they can do their best working from home with their new autonomy and flexibility. Employees who would prefer to maintain their 'new normal' working from home are opting to quit. Hence this unique challenge is creating worldwide attention and a significant challenge for employers.

Using Maslow's hierarchy of needs as a foundation, the pandemic has emphasized the top three needs of employees: (i) love and belonging; (ii) esteem; and (iii) self-actualization. At this 'turning point' workers are learning to advocate for their needs, wants and demands in return for their work, time and talents. This moment resembles one of the rare points in history: in the mid-1300s the bubonic plague (Black Death) pandemic in Europe caused millions of deaths and triggered opportunities for farmworkers to dictate their needs and wants in return for their labour. We see similar implications for the COVID-19 pandemic in the modern world and believe that this will increase the need for artificial intelligence (AI), technology, digitalization and robotics.

Conclusion

As we enter into an age of experimental positioning of work and life balance, many organizations are adapting and adjusting to the needs, wants and demands of their employees. Leaders in organizations are listening to the concerns of their employees. The crucial question is will employers provide new policies and procedures to recruit and retain employees that will meet their needs according to Maslow? What is unclear is how the work culture paradigm may shift when all the pieces fall into place, and we establish a 'new normal'. Time will answer these questions, but perhaps the only safe prediction during this pandemic is that the employers will be committed to finding solutions to the long-standing and emerging issues that have popped up due to the pandemic. The question is not the demand of the jobs or the desire of the employers to hire employees in their organizations; the real question is if the workers want to find jobs and if the jobs will meet their new needs, wants and demands. As for the corporate sector, this will be a true test of recruitment, retention and quality of work–balance, for all the hospitality and tourism industries facing immense challenges and struggles to maintain employees at this 'turning point' in the history of this global pandemic. Using the framework of Maslow as a lens to restructure work may be a useful tool to meet the future needs of employees.

Discussion Questions

1. Should employees vocalize their needs, wants and demands to their employers? Explain the rationale behind your answer.
2. What makes this situation unique as compared with economic recessions and wars the world has faced in the past? What are some of the issues your country has faced during this time?
3. It is human nature to recognize one's own needs. What are some of your needs that you have identified that you want to fulfil in this lifetime?
4. Have you given a thought to your self-actualization needs and if yes, what are they? If not, give yourself time to think about them and possibly share in your groups.
5. What are some of the problems/issues that you have faced during the COVID-19 pandemic and how do they resemble those of the workers in the industry?

Teaching Notes

The purpose of the case study is to understand the needs, expectations, motivations and demands of the workers who are choosing not to return to the hospitality and tourism industry during and after the COVID-19 pandemic. This will provide an opportunity to reflect and act with a sense of urgency to identify possible creative solutions to the long-standing issues such as poor wages and working conditions. In addition, it will open a conversation on newer challenges like the anxiety of getting viruses (COVID-19 and other viruses) at work and 'new' standards of health and hygiene at the workplace. While leadership in the hospitality industry maintains high standards of service, the back of the house has always been a fine-tuned machine that provides excellence in service to its guests. Understanding the needs of employees including motivation to work, life balance, job satisfaction, safety and sanitation standards, health and wellness, and workplace flexibility have become significant considerations for management. While society is experiencing a lack of employees during and post-pandemic, the hospitality and tourism industry has been significantly affected by the lack of potential employees. This case study highlights the many challenges in maintaining employees and finding new ones during and post-pandemic from the lens of leadership theory. Hence it is recommended that the instructor allows critical thinking discussions in a group of three to four students, where each student will have an opportunity to look at the phenomenon from different standpoints and discuss the issue with their fellow students. Once an initial discussion has been had in smaller groups, the instructor can then bring all the students together to discuss the case at length as a whole class and find out different ideas and strategies that were discussed in the small groups. This will promote group learning, help facilitate discussion and provide opportunities for each student to contribute.

Target audience

Since this topic is timely and constantly evolving, it will garner attention from universal audiences who are looking for answers from practitioners and educators in the field. This case can be used in bachelor's degree courses and postgraduate-level introductory courses in hospitality and tourism, as individual or group work.

Teaching methods and equipment

This is a case study for facilitating a classroom discussion between students of a real-world problem. The main purpose of the study is to shed light on the issue with the intention of generating and discussing practical solutions to the problem. This case avails the possible solutions along with the theory that helps by providing justification. These theories, along with other literature, will help the students to critically evaluate the scenario that is at stake and evolving. Although there is no need for any other equipment, the use of flipcharts, computers and applications that encourage classroom discussion are suggested as handy tools to facilitate a better learning environment.

Teaching instructions

Assuming that the students will contribute to the discussion, it is important to learn the new ideas that Millennial and Generation Z students will provide to solve the problem. Since the students from these generations are changing the paradigm of the world's workspace, it is important and timely to listen to the thoughts of the students who are future leaders, and the instructor can step back having initiated the conversations as a coach and facilitator. This will provide an opportunity for experiential learning where students learn by reflecting on this phenomenon and creatively find solutions to the problems. The instructor should evaluate what is learnt based on classroom discussion and understanding of the concepts and challenges of operating a human resources department. The instructor may choose to instruct the students to carry out additional research on the issue. This could help in writing future strategies that were not discussed initially and be used as a second round of discussion based on the updated information.

Teaching instruction plan

Discussion point	Time required (min)	Teaching strategies
Introduction	10	The instructor begins the case analysis by asking questions about the learners' recent visit to a restaurant or any organization related to hospitality and tourism and their experience. The instructor will randomly ask a few students to volunteer to narrate the positive and negative experiences from their visit. For a voluntary response from students, the instructor asks the students to raise their hands. The instructor writes the points on the board. The instructor asks a few questions to verify the students' understanding of reading materials supplied before the case analysis.
Briefing on long-standing issues and emerging issues with connection to Maslow's hierarchy of needs	30	The instructor introduces the concepts of Maslow's hierarchy of needs and discusses the long-standing and emerging issues related to 'The Great Resignation' with the class. The instructor asks randomly selected students to provide an example for each issue and asks students to specify the example in relation to Maslow's hierarchy of needs.
Case summary	10	The instructor briefly discusses the background information of the phenomenon and discussion questions. After summarizing the case, the instructor creates teams to answer the discussion questions.
Assignment question 1	10	**Should employees vocalize their needs, wants and demands to their employers? Explain the rationale behind your answer.** **Yes** **No** The instructor asks the students to participate in the poll using any online app or a Google Form link. After finishing the poll, the instructor shows the output as a percentage. The instructor now allows the students to discuss the question and identify the pros and cons of dictating needs, wants and demands to the employers. The instructor writes all the teams' responses on the board. Each team must defend their answers.
Assignment question 2	15	**What makes this situation unique as compared with economic recessions and wars the world has faced in the past? What are some of the issues your country has faced during this time?** The instructor allows the students to discuss and identify the economic recessions of the past and their impact on the labour market. After the discussion, the instructor asks them to present the information on the uniqueness of the current scenario. The instructor also asks them to justify the phenomenon with further clarification as and when needed. Students should be able to illustrate particular issues faced by their home country.
Assignment question 3	10	**It is human nature to recognize one's own needs. What are some of your needs that you have identified that you want to fulfil in this lifetime?** The instructor now allows the teams to share their thoughts about their own needs giving reasons. Each team must defend their choices of needs they want to fulfil in this lifetime.
Assignment question 4	10	**Have you given a thought to your self-actualization needs and if yes, what are they? If not, give yourself time to think about them and possibly share in your groups.**

Continued

Teaching instruction plan Continued.

Discussion point	Time required (min)	Teaching strategies
		The instructor now allows the students to take time and give deep thought to their self-actualization needs. This allows students to really think about Maslow's hierarchy of needs and reflect on self-actualization needs which will help them in identifying what ultimately provides fulfilment for them.
Assignment question 5	15	**What are some of the problems/issues that you have faced during the COVID-19 pandemic and how do they resemble those of the workers in the industry?**
		The instructor allows the students to think about the issues they faced during COVID-19 pandemic. These issues could be fear, inconsistent public policies, confusion and financial difficulties. This will help them understand the issues the employees faced during the COVID-19 pandemic.
Closing remarks and discussion	10	The instructor concludes the case analysis session with his/her/their analysis and answers for the discussion questions using Maslow's hierarchy of needs and analysis of the phenomenon of 'The Great Resignation'.

Duration

120 min (including reading, discussion and answering questions)

Analysis

1. Should employees vocalize their needs, wants and demands to their employers? Explain the rationale behind your answer.

Yes, employees should feel confident that their needs are respected and supported by their employer. The instructor should judge the answer to this question based on the quality of the content and the rationale provided. In general, the employees could dictate their needs for their own rights. The rationale provided should match the opinions of the students as per Maslow's hierarchy of needs.

2. What makes this situation unique as compared with economic recessions and wars the world has faced in the past? What are some of the issues your country has faced during this time?

The COVID-19 pandemic is a unique phenomenon that is lived by today's generation. Health has become more important than ever. The virus has transformed the personal and professional lives of the masses around the globe. WFH became a new normal for many who had not previously experienced it. Each country faced many unique challenges, for example in the USA people found it difficult to isolate themselves completely from their loved ones. In many European countries people experienced popular tourist destinations without travellers but many people realized for the first time that these destinations without travellers are more enjoyable and fun. In Asian countries such as India, people experienced a countrywide lockdown for the major part of the pandemic which created a major shift in the work culture. This is possibly one of the first times in several decades that employees had the opportunity to spend more time with their families and work at the same time. This created an opportunity to rethink future work-related decisions. Hence, this phenomenon is unique compared with a recession or a war, where people do not get the opportunity to indulge in knowing their own needs as mentioned in Maslow's hierarchy of needs.

This is a critical thinking question, where students are expected to discuss the points based on how the implications of the economic recession and wars are different from the implications of the COVID-19 pandemic. The major part of the answer should be that the health scare during the pandemic is the driving force behind the phenomenon of workers not coming back to work.

3. It is human nature to recognize one's own needs. What are some of your needs that you have identified that you want to fulfil in this lifetime?

Spending time with family while working for bread and butter is important for many and this pandemic provided an opportunity to rethink those priorities. Apart from that, it gave many a new chance to develop new businesses or entrepreneurial avenues that did not exist in the past or they did not get time to try them.

This is a self-reflection question, pointing out the importance of why one needs to invest time in learning about self needs as per Maslow's hierarchy of needs. No answer is correct or wrong. But the instructor should judge the students' response based on whether they understand the concept of self needs.

4. Have you given a thought to your self-actualization needs and if yes, what are they? If not, give yourself time to think about them and possibly share in your groups.

During the pandemic, WFH opportunities and schedule flexibility options reintroduced people to their inner desires. For example, individuals travelled more, pursued new hobbies, or spent more time with their families, learning to cherish and indulge in these activities.

This is a self-reflection question. In this case, the instructor should read into whether students understand the highest need as per Maslow's theory and their reflection on self-actualization.

5. What are some of the problems/issues that you have faced during the COVID-19 pandemic and how do they resemble the workers in the industry?

This is a critical thinking question where the students are expected to relate to the workers in the hospitality and tourism industry. The answers could be anywhere from anxiety about catching the virus to their preferences of working from home. The instructor should judge the students' response based on if they can relate to the workers in the industry and understand the five modules of Maslow's hierarchy of needs.

References

Bloom, N. (2021) *Hybrid is the Future of Work*. Stanford Institute for Economic Policy Research (SIEPR), Stanford, California. Available at: https://siepr.stanford.edu/publications/policy-brief/hybrid-future-work (accessed 29 November 2021).

Joblist (2021) Q2 2021 United States Job Market Report. Available at: https://www.joblist.com/jobs-reports/q2-2021-united-states-job-market-report (accessed 29 November 2021).

Johns Hopkins University (2022) COVID-19 Dashboard. Available at: https://coronavirus.jhu.edu/map.html (accessed 12 August 2022).

Maslow, A.H. (1943) A theory of human motivation. *Psychological Review* 50(4), 370–396. https://doi.org/10.1037/h0054346

Molla, R., (2021) Service workers are getting paid more than ever. It's not enough. Available at: https://www.vox.com/recode/22748448/service-food-hotel-workers-pay-raise-resignation-jobs-wages-benefits (accessed 12 August 2022).

NPR (2021a) The Great Resignation: Why People are Leaving their Jobs in Growing Numbers. NPR, 22 October. Available at: https://www.npr.org/2021/10/22/1048332481/the-great-resignation-why-people-are-leaving-their-jobs-in-growing-numbers (accessed 29 November 2021).

NPR (2021b) A $500 Sign-on Bonus to Deliver Pizzas? Here's What to Know About Hiring Incentives. Available at: https://www.npr.org/2021/07/06/1012344023/heres-what-you-should-know-about-that-eye-popping-sign-on-bonus (accessed 29 November 2021).

Wellable Labs (2021) 2021 Employee Wellness Industry Trends Report. Available at: https://www.wellable-labs.com/research/employee-wellness-industry-trends-reports/2021 (accessed 29 November 2021).

4 Last Time They Were Here, This Time They Were Not: Environmental Factors behind the Introduction or Discontinuation of Robot Hotels

Bình Nghiêm-Phú[1]* and Hằng Thị Bích Trần[2]
[1]*University of Hyōgo, Japan;* [2]*Thuongmai University, Vietnam*

Abstract
Concept hotels are hotels designed and planned to meet the needs and wants of a segment of guests based on a theme which these individuals are interested in. Due to this specialized theme, and thus the related products and services, a concept hotel's target market may be small and limited. None the less, the construction and operation of this concept hotel may require a considerable amount of financial investment and other resources. Examining the environmental factors affecting the establishment or suspension of an existing concept hotel will provide meaningful implications for the planning and management of other concept hotels, in particular, and concept-based businesses, in general, in the future. This case study illustrates the method used to examine the macro- and micro-environmental factors surrounding a concept hotel to identify the reasons behind the opening and closing of this hotel and its products or services using *Henn na Hotels* in Japan as the example. Future concept-based businesses can find some implications from this particular case.

Subjects: concept-based business, business environment, secondary data analysis

Learning Objectives

At the end of the case study, a learner will be able to:

1. Get acquainted and reacquainted with the micro- and macro-business environment factors.
2. Use these factors as the framework to analyse the reasons behind the introduction and discontinuation of certain concept-based businesses.
3. Practise information collection and analysis methods, focusing on secondary sources (e.g. newspapers, websites and user-generated content).

Case

Hotels are business units providing guests with (mostly) short-term lodging and other products and services, such as dining, drinking, sightseeing, transportation, massage and spa treatment. The earliest lodging facilities may have appeared thousands of

*binhnghiem@gmail.com

DOI: 10.1079/9781800621022.0004

years ago in the era of the Persian Empire (6th century BC). However, the oldest hotel (a hot spring inn) was established much later, 705 AD, in Japan (*Keiunkan*, Yamanashi Prefecture). Today, many types of hotels are operating worldwide, including concept hotels.

Concept hotels are hotels designed and planned to meet the needs and wants of a segment of guests based on a theme which these individuals are interested in. Due to this specialized theme, and thus the related products and services, a concept hotel's target market may be small and limited. None the less, the construction and operation of this concept hotel may require a considerable amount of financial investment and other resources. Examining the environmental factors affecting the establishment or suspension of an existing concept hotel will provide meaningful implications for the planning and management of other concept hotels in the future.

As mentioned earlier, Japan is home to the oldest hotel, *Keiunkan*. Interestingly, this hotel itself is a concept hotel (hot spring *ryokan* or traditional Japanese inn). Nowadays, Japan is also home to many other modern concept hotels, such as samurai hotels, animation hotels, robot hotels, capsule hotels and ice hotels. In particular, the first robot hotel, *Henn na Hotel Huis Ten Bosch* (Nagasaki), was opened in July 2015 with less than 100 robots of six types. Today (as of March 2022), *Henn na Hotel* has evolved into a hotel chain with 21 members in Japan, Korea and the USA. The number of robots increased to approximately 300, belonging to about 30 types.

The robot hotel concept

The *Henn* element in *Henn na Hotel* (変なホテル) means 'strange' or 'change' in Japanese. Its brand concept certainly is 'a hotel that promises to change'. The mission statement consistently reiterates this concept:

> Henn na Hotel supports a comfortable hotel stay with various advanced technologies. Our hotel is not bound by one type and promises to continue evolving according to the times and needs.
>
> (*Henn na Hotels*' websites)[1]

Henn na Hotels' founder, Chief Executive Officer (CEO) Hideo Sawada, began preparing for their inaugurations in 2013. He aimed to create the most efficient hotels globally by inventing a new concept that could compare with the low-cost carriers (LCCs) that had utterly changed the world's airline industry. The cost has been kept to a minimum by reducing energy use (30% compared with traditional hotels), providing only essential room products and services (vending machines sold the extras), and reducing room cleaning (this service was not available with less-than-a-week stays), among others.

The employment of robots (Table 4.1) was another cost-saving strategy. For example, *Henn na Hotel Osaka Shinsaibashi* (Fig. 4.1a) had two velociraptor robots and a troupe of dancing humanoid robots to greet incoming guests (Fig. 4.1b, c). In addition, the check-in and check-out procedures could be done automatically via machines (Fig. 4.1d). Certain guest rooms even had exclusive cleaning machines. In addition, artificial intelligence (AI) chatbots supported room reservations on the hotels' websites.

The use of robots has been the trend in the world hospitality industry in recent years. This trend can also help with the depopulation and labour shortage issues Japan has faced for many years.

With the help of experts in architecture, construction, engineering, environmental design and robotics, Mr Sawada launched the first *Henn na Hotel* in the largest theme park in Japan, Huis Ten Bosch (Nagasaki), in July 2015. Since then, other hotels have been continually opened or projected in other big cities in Japan and abroad, such as Sendai, Tokyo, Nagoya, Osaka, Kyoto, Nara, Fukuoka, Seoul and New York.

Generally, the hotels were designed not only for the convenience of human guests, but also for the efficiency of robots. For example, *Henn na Hotels* used face-detection locks instead of room keys. In addition, standardized walkways and steps ensured that porter robots could function efficiently.

The layoff of some robots

Unfortunately, only 3½ years after the first *Henn na Hotel* was opened (July 2015), in

Table 4.1. Examples of robots. From *Henn na Hotels'* websites.[2]

Area	Type
Reception	Humanoid robots
	Dinosaur robots
	RoBoHoN
	Light holograms
Guest room	Robot cleaning
	Futocon
	Chomecast
Others	Porter robots

January 2019, the chain informed the public that it had stopped using half of its robots and employed more human staff. Two significant incidents might have led to such a grim outcome.

Incident 1: at the reception

A couple of foreign guests arrived at the *Henn na Hotel*. Signs instructed them to use a nearby machine to scan their passports. Unfortunately, the language on display was Japanese, and the guests could not have their documents reviewed after several unsuccessful attempts. Therefore, they decided to turn to the two robots present at the reception for help. They asked the robots whether they could check into their rooms or not. The answers they got from these two were some blinks and greetings of *konnichiwa* or hello in Japanese. Of course, their problem was unaddressed. They had to call the human staff to help with the check-in procedure as a last resort.

However, their problem might not have ended here since on occasion it was found that the porter robots were unable to carry luggage to rooms that were outside the robots' predetermined courses. In a related incident, the robots in charge did not adequately clean guest-room windows since the robot could not work under bad weather conditions such as heavy rains.

Incident 2: in the guest room

The in-room robot *Churi*[3] was an assistant to the hotel guests. Unfortunately, *Churi* regularly misinterpreted the requests and underperformed their tasks. For example, a guest who snored, probably quite loudly, was repeatedly awakened by the robot since they assumed that the guest would like to ask them some questions. However, when requested (e.g. adjustment of room temperature and information about nearby attractions), *Churi* could not provide satisfactory answers due to their lack of intelligence compared with other AI assistants such as Alexa and Siri.

In addition to *Churi*, another robot, *Tapia*,[4] was employed to entertain guests. However, external security engineers publicly claimed that they could use a *Tapia* as a remote camera to survey the guest rooms. The chain responded late to this warning, and experts criticized them for putting their guests at risk.

The business environment factors

Generally, the business environment includes two large entities – the micro environment and the macro environment (Kotler and Armstrong, 2021). The micro environment incorporates six main stakeholders, including: (i) the hotel itself (e.g. its managers and staff); (ii) the guests; (iii) the suppliers (e.g. the businesses that sell goods and services to the hotel); (iv) the competitors (e.g. other hotels that serve similar products and services); (v) the intermediaries (e.g. booking sites and travel agents); and (vi) the public at large. On the other hand, the macro environment comprises the hotel's external environment factors, including the demographic, sociocultural, technological, economic, political and natural conditions of the hotel's geographical location. Among these factors, only the managers and staff of a hotel are part of the internal environment, all the other factors (both in the micro and macro environment) are external. The hotel can actively control the former yet may not take care of the latter effectively.

With a concept hotel, the introduction or discontinuation of its products and services does not only involve its guests, employees, managers and suppliers but it also affects or is affected by other stakeholders.

Fig. 4.1. Robots and machines at *Henn na Hotel Osaka Shinsaibashi*. (a) The front of the hotel. Inside guests were greeted by two velociraptor robots (b) and dancing humanoid robots (c) and automatic machines were provided for check-in and check-out procedures (d). Photos courtesy of the authors.

Therefore, a thorough examination of the business environment factors (micro-macro and internal-external) to identify the roles of the stakeholders is a necessary practice. This case study adopts the theoretical framework mentioned by Kotler and Armstrong (2021) to carry out this task.

Information about business environment factors is abundant and available in various sources, such as news articles, blogs and YouTube videos and comments. This information can be analysed using a deductive content analysis approach (Krippendorff, 2019). The business environment factors themselves serve as the scheme to categorize and assess the data.

Discussion Questions

1. What are the reasons behind the initial introduction of the robot hotels?
2. What are the reasons behind the temporary suspension of certain robots in these hotels?
3. How can we identify the reasons given the available secondary and primary sources of information?

Teaching Notes

The intention of this case study is to help learners get acquainted and reacquainted with the micro- and macro-business environment factors mentioned by Kotler and Armstrong (2021). They can alternatively categorize these factors into internal and external groups to better understand the impacts of each element. The learners will later use these factors to analyse the reasons behind the introduction and discontinuation of certain concept-based businesses, such as that of a robot hotel.

The learners will get familiar with and practise collecting secondary information available in newspapers, websites and social media platforms. Popular search engines, such as Google, can be used in the information searching process. The main keyword is the hotel's name (*Henn na Hotel* or 変なホテル). The supporting ones are the business environment factors.

After that, the learners will learn to analyse this information using the business environment factors as the analysis scheme. They will follow a deductive process, which is detailed by Krippendorff (2019), by categorizing the information into their respective elements (macro-environment factors: demography, socioculture, technology, economy, politics, nature; micro-environment factors: the company, customers, suppliers, competitors, intermediaries, the public). From there, the learners will identify the favourable conditions that lead to the establishment of a concept hotel (discussion question 1) and the unfavourable conditions that cause the discontinuation (discussion question 2). They should adopt a chronological approach (before establishment and between establishment and discontinuation) during the analysis process (discussion question 3).

This case study is suitable for a group of five to seven students. Each student may be in charge of one or two environmental factors. However, all of them need to work together by the end of the class to summarize and report the findings.

Target audience

Undergraduate and graduate students majoring in business management and marketing studies can use this case study. It is also relevant for entrepreneurs who want to start their concept-based businesses.

Teaching methods and equipment

Each student needs a personal computer and a stable Internet connection. They will search for the relevant information on their computers and can save the information in a Microsoft Word document or the like.

The instructor explains to the students the purpose of the case study and the information searching method. After that, they instruct the students to analyse the data and prepare the final report or present the key findings.

Teaching instructions

The instructor needs to spend a proper amount of time getting familiar with the case study in advance, particularly with the information sources. They may also need to prepare or recommend alternative information sources, such as interviews with experts or customers, to compensate for the lack of information or diversify the information inputs.

The case study requires three 60-min sessions. The details are shown in the teaching instruction plan.

Teaching instruction plan

Discussion point	Time required (min)	Teaching strategies
Introduction	10	The instructor begins the case study by introducing the concept hotel business. They may show students the hotels' websites and other related videos available on YouTube.
Briefing on the business environment factors	10	The instructor reminds students about the macro- and micro-business environments. They may ask students to join them in identifying and discussing the related elements.
Case summary	10	The instructor mentions the background information, the issues of the robot hotels, and the tasks. They explain to the students how to collect information about the business environment of the robot hotels.
Identification and collection of secondary information	30	The instructor helps the students to choose the search engine and the keywords. They also instruct students to identify the relevant information and store it for later use.
Analysis of secondary information	40	The instructor explains the deductive analysis method using the ten environment factors as the coding scheme. They show the students how to categorize the information into the relevant elements and the time frames (introduction versus discontinuation).

Continued

Teaching instruction plan Continued.

Discussion point	Time required (min)	Teaching strategies
Additional information collection and analysis (implemented when necessary)	20	The instructor asks the students to search for and analyse additional information to strengthen the observations. If this task is unnecessary, the instructor may lengthen the information analysis session or the findings summary session.
Summary of key findings	20	The instructor asks the students to summarize the key findings to answer two assignment questions: 1. the reasons behind the initial introduction of the robot hotels, and 2. the reasons behind the temporary suspension of certain robots in these hotels.
Presentation of key findings	30	The instructor asks the students to present their findings.
Conclusion	10	The instructor summarizes the sessions, the key findings, the implications with future concept business management, and the importance of data analysis.

Duration

180 min

Analysis

The analysis of the information related to the case study aims to identify the reasons behind: (i) the initial introduction of the robot hotels; and (ii) the temporary suspension of certain robots in these hotels. The summary of the observations is in Table 4.2.

The reasons behind the introduction of robot hotels

THE MICRO-ENVIRONMENT FACTORS

The company. *Henn na Hotels* were founded by their CEO, Hideo Sawada, as he wanted to create a new hotel model that mainly relied on robotic technologies. By reducing the human staff to a minimum (less than 10%) while raising the proportion of robot staff to a maximum (more than 90%), Mr Sawada aimed to improve the efficiency of the hotel business, in terms of both human resource management and cost management.

Specifically, robots were employed in many positions, from reception and luggage carriers to lawn mowing and window cleaning. The robots could help increase productivity, in a sense, and create a fun and exciting atmosphere, in another. Human staff only worked in areas where robots could not perform, such as bed preparation, bathroom cleaning and system maintenance. The initial cost of the robots might be high but it was less than that of the human staff. Thus, there were high expectations of cost-competitiveness.

Mr Hideo Sawada, CEO, quoted by BBC News (17 July 2015)[5] explained: 'I wanted to highlight innovation. I also wanted to do something about hotel prices going up.'

Mr Yasuhiko Hoshi, an associate director, quoted in an article by wundermanthompson.com (26 February 2018)[6], said: 'We think that the number of people who can work for hotels will be less and less because of the population decline. We reached the idea that hotel employees can be replaced by robots.'

Mr Tatsuya Fukuda, a director, was quoted by traveller.com.au (21 March 2018)[7] saying: 'We look for high productivity at our company, and the robots are good for that. We are always looking at ways of introducing the latest technology ... it is very exciting.'

In addition to the robots, the guest rooms had next-to-minimum facilities, including a bed, lighting devices, and electric and automatic devices. Thanks to special

Table 4.2. Factors affecting the operations of robot hotels.[a] Authors' own table.

Factors	Introduction (2015)	Discontinuation (2019)
Micro environment		
The company	Vision and mission: comfort and efficiency	The meaning did not change but the approach did
Customers	Preferences for more novelty and less human contact	Other market segments with different preferences also existed
Suppliers	Robot makers and AI programmers	Slow advancement of robots and limitation of current AI technologies
Competitors	Other hotels using robots	Hackers
Intermediaries	Mainly unrelated	Mainly unrelated
Public	Mainly unrelated	Mainly unrelated
Macro environment		
Demography	Ageing population and labour shortage	Unchanged
Socioculture	Preferences for 'strange' things and objects	Unchanged
Technology	Advanced automatic technologies	Unchanged
Economy	Advanced economic development	Unchanged
Politics	Governmental support and encouragement for advanced technological development	Unchanged
Nature	Usefulness of advanced technologies for disaster management	Unchanged

[a]AI, artificial intelligence.

bricks, angled roofs, reflective paint, radiation panels and body-temperature sensors, the hotels did not need air-conditioner units. Therefore, the chain could reduce the cost of electricity and other frills (e.g. approximately 30% of traditional hotels' energy costs). The robots then became attractions in and of themselves, especially with the three-star facilities at *Henn na Hotels*.

Mr Tatsuki Miura, a manager, stated:

> We initially adopted robots with the aim of holding labor costs down. But then, the robots became an attraction in their own right, adding to the entertainment and drawing popular attention. The Henn na Hotels are rated at three to three-and-a-half stars – not as high as might be hoped – so performances by these robots can add real value to customers' stays.
>
> (cited in nippon.com, 23 October 2018)[8]

One byproduct of these strategies was a reasonable price range for room services at the *Henn na Hotels* (starting from as low as US$50) with their city-centre locations necessarily being taken into account. The concept of the *Henn na Hotels*, thus, harmoniously fitted with and supported the overall operations of its mother company, H.I.S., a Japanese travel agency specializing in low-cost package tours. Concept (novel) and price (low) seemed to be the effective positioning strategies that *Henn na Hotels* adopted.

The guests (customers). The first *Henn na Hotel*, as mentioned earlier, was opened in a theme park, Huis Ten Bosch. Guests to this theme park (approximately 2.8 million in 2015), undoubtedly, were the first potential market segment of this hotel. Since Huis Ten Bosch recreated a European town in Japan, an Asian country, it provided a sense of novelty from what has typically been present here. A robot hotel in such a setting was an additional element to this difference.

Mr Tatsuki Miura stated:

> Staying at an impersonal hotel either before going to, or coming back from, an amusement park puts a damper on what is supposed to be a good time. So, we put robot dinosaurs in the facilities and robots based on fossil fish to swim in a fish tank. While some kids are scared and cry at first, soon they are smiling and having their pictures taken with these robots.
>
> (cited by nippon.com, 23 October 2018 – see note[8] again)

Moreover, business guests, many of whom wanted to avoid extra human contact, were also the first to be targeted. The hotels hoped to scale down the pressure and hassle of guest management by limiting direct guest–staff communications.

Mr Allen Lee, a manager, quoted by natgeotravel.com (10 October 2017)[9] said: 'One great advantage of this hotel is that customers cannot complain to robots.'

Mr Tatsuki Miura (nippon.com, 23 October 2018):

> It takes a surprising amount of energy to talk to people, even hotel staff. We figure this is going to be just that much harder for a businessman to take after wearing himself out in negotiations or other business meetings. And we thought, how can we leverage the characteristics of robots for business travelers as well as tourists? Now, at Henn na Hotel, you can go your whole stay, from check-in to check-out, without ever talking to another person.
>
> (cited by nippon.com, 23 October 2018 – see note [8] again)

Nickname Donikantoku, a blogger, wrote on unborderedlife.com (29 December 2019)[10]: 'A traveler really can do everything they need at this hotel without speaking to another human being.'

The suppliers or partners, the competitors and others. The *Henn na Hotels* could not open without support from robot makers, program developers, architects and designers in Japan. Examples of such notable figures and entities include architecture professors at the University of Tokyo, engineers at Yukai Engineering Inc., and developers at Qbit Robotics, among many others.

The *Henn na Hotels* were also the only one of its type in the hotel market. The noncompetitive situation has provided *Henn na Hotels* with many favourable conditions to spread its operations in Japan and other countries. However, overlooking the 'robot concept' element, competitors in the same price category (three-star hotels), especially traditional family and business hotels, could not be ignored (Fig. 4.2).

The role of the intermediaries and public factors of the micro environment in the introduction of these robot hotels was somewhat insignificant. However, these stakeholders might also support the introduction of robot hotels since the robot hotels could provide them with new products and services to diversify their offerings (e.g. for tour operators). They could also demonstrate the pride of Japanese people in their technological advancement and uniqueness (e.g. for the public).

THE MACRO-ENVIRONMENT FACTORS. All the macro-environment factors seemed to support the establishment of robot hotels in Japan. First, the ageing population and, thus, the shortage of labourers has facilitated the development of robot-based services. In 2015, Japan's population was approximately 127.1 million. This number, unfortunately, has continually decreased since then. Less than 50% were working, with people aged 65 and above accounting for approximately 12.5% of the total workforce. Machines and robots, in addition to foreign workers, thus, have been increasingly employed to deal with the labour shortage issue.

Secondly, the local culture, which has shown great interest in strange and weird things, also helped provide a promising foundation for robot hotels. Strange fashions (e.g. animated outfits), strange ice-creams (e.g. pepper flavour), and strange toilets (e.g. fully automated and fashionably designed units), for example, started in Japan before going to other countries. In addition, the pursuit and respect of personal privacy and solitude of Japanese people could further support the establishment and development of robot hotels.

Thirdly, the advancement of technology in Japan has doubtlessly boosted the development of hotels using robots. For example, *Asimo*, one of the first humanoid robots, was created by Honda, a Japanese company. *Lovot*, a pet robot of Groove X Inc., was also very popular in this country.

In addition, the fourth factor of natural conditions has also added extra force to the development of robots and the robot-based business sector. For example, robots and

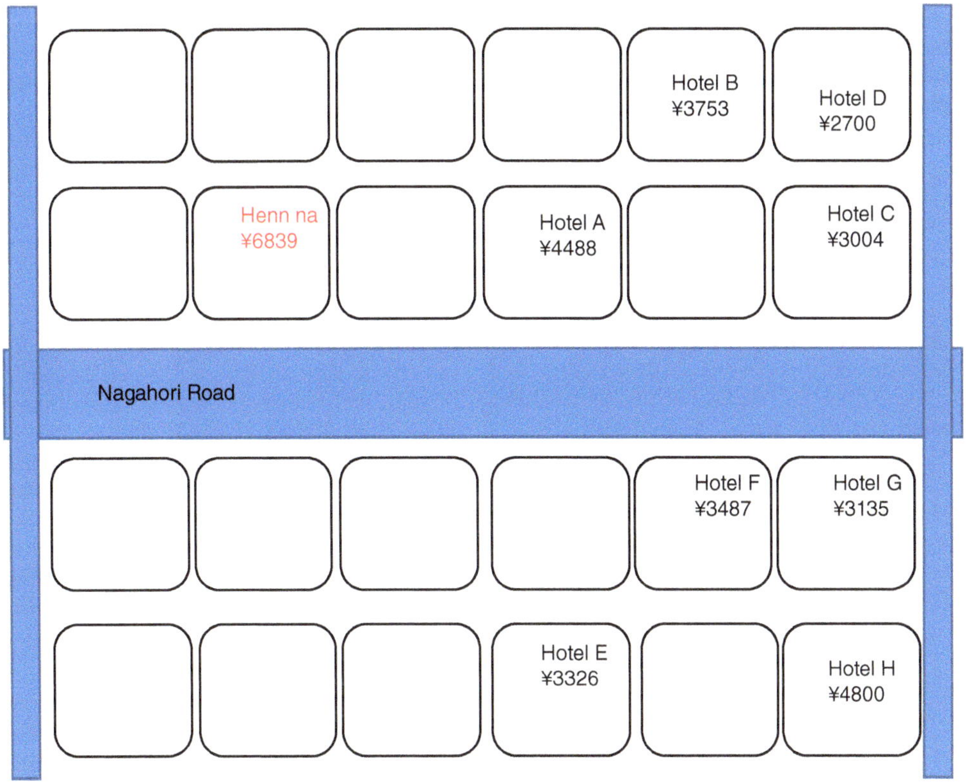

Fig. 4.2. *Henn na Hotel Osaka Shinsaibashi* and other hotels nearby. (Information source: Google Maps, accessed on 1 November 2021; the prices are displayed in Yen, 1 Yen = approximately US$0.009.) Authors' own figure.

robotic technologies have been effective in disaster-hit areas, such as the Fukushima nuclear power plants.

Finally, the fifth and sixth factors, the power of the third biggest economy in the world (US$4.389 trillion in 2015) and the government's commitment have certainly provided other incentives for the inauguration of robot hotels. Since the population was decreasing and the economy was struggling, the Japanese Government has considered robotics one of the pillars of its economic growth strategies.

The reasons behind the discontinuation of certain services at robot hotels

Only 3½ years after the first *Henn na Hotel* was opened (July 2015), in January 2019, the chain informed the public that it had stopped using half of its robots and employed more human staff. The original robot hotel concept, thus, had failed significantly.

THE CHANGES IN THE MICRO ENVIRONMENT. While the macro-environment factors had mainly remained stable, other factors in the micro environment unexpectedly but unsurprisingly emerged, which were unobserved at the time of introduction (July 2015). The guests, for example, did not only include those who wanted to avoid human contact or who had great enthusiasm and patience with robots. Many of the robots were not advanced and intelligent enough. As a result, the guests could not enjoy the comfort and efficiency that the hotels had promised to deliver.

Mr Taishi Mito, a guest, quoted by cnet.com (21 January 2019)[11] stated: 'They [the robots] were really slow and noisy, and would get stuck trying to go past each other.'

Mr Yukio Nagai, a manager, as quoted in the UK's *Daily Mail* (15 January 2019),[12] said: 'We have not quite figured out when exactly the guests want to be served by people, and when it is okay to be served by robots.'

In addition, the hotels had not adequately undertaken the maintenance and enhancement of the robots and the operating programs. Perhaps, since the hotels had expected a long time of service for these robots due to the initial high costs, they might have neglected further investments for proper upgrading. Or perhaps, since the makers and programmers have reached their limits, further advancement could not be expected. Moreover, the hotels had underestimated the security threats about electronic and Internet-based machines and robots. Risks for hotel guests, including personal data and sensitive images, were apparent.

Mr James Hacon, a marketing expert, talked to *Design & Build Review* and said:

> 'Unfortunately, some hoteliers are turning to technology to give them a differentiating factor, creating a PR story ... You only need to see videos coming out of Japan – where human-mimicking robots guide guests to rooms or deliver room service – to see how slow and clunky they are. I would suggest they are actually detracting from the customer experience.'
>
> (cited by designbuild.nridigital.com in March 2019)[13]

To deal with these issues, *Henn na Hotels* have discontinued use of half of their robots since January 2019, starting with the in-room robots *Churi*. They have also employed more human receptionists, in addition to human porters. The role of the robots became that of entertainers rather than that of actual staff. The concept of robot hotels, thus, has been significantly altered.

Conclusion

The *Henn na Hotels* were opened after long and careful preparation, with lots of internal and external support. Unfortunately, the original concept did not last long, only 3½ years. Although the mission of creating comfort for guests and efficiency for those running hotels remained unchanged, they significantly reduced the participation of robots in this process.

The owners and managers of the *Henn na Hotels* seemed to overestimate the prowess of the robots, especially their potential to fully and effectively replace human staff. They also miscalculated and misinterpreted their guests' preferences and characteristics. The misunderstanding led to the definition of an impractically high goal. The reality, however, reset the goal in a sense and broke the original concept in another. The failure of the robot hotel concept, thus, seemed to be the result of internal mismanagement rather than that of external forces.

The *Henn na Hotel* case again illustrates the vital lesson of understanding oneself and one's customers. A trial of the new business model should also be carried out before expansion, primarily when that might involve a significant change from a niche operation (concept development) to a mass one (concept promotion). These strategies would help prevent extra losses in terms of financial investment caused by the disuse of the irrelevant robots. They would also help reduce the damage done to the concept and the image and reputation of the hotels and their mother company. Other concept-based businesses can refer to the *Henn na Hotel* case to project their operations and avoid similar potential failures in the future.

Notes

1. Available at: https://www.hennnahotel.com/maihama/concept/ (accessed 25 September 2021).
2. Available at: https://www.hennnahotel.com/ (accessed 25 September 2021).
3. An image of *Churi* can be found at: https://www.hisgo.com/us/destination-japan/nagasaki/henn-na_hotel.html (accessed 25 September 2021).
4. An image of *Tapia* can be found at: https://mjirobotics.co.jp/en/ (accessed 25 September 2021).
5. BBC News (2015) Five things about Japan's robot hotel. 17 July. Available at: https://www.bbc.com/news/world-asia-33562368 (accessed 23 September 2021).

[6] Yee, C.M. (2018) Japan's robot hotels. Insight, Wunderman Thompson, 26 February. Available at: https://www.wundermanthompson.com/insight/japans-robot-hotels (accessed 23 September 2021).
[7] Miller, J. (2018) Tokyo's robot hotel, Japan. What it's like to stay in Henn-na Hotel in Ginza. *Traveller*, 21 March. Available at: https://www.traveller.com.au/tokyos-robot-hotel-japan-whats-it-like-to-stay-in-henn-na-hotel-in-ginza-h0xq2m (accessed 23 September 2021).
[8] Fukiko, M. (2018) World's First Robot-staffed Hotels Make Business Travel Inroads. *Japan in Video*, 23 October [originally written in Japanese]. Available at: https://www.nippon.com/en/guide-to-japan/gu900045/ (accessed 23 September 2021).
[9] Sunder, K. (2017) Check into the Future at the World's First Robot Hotel in Japan. *Traveller*, 10 October. Available at: https://natgeotraveller.in/check-into-the-future-at-the-worlds-first-robot-hotel-in-japan/ (accessed 23 September 2021).
[10] Donikantoku (2019) Henn Na Hotel: a Stay at the World's First Robot Hotel. *Unbordered Life*, 29 December. Available at: https://unborderedlife.com/henn-na-hotel-a-stay-at-the-worlds-first-robot-hotel/ (accessed 25 September 2021).
[11] Burton, B. (2019) Hotel fires half its robot staff for sucking at their jobs. *CNET*, 21 January. Available at: https://www.cnet.com/news/hotel-in-japan-fires-half-its-robot-staff-for-being-bad-at-their-jobs/ (accessed 25 September 2021).
[12] Prigg, M. (2019) Japanese hotel staffed by 243 robots fires more than hald the bots. *Daily Mail*, 15 January. Available at: https://www.dailymail.co.uk/sciencetech/article-6594853/Japanese-hotel-staffed-ROBOTS-forced-fire-half-mechanical-workers-complaints.html (accessed 25 September 2021).
[13] Davies, R. (2019) Rage Against the Machine: Lessons from the World's First Robot Hotel. *DBR Hotels*, March. Available at: https://designbuild.nridigital.com/dbr_hotels_mar19/rage_against_the_machine_lessons_from_the_worlds_first_robot_hotel (accessed 25 September 2021).

References

Kotler, P. and Armstrong, G. (2021) *Principles of Marketing*. Pearson, Harlow, UK.
Krippendorff, K. (2019) *Content Analysis: an Introduction to its Methodology*. Sage, Los Angeles, California.

5 The Tourism Pursuit App – Can Tourism Destinations be Gamified?

Ranjana Tiwari*
Jain (Deemed to be University), Bangalore

Abstract
Tourism officials from the town of South Dundas in Ontario, Canada have launched a new app that utilizes gamification and interactive elements to promote tourism for both tourists and residents. This brand-new app, called the Tourism Pursuit, is a fresh way to learn about and locate South Dundas' areas of interest. Interesting to note is the app was developed with the aid of students from a local college. The purpose of the case study is to highlight the use of gamification, a trending medium, as an interactive tool to boost tourism in a region in the wake of the coronavirus disease 2019 (COVID-19) pandemic.

Subjects: information and communications technology, marketing, destination management, contemporary issues and trends

Learning Objectives

At the end of the case study, a learner will be able to:

1. Critically analyse the effectiveness of the Tourism Pursuit gaming app.
2. Explain whether gamification can be an answer to the revival of tourism destinations after the coronavirus disease 2019 (COVID-19) pandemic.
3. Analyse the trends in using digital technologies for travel destination promotion.

Case

'There's no doubt about it that something like this is unique. It's good for South Dundas.'
(Steven Byvelds, Mayor, South Dundas, 2021 cited in Haworth, 2021)

'This initiative will provide key areas of support ... Being able to launch this app just as the province begins its stages of recovery is fitting.'
(Jim McDonell, the Member of Provincial Parliament, Stormont-Dundas-South Glengarry, 2021 cited in Haworth, 2021)

*ranjanatiw88@gmail.com

 Teaching Cases in Tourism, Hospitality and Events (eds. S.K. Dixit and S. Piramanayagam)
DOI: 10.1079/9781800621022.0005

On 23 June 2021, Wednesday morning was a regular morning for the people at South Dundas Province, Canada, until Norene Hyatt-Gervais, South Dundas Tourism Committee Chair announced, 'Let the games begin!' (Comfort, 2021a). It was the time when the South Dundas Tourism Committee, in partnership with St. Lawrence College and the Municipality of South Dundas, launched the South Dundas Tourism Pursuit app to promote South Dundas' various tourist attractions and events. During an event at Stone Crop Acres Winery and Vineyard, South Dundas Deputy Mayor Kirsten Gardner introduced the municipality's new tourism promotion tool, which was developed in collaboration with the college. She said: 'Who would have thought that a brainstorming session by our tourism committee in September of last year, during a pandemic nonetheless, would have led to us being here today' (cited in Morin, 2021). During the launch, Mayor Steven Byvelds praised the partnership between the college and the municipality for their collaboration on a new tourism tool. He further added that he hoped that the app will catch on and that it would help boost tourism in the area after the lockdowns due to the COVID-19 pandemic.

The moment was sure to be a proud one for Glenn Vollebregt, President and Chief Executive Officer (CEO) of St. Lawrence College: 'This app demonstrates the power of collaboration and the power of connection ...' (cited in Morin, 2021). Vollebregt said that he hoped the event would help create more connections between the college and its communities.

The Tourism Pursuit app is the brainchild of James Dupuis, Professor/Coordinator of Game Programming at St. Lawrence College, who along with his nine students from the college built the app in 1½ years. During the launch of the app, he demonstrated its features including badges collected for visiting locations across South Dundas, mini-games and rewards.

Background

Tourism is a vital part of the South Dundas economy. It accounts for over 20% of the municipality's workforce and supports over 200 jobs. There are more than enough jobs in South Dundas to keep everyone happy. From the shops and restaurants to the campsites and the water parks, there are plenty of reasons to visit South Dundas. The concept of South Dundas Tourism Pursuit was conceptualized in 2020 during a brainstorming session of the South Dundas Tourism Advisory Committee which was held in response to the impact of the pandemic on the tourism industry.

While recognizing the impact of the pandemic on the tourism industry, the session focused on being innovative and doing something that hadn't been done in the past in the United Counties of Stormont, Dundas and Glengarry. One of the things that stood out from the rest was the beautiful location of the community along the St. Lawrence Seaway. The history and culture of Canada's original highway and its original tourist route are very deep connections to the region's agricultural heritage. Moreover, the committee pondered, no longer are tourists required to use tourist booths to find information, they are now using their mobile phones to find it. This is how the county's tourism committee came up with the idea of creating an app that will allow visitors to find information about various attractions in the area.

According to Kirsten Gardner, the committee's goal was to create something fun and innovative for all ages:

> We didn't want to play to just one type of visitor. We wanted it to be for our residents, for people across SDG who are familiar with South Dundas, but maybe haven't visited for a while, and for any folks who have never been to South Dundas ...
>
> (Kirsten Gardner cited in Comfort, 2021b)

The briefing of the tourism app was then given to Rob Hunter, South Dundas Economic Development Officer, who thought it can be very costly to develop an app. So, they decided to approach St. Lawrence College:

> so, I thought, you know, why don't we approach St. Lawrence College and see if the gaming program and the students could actually take this real-world project, so they will actually gain work skills that will help them when they go to seek a job after graduation ...
>
> (Rob Hunter, cited in Vandermeer, 2021)

After consultation, James Dupuis, Professor of Game Programming, St. Lawrence College along with his nine students, Daniel Repko, Pedro Civa Wiltusching, William Lafleur, Brigitte Demers, Aidan McDonald, Rory Lacayo, Nick Voss, Johan Carette and Serhii Marchenko, took charge of developing the app. For Dupuis, being involved with developing the app was a great opportunity for his students to get hands-on experience with real-world issues. For their last semester, these nine students were allowed to work on a project for a client. According to instructor Dupuis, this gave them a unique opportunity to improve their skills and résumé. The team took almost 14 weeks of classes to develop the app.

Gamification in tourism

Gamification is the process of adding game mechanics to non-game environments, such as websites, social media platforms and intranets. It is a strategy that aims to enhance the systems, services and activities of an organization or individual to create an engaging and motivating experience for users. Using games to engage people is not new in tourism. Before the rise of handheld devices, many establishments that offer attractions had already incorporated gaming into their offerings. Gamification in tourism is a concept that aims to add elements of real life to the gaming experience of mobile platforms. The goal is to make history more accessible to today's digitally immersed travellers by combining the physical and virtual worlds.

Some examples of gamification in tourism

Questo is a start-up that offers a travel-themed game platform that enables players to experience the cities while they are playing a game. The users discover new places and meet new people with this game. Each level comes with various challenges and clues to guide users through the game. Users can also customize their quests and locations. Blending fiction and reality, Escape From the Past lets users play as the main character of the story while solving challenging tasks related to the immediate environment. In CollezionAlghero, players collect digital and physical stickers to help solve games. They can also check-in to specific places using their mobile app. With special augmented reality (AR) tech, players can challenge themselves to win or lose stickers based on historical characters and locations.

How does the Tourism Pursuit app work?

In the next 14 weeks from the decision, the team of Professor Dupuis and his students came up with the app Tourism Pursuit, which showcased 19 locations and one hidden attraction that was not listed in the app, in South Dundas.

> It gave them kind of that real-world experience, having these commitments that they had to kind of have milestones hit, and then it kind of also gave them that creative freedom to make these mini games on the side as well. So, it was a kind of perfect mix of both worlds ...
>
> (James Dupuis cited in Vandermeer, 2021)

To acquire all of the badges, visitors need to visit all 19 participating South Dundas attractions. The signboards with QR codes will be found at all the places. The visitors simply need to launch the South Dundas Tourism Pursuit app and tap the QR scanner in the lower right corner. Once the QR code is scanned, the visitors can unlock their badge and get access to the history of the attraction. Also, some of the badges unlock mini-games that will keep visitors entertained for hours. When all the badges are scanned, that will lead to the completion of the game, participants can collect all the prizes and share the badges with their friends and family.

'You collect your badges and it's something that all ages can enjoy in the car on a rainy day on vacation. It just brings everybody together,' said Norene Hyatt-Gervais, Chair of the Tourism Advisory Committee for South Dundas on the launch day of the app (Vandermeer, 2021).

The locations shown in Box 5.1 are included in the app.

Referring to the tourist attractions included in the app, Hyatt-Gervais noted, 'We

Box 5.1. Tourist attractions included in the app. Author's own data.

Morrisburg Beach	Stone Crop Acres Winery and Vineyard	Smyth's Orchard	Upper Canada Village	Two Creeks Conservation Area
Iroquois Beach	Upper Canada Playhouse	Upper Canada Creamery	Dr. Locke Mural	Iroquois Plaza
Morrisburg Plaza	Holy Trinity Anglican Church	The New Canadians Mural at H & I Supply	Morrisburg Waterfront	Iroquois Seaway Lock
McIntosh Apple Plaque	Dentz Orchards and Berry Farm	Loyalist Park	Robert Graham Conservation Area	

kind of like to call them hidden gems or nooks and crannies in our township that even our own residents sometimes are not aware of ...' (cited in Vandermeer, 2021). Indeed, Dupuis said it was the students who looked for the attractions to be included in the app, and although he lived in a nearby area and had been to a few of these places, he was not aware that there are many hidden gems out there.

The tourism officials of the region mentioned that they were also doing their best to make sure that the residents are educated about everything that is happening in their community.

The tourism officials were impressed by the app, which they claimed was the first gaming application that's specifically geared towards tourism in Ontario. They were able to meet several people who were playing the game and they thought it was phenomenal.

Results and reactions

Hyatt-Gervais said the app's success had been great so far, and she hoped that more people would download it now that COVID-19 restrictions are dropping. She said: 'I think this is going to be an amazing thing for our region. It's innovative ...' (cited in Vandermeer, 2021).

South Dundas' Mayor Stephen Byvelds said that tourism is extremely important to the area's economy. He noted that the council and various tourism businesses work together to make sure that visitors get the most out of their stay in the region. He also said the app would help with tourism in South Dundas and that the committee had worked extremely hard on the project.

The residents and visitors welcomed the app enthusiastically. As one person said: 'It seems like a great app for families to use during a day out, as it has a treasure hunt quality that I think kids (and even adults) will enjoy ...' (cited in Paquet, 2022).

> 'My one wee quibble is that the link to the map for each destination is tucked away in the top right-hand corner, where I didn't see it at first—but that's a very minor quibble ... On the bright side, I discovered all sorts of cool places I'd never been to before. In fact, I didn't even make it to all of the destinations. It could be useful to plan two separate days out—one to discover the sites east of County Road 31 (AKA Bank Street) and one to discover the sites to the west ...'
>
> (cited in Paquet, 2022)

South Dundas' tourism director encouraged all residents and visitors to download the free app to get in touch with the area. Steven Byvelds, Mayor, South Dundas, seemed hopeful and said the app would help promote South Dundas and its tourism industry.

> 'As we start to come out of these lockdowns and restrictions this summer, we're all looking for things to do that are safe and enjoyable. What better way than to explore a beautiful area like South Dundas?'
>
> (Mayor Steven Byvelds cited in Paquet, 2022)

Further, according to Miles Harrison, 'it's fun, and as a resident for a little over a year, I am learning a lot about our community, it history, and local businesses. ...' (cited on South Dundas Tourism Facebook page, 2021).

Challenges

The Tourism Pursuit app is a promising technology that can enhance various aspects of tourism, such as education, interpretation and entertainment. It also provides a variety of experiences and knowledge that are not available in other areas. However, there are a few challenges that need to be addressed, such as while visiting a destination, due to time constraints, some people might not want to engage in a gamified visitor guide. Also, when people visit a destination, they tend to look at the historic sites or scenes on their mobiles instead of playing games on their mobile devices. Moreover, as the scanning of QR codes and playing of games in the app requires a data connection, the uninterrupted Internet availability and battery time are some constraints.

The implications of the Tourism Pursuit app are yet to be seen, as the region is recovering from the effects of the COVID-19 pandemic, as is the entire world. However, to engage visitors, South Dundas tourism launched a travel photo contest on the app on 20 October 2021. Participants had to take a photo while they were in South Dundas and enter to win one of three great prizes. All entries had to be submitted by 3 November 2021. To participate, the visitors had to download the free app, pick one of the 19 destinations showcased in three categories in the app, take a selfie there and post it to the South Dundas Facebook Group. The first, second and third prizes were worth US$300, US$200 and US$100, respectively.

Discussion Questions

1. Explain the concept of 'gamification' which can be used in the tourism industry, using the present case study as a reference.
2. Identify the pros and cons of gamification in the tourism and hospitality industry.
3. With reference to the case study outlined here, explain how gamification can be used for tourist engagement in a destination.
4. Discuss the innovation in digital technologies for tourism destination promotion using evidence from contemporary studies.

Teaching Notes

The case is designed to familiarize learners with the concepts of gamification, and its use within the proximity of the tourism and hospitality industry. The main aim of the case study is to help graduate and postgraduate students of travel and tourism courses to understand how destination management organizations or government tourism organizations can effectively leverage digital technology and gamification in tourism engagement. As most travel destinations and companies are trying to revive themselves after the global COVID-19 pandemic, students can further find out if using gamification in the tourism industry could assist in the revival of travel and tourism. This can be achieved by further discussing the pros and cons of using gamification in different industries and comparing it with travel and tourism, so that students have a comprehensive understanding of the gamification tool. Moreover, students will gain perspectives of the trends and challenges in using digital technologies for travel destination promotion. It is recommended that the facilitator discusses the case study, games and technology, and destination marketing and promotion techniques using gamification.

The case study is designed for a minimum of four or five students. The instructor of the case study is expected to take up the role of a member of the destination management organization and pose the case to the students.

Target audience

The case study is meant for graduate and postgraduate students of travel and tourism management, as part of their innovation and technology curriculum. It can also be used in the destination management curriculum.

Teaching methods and equipment

Group discussions can be initiated by analysing the performance of the tourism industry in South Dundas over the past couple of years. Then, the case study can be activated using scenario analysis, and the emergence of the need to develop a tourism gamification app. The present case study is perfect for 'Think-Pair-Share' activities, collaborative active learning and quizzes. Although there is no need for other equipment, the use of projectors and computers is highly encouraged to showcase the launch video of the app and the working of the app using the official Facebook page of South Dundas Tourism (available at: https://www.facebook.com/groups/southdundastourism (accessed 9 August 2022).

Teaching instructions

The case can be discussed in the classroom effectively. It can be taught in 2 hours with the help of a teaching instruction plan. The instructor should interact with the students and evaluate how familiar they are with the concept. For instance, if the students are not familiar with the concept of gamification, the tutor could ease them by asking them to describe their experience using physical and digital games. The instructor can then start the discussion with a brief introduction of games, gamification of destinations, and South Dundas. This can be followed by discussion of various advantages and disadvantages of the gamification technique. Later, the instructor can discuss the post-pandemic revival strategies of South Dundas Tourism. He/she can start the discussion by asking the students what they think was the right course of action to revive tourism post-pandemic.

Teaching instruction plan

Discussion point	Time required (min)	Teaching strategies
Introduction	15	The instructor can start the class by asking students to play a quick 3–5-min basic game on their mobile phones. And then, asking questions about their feelings about the game and excitement level while playing the game. Further, he/she can pose questions (e.g. Will they be happy if they get to play games in a tourist destination?).
Briefing on concepts related to innovations and technologies in tourism	15	Next, the facilitator introduces the concept of gamification in the tourism industry and how different departments are effectively using gamification in customer engagement. He/she can further probe students to think of some current examples of the same.
Summary of the case study	10	Then, the instructor can briefly discuss the scenario of the case study, the background and the idea behind the creation of the app. The instructor can discuss the reactions of the destination management organization (DMO) and government tourism officials involved in the app and tourism development.
Discussion question 1	15	**Concept of 'Gamification'** Now, the instructor can keep the discussion forum open and ask students about the various kinds of games that can be played while a tourist is in a tourist destination. They can discuss the physical as well as digital games for starters. They can further discuss the games such as Pokémon GO, which was extremely popular among students.

Continued

Teaching instruction plan Continued.

Discussion point	Time required (min)	Teaching strategies
Discussion question 2	20	**Pros and cons of gamification** The instructor then can divide the class into four groups. Two groups will discuss the pros and cons of gamification from the perspective of service providers. The other two groups will discuss the pros and cons of gamification from the perspective of a tourist destination.
Discussion question 3	20	**How gamification can be used for tourist engagement** Following this, the instructor can initiate the discussion with the latest types of gamification techniques used around the world. Students can be asked to find out more gamification tools and techniques for tourist engagement. The instructor can then ask: If the students were to design some games for a tourist destination, what kind of games would they propose?
Discussion question 4	15	**Innovation in digital technologies for tourism destination promotion** Following this, the instructor can ask students about the technological innovations which can lead to gamification and thereby engage tourists. They can discuss augmented technology, virtual reality (VR) and artificial intelligence.
Closing remarks and discussion	10	The instructor then can finish the case analysis session by summarizing the overall outcomes of the students' answers to the discussion questions.

Duration

120 min

Analysis

1. Explain the concept of 'gamification' which can be used in the tourism industry, using the present case as a reference.

Gamification is a technique used by designers to add gameplay elements to non-gaming settings. Further, gamification is the process of deriving the addictive elements from games and applying them to real-world activities. It is a process that is focused on optimizing the human experience. Games and their elements have been used in education and entertainment for thousands of years. Some of these include points, badges and leaderboards. A point is used as a visual identifier of achievement in sports, reward cards and video games. Achievement badges are also used to show achievement in various ways. While badges and leaderboards are often part of gamification, they do not capture the essence of the game. Gamification is about engaging and motivating people through their core drives. Core drives are the core elements of gamification, they start with the core values, and they are the ones that truly motivate people.

2. Identify the pros and cons of gamification in the tourism and hospitality industry.

The pros and cons of using gamification for training tourism and hospitality employees are shown in Table 5.1 while the pros and cons for the tourism destination are shown in Table 5.2.

3. With reference to the case study outlined here, explain how gamification can be used for tourist engagement in a destination.

There are several ways to engage tourists or guests in a destination using gamification.

These include:

- Augmented reality (AR) games – With the emergence of smartphones, more and more people will be able to experience the attractions of a city or region simply by playing games on their devices. This is already happening in some of the world's most famous cities such

Table 5.1. Pros and cons of using gamification for training tourism and hospitality employees. Author's own data.

Pros	Cons
Entertaining – Not every training resource is created equal, and there are many reasons why training aids and guides are not being used by sales representatives including that gaming is more entertaining.	Lack of strategic connection – It is important that employers thoroughly assess their objectives and determine if they should gamify the training activity. If they cannot make these connections, then stop playing and get back to the drawing board.
Portable – In earlier times, training manuals from training sessions were mostly the source of acquiring generic and job-specific skills. Now, organizations are using gamification techniques to instil transferable skills into its employees.	Frequency – Employers would want the game to be focused on the exercise itself and not on the people around it. If the game is too complicated to play and win, then it will distract and take away from the purpose of the activity.
Competitive – There's a more competitive role in the workplace for sales or marketing people. Games typically have a winner and loser, and they can be used to attract more sales reps. By having a competitive atmosphere, games can be used to motivate and attract new players.	Quality – When a website has not been updated for a long time, it shows that something is not being done right. This is very common in the gaming industry as well. In this case, employees will not take the company or employer seriously. Instead, they will start acting as if they are not valued enough.
Managerial insights – When employees play in a learning experience, their managers can easily identify who is struggling and who is succeeding at it. They can also monitor how many times they play and what their pace is.	Budget and resources – This is an extension of quality, but it is of great importance. If a company has an IT department they should be allocated time to use their expertise to create a game that is both effective and memorable. Allocation of a proper budget and resources is the key to success.
Reward – The only thing that employees love more than a good competition is being rewarded. This is the only thing that makes a game great.	

Table 5.2. Pros and cons of using gamification for the tourism destination. Author's own data.

Pros	Cons
Self-tour package – Gamified walking tours are different from standard tourist attractions. They allow players to experience a city on a scavenger hunt adventure through a mobile app. These tours are usually themed differently and can be self-customized by the tourists to fit their interests.	Over-tourism – One of the cons associated with the gamification of a destination is over-tourism. With the popularity of the game, the destination or attraction could be flooded with the players.
Connection with the unexplored – Gaming in cultural heritage sites provides an immersive experience that lets visitors experience the past in a way that is familiar to them.	Accidents – As happened in the case of the Pokémon GO game. Players were so immersed in the games that often they met with accidents.
Learning – These games are more engaging and effective than information brochures, tourist guidebooks and maps in terms of teaching and learning. Through these advancements, learning methods have become more effective and memorable.	Hurting community sentiments – Sometimes, the games are designed by a team that is unaware of the local community's religion, beliefs and values. Thus, having such an uninvited element in the game might create repulsion among those in the local community at the tourist destination.
For storytelling – Players can alter the story at any time, and they can also create their versions of the game at any time. The game is so immersive that players can't wait to get started.	

as London and Berlin. Interactive games that are based on real-world locations bring history to life and make the tourist experience memorable.

- Gamified travel tours – Tour guides are now more game-like with the integration of various game elements, such as Foursquare. Players can now collect points and badges to use in their favourite cafés. On a scavenger hunt adventure, the participant will explore the city on a mobile app. The goal is to complete the scavenger hunt and visit various locations in the city. Gamified walking tours can introduce guests to discoveries and uncover hidden secrets. They can also create their own tours and discover hidden attractions.
- Augmented experiences in theme parks – Adventure parks are ideal places to implement games as they already have playgrounds that are designed to be played on their own. The Walt Disney Company started using AR for enhancing park attractions early on. The next level of use is developing games that tell stories and engage guests through their theme.
- Gamified experiences in cultural heritage – Gaming in cultural heritage sites provides an immersive experience that draws visitors back to the past. Through historical quests and missions, players can experience what it was like to live at these sites as if it were in the present. New forms of learning are becoming more prevalent in cultural heritage sites, such as video games and gamification. Through the use of mobile platforms, these games provide new opportunities to introduce education and culture to the public.
- Gamification in storytelling – Gaming can be used in various ways such as storytelling. This can be done through various means such as social media platforms and gaming. Players can alter the game's story at any time, and they can also customize their challenges. The game's designers help players develop their storylines and gameplay mechanics. Transmedia storytelling is a process that involves using various media platforms to transfer a story.
- Gamified virtual experience – Games are often versatile but can also be adapted to other contexts such as virtual travelling. Google is currently showing how virtual travel can look like in Google Earth. The immense influence of computer games in creating virtual worlds has remained a niche area. Despite this, the potential of this area is enormous. Through technology, people can now travel in time and experience the world through their virtual characters. It allows them to interact with ancient rituals and foreign traditions. The advancements in game design support the development of virtual cultural heritage travel. This concept could serve as a way to protect the cultural heritage sites of our country while still allowing mass tourism to thrive.

4. Discuss the innovation in digital technologies for tourism destination promotion using evidence from contemporary studies.

Digital technologies are often defined as electronic tools and systems that are used to generate, store or process data such as mobile phones, multimedia, social media and online games. The rapid emergence and evolution of digital technologies are transforming how people live and work. Much attention has been focused on digital marketing and E-commerce as a way of building a brand and reaching new markets. AR, virtual reality (VR), artificial intelligence, geotagging and other cutting-edge technologies are creating new ways to enhance and improve the experiences of tourists. With the rapid emergence and evolution of digital technologies, tourism has become more complex and unpredictable. This is a key challenge for the sector as a whole.

VR is expected to gain widespread acceptance in the near future. The increasing number of customers wanting to book their stay online could benefit the hotel industry by allowing customers to get a better understanding of a hotel's environment. Some companies are even exploring the possibility of using VR to enhance the customer's experience. Another example is the possibility of allowing people to experience the rooms of hotels through VR. Using VR, hotels are currently offering guests the ability to book their rooms and attractions through their smartphones.

AR is a conceptual technology that uses digital technology to alter the perception of real-world objects in the eyes of people. It is a type of technology that enables people to experience real-world locations through their smartphones. This technology can enhance the experience of living in real-world environments. It can provide information on various topics that are most relevant to the tourism industry. In terms of the tourism industry, gamification has huge potential. For instance, AR could allow hotel guests to take photos of themselves with virtual characters or even re-create famous scenes in their hotel room. AR can also enhance the appearance of other areas outside the hotel. For instance, it could allow people to point their phones at restaurants and find reviews, and it can also help them visualize how a view would have looked in the past, including historical buildings that were once prominent but are now just ruins or no longer there.

Through the use of AR and VR technologies, designers can create an immersive environment that is both safe and enjoyable for the people who are in it. This is very important since it allows visitors to easily find the information they need, without relying on a local contact. As a result, the tourism industry has embraced the digital age.

References

Comfort, R. (2021a) South Dundas Tourism Pursuit – Let the games begin! *The Leader*, 23 June. Available at: https://www.morrisburgleader.ca/2021/06/23/south-dundas-tourism-pursuit-let-the-games-begin/ (accessed 11 October 2021).

Comfort, R. (2021b) South Dundas and St. Lawrence College team up for new app. *The Leader*, 22 April. Available at: https://www.morrisburgleader.ca/2021/04/22/municipality-and-st-lawrence-college-team-up-for-new-app/ (accessed 11 October 2021).

Haworth, J. (2021) South Dundas' Tourism Pursuit app launches, brings visitors to popular scenic spots. *Standard Freeholder*, 23 June 2021. Available at: https://www.standard-freeholder.com/news/local-news/south-dundas-tourism-pursuit-app-launches-brings-visitors-to-popular-scenic-spots?r (accessed 11 October 2021).

Morin, J. (2021) South Dundas Tourist Pursuit app launched. *The Chesterville Record*, 30 June. Available at: https://chestervillerecord.com/2021/06/30/south-dundas-tourist-pursuit-app-launched/ (accessed 12 October 2021).

Paquet, L.B. (2022) Tips for day trips to Morrisurg and Iroquois, Ontario. Available at: https://ottawaroadtrips.com/2021/07/12/tips-day-trips-morrisburg-iroquois-south-dundas-ontario/ (accessed 12 August 2022).

South Dundas Tourism Facebook page (2021) Available at: https://www.facebook.com/photo?fbid=10157908150401816&set=g.1423396667850070 (accessed 12 August 2022).

Vandermeer, N. (2021) New tourism app highlights 'hidden gems' in South Dundas. *Ottawa News*, 2 July. Available at: https://ottawa.ctvnews.ca/new-tourism-app-highlights-hidden-gems-in-south-dundas-1.5494845 (accessed 12 August 2022).

Further Reading

Xu, F. and Buhalis, D. (2021) *Gamification for Tourism*. Channel View Publications, Bristol, UK.

6 The Role of Locational Factors in Influencing the Overall Rating Experience of Hotel Visitors: an Application of Tripadvisor Data in Barcelona

Sushant M. Desale,[1]* Andres Artul-Tur[1] and Maria Isabel Osorio-Caballero[2]
[1]*Technical University of Cartagena, Spain;* [2]*Universidad Nacional Autónoma de México, México*

Abstract
The analysis of data provided by tourism platforms through Web 2.0 opens new ways to investigate how to improve trip experiences from a destination management organization's (DMO) perspective. This case study builds on Tripadvisor users' generated content to analyse the lodging experience and overall rating by around 70,000 hotel visitors in the city of Barcelona. The online reviews are processed using text mining and depuration techniques. The overall rating of the hotel stay is then explained by a set of covariates, mostly including locational and services rates' perception (room quality, cleanliness, room service, etc.) by means of a regression analysis using an ordered probit model. Through the case, we will learn to: (i) prepare the data set for the experiment; and (ii) run the model and discuss the main findings.

Subjects: business model, managing service quality, site or location choice, big data analysis, destination management, social networks' analysis, quality rating, text mining

Learning Objectives

At the end of the case study, the learner will be able to:

1. Identify how user generated content (UGC) can help taking decisions to improve the tourist management and marketing policies at destinations.
2. Explain how the impact of location plays on the tourist's stay ratings.
3. Identify how hotel managers can contribute in a methodological way to improve the use of Web 2.0 data for research analysis in the tourism and hospitality field.

Case

Experiences are the most important factor to consider when attempting to understand consumer/visitor behaviour, especially in the tourism industry simply because of the intangible service nature of the industry. Accordingly, this study analyses the dimensions

*sushantmadhukar.desale@edu.upct.es

 Teaching Cases in Tourism, Hospitality and Events (eds. S.K. Dixit and S. Piramanayagam)
DOI: 10.1079/9781800621022.0006

of visitor experiences in the context of hotels. An understanding of the dimensions of hotel visitor experience will help to better manage the total experience of a visit to a hotel. The study was multidisciplinary involving tourism and hospitality, computer science and information systems. The attributes which affect Tripadvisor users' experiences are investigated by using online reviews of those who were staying in the hotels of Barcelona (Spain). Text mining and an ordered probit regression (oprobit) model are applied in this study to help analysing the online reviews and determinants. Text mining is also used to divide between positive and negative reviews from the guests at each hotel through their opinions. While oprobit is used to identify the key variables, the key variables then indicate the keyword which often appears based on the number of occurrences from both negative and positive reviews.

User-generated content

The hotel market has become quite competitive over the years, hotels attempting to be distinguished from others by their quality of service and the facilities they offer. In order to select the best option for accommodation, tourists utilize rating systems of various booking sites. Understanding the behaviour of consumers by means of examining their experiences helps researchers realize visitor preferences with regard to the services provided or experiences encountered. Mostly they share their experiences via well-known social media platforms. This information is likely to have an impact on the number of room reservations at a hotel and for the hotel to raise their service quality and staff service provision, inter alia. This case study purposes to investigate various elements associated with hotel visitor experiences. Various techniques are involved for online review analysis (i.e. data mining, sentiment analysis, comparative analysis). In this study the methodology consists of determining the keywords and phrases/remarks in reviews on various visible features of hotel services and facilities including its location. The most frequent keywords were then categorized as the main customer concerns.

The latest studies about the analysis of hotel reviews are quite diverse and they use social media as the resource, which deals with a large number of consumer reviews (see Xiang *et al.*, 2015). Authors construct a model from hotel guests' experience and relate it to the satisfaction rating, and also run a comparative analysis from the major social media platforms in the field of hospitality and tourism. For example, research results from a study by Xiang and Gretzel (2010) on the role of social media in online travel information search, indicated that most tourists in the USA use social media to collect information about the destinations to which they plan to travel. The analysis revealed that the search engine results concerning travel destinations came from social media websites. It was also evident that most of the visitors to these sites shared their experiences through social media sites after travelling to some of the destinations (Xiang and Gretzel, 2010). According to Google Analytics 2015, Tripadvisor consists of 23 other travel media brands reaching 375 million monthly visitors, and more than 250 million reviews and opinions covering more than 5.2 million accommodation, restaurants and attractions (Google Analytics, 2015). The unique feature of Tripadvisor is that it is a travel community website; as a result, tourists have the power to interact with each other, becoming part of the place-making and experience-evaluating process.

The insights provided by visitors on Tripadvisor could be regarded as a combination of factual information, impressions, sentiments and empathy. These sentiments and empathy influence the visitor's experience and their level of gratification. Specifically, there are various aspects that affect the perceived empirical value of hotel visitors in Barcelona. The fundamental dimensions of this empirical value encompass: (i) location; (ii) neighbourhood; (iii) room; (iv) service; (v) sleep quality; and (vi) cleanliness. And the final outcome demonstrates that location seems to be a key driver of the overall experience. However, neighbourhood and hotel services also play a decisive role in affecting the visitor's reviews. For example, location of the hotel in terms of neighbourhood (i.e. the closeness to the city centre) is a key factor influencing the lodging experience as declared by customers. The rest of the services provided by the hotel also seem to affect the probability of obtaining a higher rating for the experience.

Steps 1 and 2: Downloading data from the online platform, and preparing the data set

Data in the form of travellers' ratings of their experience of hotels in Barcelona was collected from the online Tripadvisor.com platform. The study focused on two main determinants: (i) location of the hotel in terms of neighbourhood (i.e. the closeness to the city centre) as the main variable of interest; and (ii) quality of services provided from a short list given by Tripadvisor.com website's users.

Once all the big data from users was downloaded, we selected the target of the study through *Glogg software*. Glogg software is an open source, small but powerful platform that lets you browse and search through long and complex log files on the computer system. Since the log files can be massive and complicated to read, software like Glogg helps with handling a large amount of data to make it easier for programmers and system administrators to use in order to select smaller blocks of reviews or information. It is a user-friendly software providing graphical and interactive tools to handle in different operating system environments.

Initially we used the text mining and encoding option in order to locate the data, particularly hotels in Barcelona. Later on we transferred the selected data in txt. format. Then this raw file was imported in MicroSoft Excel (.xls) format and the separation was carried out through a delimiter function in order to convert the data (i.e. numbers, values and text) into columns for further analysis. This raw data set was then converted to the final data set (i.e. in variables which were annotated to the attributes as illustrated below) and finally it was imported into STATA software to run regression analysis. In sum, we captured the data sets identified by the keywords 'Barcelona' and 'Hotels', then transferred the .csv files provided as output by Glogg into Excel sheets to depurate errors in the data and then transferred them to STATA .dta files for the final statistical analysis. As a result, a number of variables were identified, as follows:

- *Location*: Tripadvisor defines the hotel location as part of the information provided in their reviews, that is, if the hotel is located in the capital city, in a town or if it is a mountain resort. In this case, we worked with locations based in neighbourhoods inside the city of Barcelona in Spain for hotel visitors (see next variable).
- *Neighbourhood*: This denotes the neighbourhood zone (different neighbourhood zones exist based on their distance from the city centre).
- *Room*: This is a numerical rating of room conditions based on size, comfort, etc. (scale ranges from 1 for worst rating to 5 for best).
- *Service*: This is a numerical value based on available service facilities (scale ranges from 1 for worst rating to 5 for best).
- *Sleep quality*: This is a numerical value based on bed quality in terms of size, comfort, etc. (scale ranges from 1 for worst rating to 5 for best).
- *Cleanliness*: This is a numerical value based on surroundings as well as the room (scale ranges from 1 for worst rating to 5 for best).

This case study employs data from a sample collected between April and August 2021. Data was gathered from Tripadvisor.com for the hotels situated in Barcelona (Spain). The hotels were grouped based on the neighbourhood location (e.g. Gracia, Sant Marti, Sarria, Sant, Gervasi, Les corts, etc.) and the overall rating of the experience.

Approximately 223 hotels in Barcelona were listed on Tripadvisor.com. The hotels were classified on the basis of their location throughout the city space. Further information illustrated the categories in which the gathered data was classified. From locations provided we could clearly recognize that hotels that visitors preferred were those in the city centre. Some of the individual hotels and resorts which were rated on Tripadvisor were far from city centre area. Finally, the categorization of all of them was done mainly on the basis of their neighbourhood as indicated previously. The objectives were to analyse the public review towards their staying in the hotel in terms of rating of the overall experience at the hotel, and additional service quality for other items, also mining some of the frequent words which appeared in the public reviews for sentiment analysis of users.

From the reviews on Tripadvisor, it was noticed that most reviews were posted by visitors to hotels situated in the city centre or downtown of Barcelona. By contrast, fewer reviews were posted by visitors to hotels situated more on the outskirts. However, visitors specifically indicated in their comments that they enjoyed the trip since they got the chance to visit most of the tourist attractions, which ultimately reflects the location of the hotel and its neighbourhood location.

Tripadvisor provided total raw data of more than 1 million registered reviews in 2019, out of which 70,000 registered reviews for hotels in Barcelona were selected and extracted using Glogg software. We employed a 'keyword search system' in Glogg software to differentiate positive and negative reviews based on various determinants. The data was further analysed for frequency of occurrence location wise, which resulted in 53,715 reviews (75% approx.) being selected from the visitors of hotels situated in the historical centre of Barcelona and the remaining from the outskirts. Additionally, information on the state of cleanliness, condition of the room, service facilities, and sleep quality were also recorded for the final data set in this case study.

Steps 3 and 4: Running the statistical test on data, and getting the main conclusions from the analysis

After collection of the defined data set, we uploaded it into the statistical software for the purpose of subsequent analysis. There are various software packages that could help us in this statistical analysis, including the simplest one, MicroSoft Excel, or more complex ones such as STATA (StataCorp) package, SPSS (IBM) or EViews (QMS). For this study we used STATA V.16 software to run our regression analysis. The data set includes the variables for analysis described in Table 6.1.

Table 6.1. Variables in the analysis. Authors' own table.

Variable name	Label	Explanation
Neighbourhood	neighnumb	This denotes the neighbourhood zone (different neighbourhood zones exist based on their distance from the city centre of Barcelona)
Location	locat_rate	Numerical value based on the visitor's rating of the hotel's specific neighbourhood location in Barcelona (ranges from 1 for worst to 5 for best)
Positive reviews	numb_positrev	Positive hotel qualities based on a customer's perception/ experience
Cleanliness	clean_rate	Numerical value based on surroundings as well as room (ranges from 1 to 5)
Room	room_rate	Numerical value based on room conditions based on size, comfort, etc. (ranges from 1 to 5)
Service	serv_rate	Numerical value based on available service facilities (ranges from 1 to 5)
Sleep quality	sleep_quality_rate	Numerical value based on bed quality in terms of size, comfort and sleep (ranges from 1 to 5)

Ordered probit regression model

The ordered probit regression (oprobit) model is a method of ordinal regression which provides the fit to ordered ranked data where the dependent variable follows a hierarchy (ranking) of numbers or ratings in this case. After that, a number of explanatory factors are defined in the model. The oprobit model is a probability one, so the user is modelling the probability of an event to occur, in this case, how probable are the ratings in the scale employed by Tripadvisor to be chosen. In brief, the oprobit model utilized here is to evaluate the link between the *ordinal dependent variable (i.e. overall rating)* and a set of *independent variables (in this case geo-location (neighbourhood), positive reviews, cleanliness rates, room rates, service rates, sleep quality rates)*. Ordinal variables include a rating of values where the user can choose one of them, so the dependent variable is defined as a categorical and ordered one. Most common examples are rating systems (poor, fair, good, excellent) opinion surveys, rankings, grades. In oprobit, the fundamental score is calculated as a linear function of the independent variables and set of cut points.

Analysis and interpretation of results

The regression model to be estimated is of the type:

$$Y_i = \alpha + \beta_i X_i + u_i$$

where:

Y_i = Overall rating given by visitor at Tripadvisor website

X_i = Set of explanatory variables: see Table 6.1

α, β and u are parameters to be estimated.

- Probit is a kind of econometric model, it has been used to model the probability (i.e. to calculate the probability from 0% to 100%) of customer reviews and what variables are working here.
- The model allows for values of estimated coefficients (i.e. the increase/decrease in the probability) showing their effects on the dependent variable of overall ratings.
- All these help to increase the check for robustness of the model through usual tests in a regression framework.

General result

The data distribution for locations of reviews on Tripadvisor corresponded with neighbourhood #1 being at the core centre of the city of Barcelona, while neighbourhood #5 was located farthest away from the city centre. As expected, most of the visits accumulated in the centre, as Fig. 6.1 shows. In particular, 55% of the visitors' sample was in the city centre, and 21% is in the next neighbourhood (#2).

Then we ran the oprobit regression, obtaining the results shown in Fig. 6.2 about the main variables (including neighbourhood, number of positive reviews, etc.) explaining

Neighbourhood	Frequency	Percent	Cum.
1	38 840	55.73	55.73
2	14 875	21.34	77.07
3	3 275	4.70	81.77
4	7 231	10.37	92.14
5	5 477	7.86	100.00
Total	69 698	100.00	

Fig 6.1. Frequency of visits and location of neighbourhoods in Barcelona (where 1 is the core of the city centre and 5 is farthest away from the city centre) according to Tripadvisor reviews between April and August 2021. Authors' own elaboration.

Ordered probit regression			Number of obs	69 698
			Wald chi2 (7)	17018.94
			Prob > chi 2 =	0.0000
Log pseudolikelihood = –11257.405			Pseudo R2 =	0.4281
ov_rating	**Coef**	**obust Std. Er**	**z**	**P> z**
neighbourhood	–0.0125	0.0016	–7.38	0.0000
numb postive reviews	–0.1012	0.0007	–129.75	0.0000
location rate	0.0356	0.0132	2.69	0.0071
cleanliness rate	0.0396	0.0174	2.27	0.0232
room rate	0.4082	0.0165	2.46	0.0145
service rate	0.0368	0.0139	2.65	0.0081
sleep quality rate	0.4215	0.0143	2.94	0.0033
/ cut 1	–4.0283	0.0768		
/ cut—2	–1.8846	0.0732		

Fig 6.2. Variables explaining the overall rating of customer satisfaction with the visit to Barcelona. These variables include rating of the neighbourhood, number of positive reviews, rating of cleanliness, room condition, etc. Authors' own elaboration.

the overall rating of customers' satisfaction of their visit to Barcelona. The following observations were made:

- There was a positive correlation between neighbourhood number and number of positive reviews (which was 100% statistically significant).
- There was a positive correlation between service and sleep quality rates (approx. 100% statistically significant).
- There was a positive correlation between cleanliness and room condition rates (significant at the 99% level).
- If the neighbourhood number variable shows a negative sign, this indicates that if we go farther away from the city centre, the overall rating will be reduced.
- Similarly, the positive coefficient of other independent variables indicates that the increase in the ratings of location, cleanliness, room, service and sleep quality, will automatically enhance the overall rating of the visit.

The value of R^2 is 0.4281, which demonstrates that this model with UGC data provides a good fit (approx. 50%) for analysing the overall rating of customer satisfaction. Next, we compute the marginal effects from oprobit to get the real computed effects of explanatories over the dependent variable in the model.

Marginal effects of covariates

There is a 48% probability of the marginal effect of covariates having an overall rating of 5. In the case of the marginal effect for this overall rating of excellent = 5, in Fig. 6.3 the main covariates affecting this are those of the neighbourhood location close to the city centre, and all rates from services and attributes of the room (cleanliness, room, service rate, sleep quality, etc.) all with a marginal effect around 1–2%, versus 4% of neighbourhood location (see column 'dy/dx' in Fig.6.3). So, if the location is taken out of the city centre by one level to the next urban ring, the probability of getting an overall rating of 5 drops by 4.9%. Positive reviews, with a negative sign, would reduce the overall rating in a counterintuitive result.

Marginal effects after oprobit				
Y = Pr (ov rating = 5) (predict, p outcome = 0.4812)				
Variable	dy/dx	Std. Err.	z	P>I z I
Neighbourhood numbe	–0.0490	0.0006	–7.38	0.0000
Numb postive reviews	–0.0403	0.0003	–129.16	0.0000
Location rate	0.0141	0.0052	2.69	0.0072
Cleanliness rate	0.0157	0.0069	2.27	0.0231
Room rate	0.0162	0.0066	2.46	0.0143
Service rate	0.0146	0.0055	2.65	0.0087
Sleep quality rate	0.0168	0.0057	2.94	0.0035

Fig. 6.3. Marginal effects of variables in the ordered probit regression (oprobit) model for overall rating. Column 'dy/dx' shows how much a change in each variable (x) would affect the overall rating of the visitor (dy). Column 'P > Izl' shows the level of significance of the coefficients, with values below 0.05 showing significance higher than 95%. Authors' own elaboration.

Oprobit regression by locational choices

Here the regression model is executed on the basis of the location of the neighbourhood for the hotels and as Table 6.2 shows its effect on the overall rating given to the experience by customers:

- *neighnumb #1 – core of city centre*: positive coefficients and the *p*-values of variables (not shown) indicate that the overall rating of customer satisfaction will increase if the customers are in the core of the city (+6.8% – i.e. raises the rating by 6.8%).
- *neighnumb #2 – very close to city centre*: results indicate that being close to the city centre also increases the customer's satisfaction (+5.4%).
- *neighnumb #3 – near to outskirts area*: results indicate that being farther away from the city centre reduces the contribution to the overall rating of the hotel experience, despite still being positive (+2.2%).
- *neighnumb #4 – far from the city centre*: in this case, being located far away from the city centre reduces the overall rating by 12.8%.

Summary

Summarizing the results of this case study, the current research seeks to identify how employing UGC data can help managers and destination management organizations (DMO) to improve the tourist management and marketing policies at destinations. In doing so, we have downloaded data from Tripadvisor.com and selected hotel reviews about Barcelona city. Then we have run an ordered probit model to identify how explanatory variables in the model help to identify factors influencing the overall rating obtained by the experience, posted by the customer on the Tripadvisor website after the stay. The primary data was around 80,000 reviews by hotel visitors to Barcelona, which were processed using text mining and depuration techniques; after depuration around 70,000 reviews were usable. Once the modelling procedure was applied, the main results have shown that location of the neighbourhood variable is the main factor affecting the overall rating in this case. Additional variables, including the state of cleanliness, condition of the room, service facilities and sleep quality, will also influence the overall rating, although to a lesser extent. Finally, we ran a model segmented by location of the hotel in four concentric rings from the city centre outwards towards the periphery of the city. Results indicate that location continues to play a key role in influencing the overall rating by the customer, showing that as the customer goes away from city centre (i.e. near to the outskirts of Barcelona), the overall rating

Table 6.2. Locational choice – neighbourhood groups from the city centre to the outskirts of Barcelona.[a] Authors' own table.

Neighbourhood 1		Neighbourhood 2		Neighbourhood 3		Neighbourhood 4	
Variables	Coefficient	Variables	Coefficient	Variables	Coefficient	Variables	Coefficient
neighnumb	0.068	neighnumb	0.054	neighnumb	–0.097	neighnumb	–0.12
locat_rate	0.007	locat_rate	0.062	locat_rate	0.072	locat_rate	0.01
clean_rate	0.061	clean_rate	0.004	clean_rate	0.083	clean_rate	–0.002
room_rate	0.051	room_rate	0.001	room_rate	–0.034	room_rate	0.07
serv_rate	0.051	serv_rate	–0.0002	serv_rate	–0.012	serv_rate	0.08
sleep_quality_rate	0.053	sleep_quality_rate	0.034	sleep_quality_rate	0.072	sleep_quality_rate	–0.0006

[a]Neighbourhood 1 (neighnumb #1) – core of city centre; Neighbourhood 2 (neighnumb #2) – very close to city centre; Neighbourhood 3 (neighnumb #3) – near to outskirts area; Neighbourhood 4 (neighnumb #4) – far from the city centre.

decreases, turning negative when the hotel is located so far away from the city centre. All these results provide more information to hotel managers in Barcelona city and contribute in a methodological way to improve the use of Web 2.0 data for research analysis in the tourism and hospitality field. For students it is a simple way of getting in touch with applied economics methods for training and gaining future professional skills in a smart activity.

Discussion Questions

1. How could this research help DMOs improve the tourist management and marketing policies at destinations?
2. How could this study help hotel managers in Barcelona attract more tourists and raise their satisfaction rates?
3. Identify two additional tourist destinations in Spain that can be considered for tourism analysis in a further extension of this research.
4. Give your opinion on how tourists' behaviour could be affected in choosing the city centres after the coronavirus disease 2019 (COVID-19) pandemic. Can you identify any changes between tourists' behaviour now, post-pandemic, from those findings of the current case study?

Teaching Notes

The aim of this case study is to familiarize students about how locational factors could play an important role in the most-visited European Union (EU) and world destinations with the help of a simple regression analysis. The main objective is to learn and experience how a much-visited tourist destination could also be impacted by socio-economic variables such as hotel location or in-hotel quality of services. The case is designed for students and professors to work in teams in class and promote discussions among groups of students to identify the main takeaway messages of the current case study.

Target audience

This case study is for students at undergraduate and master's degree level in subjects of business, economics, tourism and marketing. However, it could also attract the interest of students in the initial stages of their PhD. In order to follow this case study, it is recommended that students have: (i) at least a basic knowledge of statistics (i.e. introductory level) and regression theory; and (ii) some knowledge of data extraction and its analysis including online reviews assessment and customers' experience value theory.

Teaching methods and equipment

This case study is well suited for in-class group work. As a teaching methodology the information is provided to collect data sets from the Tripadvisor.com platform, thereafter specific data on hotels in Barcelona is extracted from the raw data set provided by the platform. Then students are required to design and run the data analysis, in order to obtain robust conclusions on other factors and specifically how location would make an impact on overall ratings as we move away from the city centre to the outskirts of Barcelona. This analysis would require use of statistical software wherein statistical methods can be performed.

Teaching instructions

- *Step 1.* Defining theoretical foundations for data gathering, cleaning and preparation, discussing and understanding the important concept to be learned with students.
- *Step 2.* Featuring and downloading the data from the Tripadvisor.com platform with the students. Using a data visualization tool (i.e. Glogg software) specific data is selected and extracted for the purpose of data analysis.
- *Step 3.* Once the final data set is created then it is imported and processed through the STATA software for statistical analysis.

It is important at this stage to explain to the students the procedures and different methods used for the analysis, and thereafter to review the findings extracted from the statistical analysis with them.

- *Step 4.* Once the findings are concluded, the groups of students should try to identify the main takeaway messages of the case study by analysing the outcomes and noting down the main conclusions which have been reached.

Duration

The duration of the case execution could be about four sessions each of 2 hours. The first three sessions could include proceeding through steps 1–4 detailed above (e.g. the initial session would include data gathering and preparation; session 2 could be devoted to data analysis with statistical software, while session 3 could be devoted to discussion of the main findings in class). Session 4 could be for further extensions and group brainstorming, including discussion questions 3 and 4.

Analysis for discussion questions

The following answers to the discussion questions are to help the teacher/instructor to develop and guide the in-class discussion with students in order to settle the main outcomes and takeaway messages arising from the case study.

1. How could this research help DMOs improve the tourist management and marketing policies at destinations?

According to the results of the whole case study, we have seen that location matters for overall rating of the tourist's stay, and accordingly that central locations of city destinations (Barcelona in this particular case) are going to be very crowded because tourists prefer to stay in or close to the city centre. There are two ways of planning the destination for the future. First, it is important to provide information to international tourists on other opportunities that they can find in more peripheral neighbourhoods (e.g. those close to Sagrada Familia Church) or giving them the capacity to select accommodation a bit farther away from the centre, so they can experience the real city life, and not just another tourist-crowded area. In this way, more marketing efforts could be devoted to open up the mind of the tourist according to the advantages that are located a bit farther away from the centre, finding lower-priced accommodation and good urban public transport communication ways to reach the city centre when needed. Secondly, it would be helpful to put in action an urban planning effort between all DMO actors where they can discuss the load capacity of each area in the city, being conscious of the trade-off between tourism revenues for the city and quality of life for residents of the city-centre areas. In fact, this view is being implemented by the city council, given the increasing problems of the tourist-crowded city centre.

2. How could this study help hotel managers in Barcelona attract more tourists and raise their satisfaction rates?

Along the same lines as the previous question, some serious debate should be raised on how to plan the city centres of crowded destinations, like Barcelona city for example. Marketing efforts should be made to better explain to tourists what they can find in places a bit farther away from the city centre. In fact it is important to develop these middle areas of the city that are not located in the very centre but have good access to it. In this case, the visitors would gain many advantages, such as lower prices and quieter and more peaceful neighbourhoods, where tourists could rest after each hard day of walking and visiting venues.

3. Identify two additional tourist destinations in Spain that can be considered for tourism analysis in a further extension of this research.

In order to compare results, students could choose other tourism cities, such as Madrid or Sevilla, and compare these with less-developed destinations like some inland places such as Burgos and Salamanca, where cultural, quiet and slow tourism dominates. Or they could select two different cities in their own countries and analyse how crowdedness could affect the tourists' satisfaction of their visiting experience.

4. Give your opinion on how tourists' behaviour could be affected in choosing the city centres after the COVID-19 pandemic. Can you identify any changes between tourists' behaviour now, post-pandemic, from those findings of the current case study?

The teacher/instructor could split the class into two groups and ask the students to think of the pros and cons of being located in a more crowded area in the city centre in terms of the pandemic or post-pandemic. One group would have to defend the positive benefits of being located close to the city's more important monuments and museums in the central area. The other group would have to argue that crowdedness is not well perceived in a post-COVID-19 scenario for some people, for example older people who fear crowds where virus transmission may be more likely.

References

Google Analytics (2015) Available at: https://analytics.google.com (accessed 18 August 2022).

Xiang, Z. and Gretzel, U. (2010) Role of social media in online travel information search. *Tourism Management* 31, 179–188.

Xiang, Z., Schwartz, Z., Gerdes, J. and Uysal, M. (2015) What can big data and text analytics tell us about hotel guest experience and satisfaction? *International Journal of Hospitality Management* 44, 120–130.

Further Reading

Dina, N.Z. (2020) Tourist sentiment analysis on Tripadvisor using text mining: a case study using hotels in Ubud, Bali. *African Journal of Hospitality, Tourism and Leisure* 9(2), 1–10.

Geetha, M., Singha, P. and Sinha, S. (2017) Relationship between customer sentiment and online customer ratings for hotels – an empirical analysis. *Tourism Management* 61, 43–54.

Kotler, P., Bowen, J.T., Makens, J. and Baloglu, S. (2018) *Marketing for Hospitality and Tourism*. Pearson Publishers, Chennai, India.

Mellinas, J.P. and Reino, S. (2019) eWOM: the importance of reviews and ratings in tourism marketing. In: Camilleri, M.A. (ed.) *Strategic Perspectives in Destination Marketing*. IGI Global, Hersey, Pennsylvania, pp. 143–173.

Sjöblom, S. (2013) Value creation through online travel reviews on Tripadvisor: case of hotels in Helsinki. Master's thesis, University of Helsinki, Finland.

7 Convert 'Good' to 'Excellent' Tourism Governance: Improve Indigenous Engagement

Joo-Ee Gan*
Sunway University, Malaysia

Abstract
Unlike most extant literature that documents poor tour ism governance, the Royal Belum case examines a successful tourism governance framework. Applying the institutional theory, the regulative, normative and cultural-cognitive factors of the governance framework are analysed to unveil its constructs. Of these, the presence of an empowered park authority with a clear mission is a decisive element. In terms of environmental governance, the framework facilitates a collaborative network between diverse stakeholders to support conservation and anti-poaching operations. Royal Belum's tourism governance is equally effective. Accompanied visitor entry is monitored closely through a permit system. A dual-checkpoint system deters the violation of permit rules. To prevent an over-supply of tourism services, a moratorium on houseboat licences was imposed. These measures ensure that Royal Belum remains a bespoke ecotourism destination. What is lacking is the meaningful inclusion of the indigenous people in the tourism offerings – the Achilles heel of Royal Belum's track record. Arguably, the conservation goals of Royal Belum should be re-examined to promote indigenous participation. The Royal Belum case offers lessons on effective environmental management and identifies areas for improvement.

Subject: indigenous community, institutional theory, irotected areas, tourism governance

Learning Objectives

At the end of the case study, a learner will be able to:

1. Identify the decisive characteristics of a good ecotourism governance framework.
2. Critically evaluate the acceptable growth for a touristic protected area.
3. Identify and analyse the strategies to enhance indigenous engagement in touristic protected areas.

Case

Dahlia Hidayat is an environmentalist who works with the Ministry of Natural Resources and Environment, Malaysia. She is part of a team empanelled to study the impact of tourism on the protected areas in Malaysia. In particular, the team is tasked with examining the tourism governance frameworks of various ecotourism destinations, with the view of recommending an effective

*jegan@sunway.edu.my

DOI: 10.1079/9781800621022.0007

model for adoption by the parks' authorities. Dahlia finds numerous examples of flawed governance frameworks – these seem to be the staple fare in the sustainability literature. The team wishes to focus on the building blocks of good tourism governance in order to unveil its constructs. Dahlia examines the Royal Belum State Park (Royal Belum) in Perak, which is reputed to be a well-managed ecotourism destination. It is immediately apparent to her that the custodian of Royal Belum, the Perak State Parks Corporation (PSPC) is instrumental in creating collaborative linkages between the stakeholders, be they state agencies, non-governmental organizations (NGOs) or local communities. Dahlia is fascinated by the dynamics between these stakeholders.

The main tasks of Dahlia's team were to:

1. Analyse the regulative, normative and cultural-cognitive factors in the ecotourism governance framework of Royal Belum.
2. Identify and evaluate the constructs that enhance Royal Belum's sustainability.
3. Identify the weaknesses in the ecotourism governance framework of Royal Belum and recommend improvement strategies.

The Royal Belum State Park

Royal Belum is part of the Belum-Temengor Forest Complex (BTFC), one of the oldest rainforests in the world (over 130 million years). It is contiguous with Thailand's Bang Lang National Park to the north, and the Hala-Bala Wildlife Sanctuary in Narathiwat Province to the north-east. In 1971, approximately 300,000 ha of BTFC was designated as the Belum-Temengor Forest Reserve (see Fig. 7.1). To further strengthen biodiversity conservation, 117,500 ha of this forest was designated as a state park in 2007 (PSPC, 2020). Creation of hydroelectric dams caused the submergence of some forest areas. This resulted in man-made lakes and islands that were once mountain peaks (Wonderful Malaysia, 2019).

The gateway to Royal Belum is Banding Island, which is connected to the nearest municipality by bridges. From there, the state park is only accessible by boat via the public jetty. The BTFC is home to the indigenous Orang Asli. While the creation of hydroelectric dams caused many Orang Asli to accept relocation to nearby settlements, some Orang Asli adapted to the altered geographical terrain and remained in the forest reserve.

Royal Belum is primarily an ecotourism destination, where the tourism activities revolve around the natural landscape of the BTFC (Wonderful Malaysia, 2019). While visits to Orang Asli villages are featured in tour brochures, tourist–community encounters are cursory in nature and limited in duration. Moreover, the Orang Asli communities currently derive limited monetary benefit from tourist visitation (Gan, 2020).

Royal Belum's governance framework

When the PSPC was established pursuant to the Perak State Parks Corporation Enactment 2001, its mandate was to protect the environmental heritage of Perak state. The PSPC's vision is '[t]o become an organisation highly capable of maintaining and managing various eco systems for mutual harmony,' and to this end, its mission is '[t]o advance positively (efficient and effective) in sustainable management of various eco systems of state parks.' (PSPC, 2020). It can be seen that the PSPC's foremost role is environmental governance, whereas ecotourism governance is merely part of its overall responsibilities. Strong, decisive, clear-sighted leadership is indispensable for good governance of protected areas (Battaglia *et al.*, 2012; Gan *et al.*, 2019). The PSPC's governance strategies demonstrate these traits. To set the backdrop for the discussion, the stakeholders in Royal Belum's governance are briefly introduced.

The prime actors in the environmental management of the BTFC are stakeholders in Royal Belum's governance – by virtue of the state park's locality. The Department of Forestry, the Department of Environment and the Department of Wildlife and National Parks (DWNP) are therefore the PSPC's

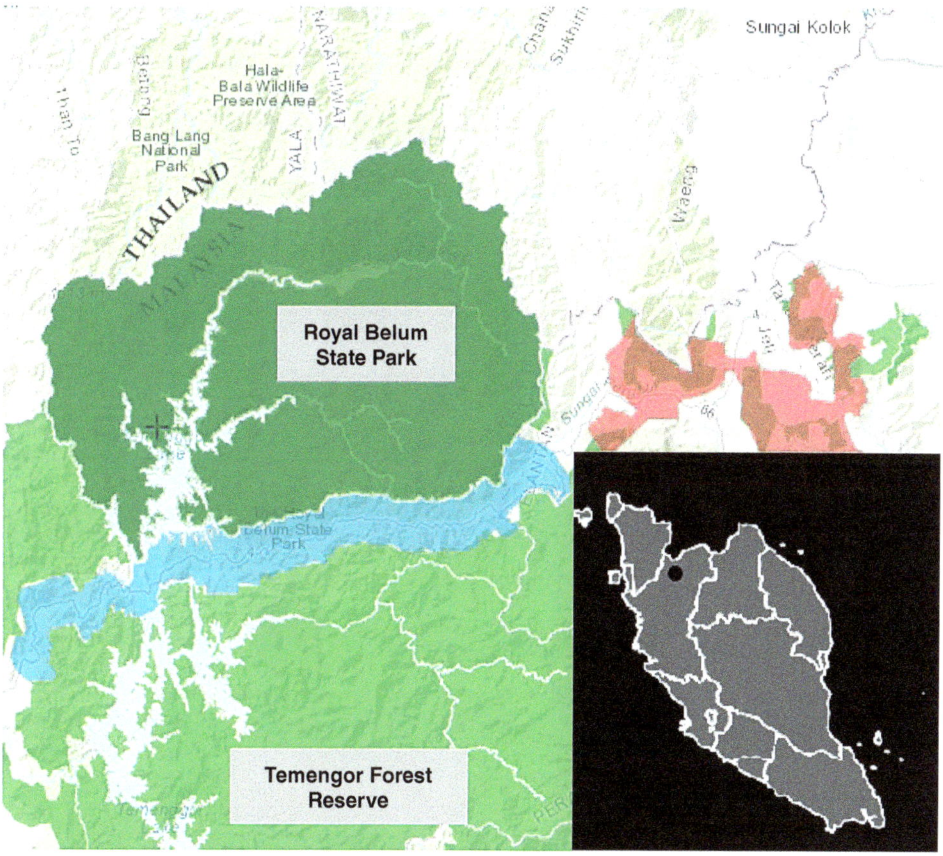

Fig. 7.1. Location of Royal Belum State Park which is part of Belum-Temengor Forest Reserve in Malaysia. Adapted from Improving Connectivity in the Central Forest Spine (IC-CFS), nd.

collaborators. In particular, the PSPC is part of a multi-agency enforcement team that combats illegal logging and poaching. In 2010, the Perak state government launched the Belum-Temengor Joint Enforcement Taskforce (BT-JETF) that comprised various government agencies and NGOs (see Table 7.1). The PSPC was an important stakeholder in this enforcement team (WWF-Malaysia, 2007). More recent collaborations involve the Malaysian Armed Forces, pursuant to the 1 Malaysia Biodiversity Enforcement Operation Network (1M-BEON) (*The Sun Daily*, 2016). This is an example where environmental governance and tourism governance intersect, since the enforcement network preserves the environmental assets central to ecotourism (Wearing *et al.*, 2020).

Where ecotourism is concerned, the Ministry of Tourism and Culture (MOTAC) and Tourism Perak (state agency) are responsible for ecotourism planning and development. However, the PSPC is invested in the tourism decision making of Royal Belum and its approval is a must. Additionally, the PSPC coordinates with the Department of Orang Asli Development (*Jabatan Kemajuan Orang Asli* (JAKOA)) in the planning of indigenous tourism. Among the private actors, the Temengor Tourism Entrepreneurs Association (*Persatuan Pengusaha Pelancongan Temengor* (PPPT)) plays a crucial role in the implementation of visitor entry control. Moreover, various NGOs contribute to the conservation and community projects in Royal Belum (see Table 7.1).

Table 7.1. Stakeholders in the tourism governance of Royal Belum.[a] Author's own table.

Agencies	Status	Role
PSPC (Perak State Parks Corporation)	Statutory body	Lead institution: mandate in biodiversity conservation, ecotourism as an ancillary goal
Department of Forestry	State department of federal agency	Environmental protection
DWNP (Department of Wildlife and National Parks)	State department of federal agency	Environmental protection
Malaysian Armed Forces	Federal agency	Environmental protection
Gerik Municipal Council	Local authority	Environmental protection, tourism licensing, waste management
WWF-Malaysia	NGO	Environmental protection, community engagement
MNS (Malaysian Nature Society)	NGO	Environmental protection, community engagement
PBF (Pulau Banding Foundation)	NGO	Environmental protection, community engagement
RIMAU	NGO	Environmental protection
Orang Asli villages	Indigenous communities	Tourism locale, environmental protection
PPPT (*Persatuan Pengusaha Pelancongan Temengor*)	Trade association	Tourism service providers, environmental protection, waste management
Ministry of Tourism and Culture (MOTAC) state office	State department of federal agency	Tourism planning and development
Tourism Malaysia Perak	State agency	Tourism planning and development
NCIA (Northern Corridor Implementation Authority)	Special federal agency established pursuant to NCER (Northern Corridor Economic Region)	Regional economic development
JAKOA (Department of Orang Asli Development)	State department of federal agency	Indigenous welfare, indigenous tourism

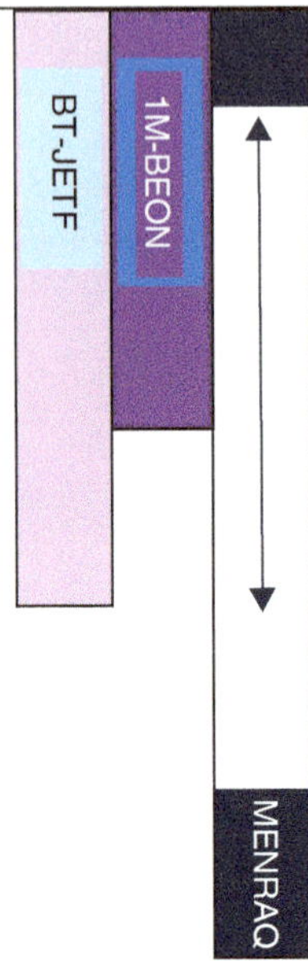

[a]BT-JETF, Belum-Temengor Joint Enforcement Taskforce; JAKOA, *Jabatan Kemajuan Orang Asli*; 1M-BEON, 1 Malaysia Biodiversity Enforcement Operation Network; NGO, non-governmental organization; Menraq, a community-based wildlife protection patrol unit consisting of Orang Asli of the Jahai tribe; RIMAU, Tiger Protection Society of Malaysia.

The PSPC also engages with the Northern Corridor Implementation Authority (NCIA), an agency set up in 2007 to implement Malaysia's Northern Corridor Economic Region (NCER, 2021). The reconstruction of the public jetty at Banding Island (completed in April 2017) was funded by the NCIA but planned in consultation with the PSPC and other stakeholders. Table 7.1 summarizes the roles of the stakeholders.

Governance strategies

In this section, the PSPC's governance strategies are appraised through the institutional theory, which proceeds on the assumption that an institution affects rules, norms and culture, which ultimately influence the socio-economic spheres (Scott, 2008; Toulmin, 2009; Lapeyre, 2011). In the Royal Belum context, the regulative aspects are the conservation regulations and the agencies involved in its governance. Normative aspects include the organizational norms that permeate the governance framework, whereas cultural-cognitive aspects refer to the stakeholders' assumptions or beliefs. As highlighted above, the PSPC's primary mission is the conservation of biodiversity in the state parks. Ecotourism activities are ancillary and the PSPC's priority is to ensure that ecotourism is conducted sustainably in Royal Belum. These objectives are the yardstick through which the PSPC's measures are assessed.

Environmental governance

The PSPC's effective leadership in environmental protection is the result of strong regulative factors. The PSPC has a clear statutory mandate (derived from the Perak State Parks Corporation Enactment 2001) to protect the environmental assets of Perak state. In short, the state ordinance empowers the PSPC. Coupled with state funding, the PSPC has the resources to carry out its responsibilities. These are important pre-conditions to its effective engagement with other stakeholders. Further, the PSPC clearly identifies the conservation challenges, and the measures in deterrence, detection and enforcement required to protect the Royal Belum. Unlike some state park authorities that fail on the procedural actualization of policies (Stoffelen *et al.*, 2020), the PSPC is able to operationalize its mandate and the governing regulations.

The normative factors are equally important in Royal Belum's governance. A multi-actor interdependence approach characterizes the stakeholders' relationship (Pellis *et al.*, 2015). Similarly, the engagement ethos of consultation and participation are observable in the PSPC's relationship with other stakeholders (Shields *et al.*, 2016). This is evident from the PSPC's involvement in the BT-JETF in 2010 and other anti-poaching operations. Effective environmental management requires the bridging of jurisdictions (federal, state or municipal) and the inclusion of private entities (NGOs and communities). A hierarchical norm would have been counter-productive, and providently, this is not the PSPC's approach. Rather, transparency and inclusivity characterize the ethos of stakeholder engagement. Good communication between the stakeholders contributes to planning and implementation, which ultimately fosters a sense of ownership towards the enforcement network (Lauber *et al.*, 2012). Despite the co-equal norm among the stakeholders, the PSPC's role as the lead institution is indispensable. This is because a structure of shared responsibility can grind to a halt where no stakeholder is willing to take the final responsibility (Stoffelen *et al.*, 2020). The PSPC's willingness to assume the mantle of leadership facilitates the smooth functioning of the network.

The PSPC also demonstrates managerial flexibility (TenDam, 1987). In terms of environmental governance, this means the making of strategic decisions to capitalize on opportunities or address challenges. Hence, in 2014, when federal funding became available for an enforcement operation through 1M-BEON (*The Sun Daily*, 2016), the PSPC seized the opportunity to invigorate the anti-poaching activities through a smaller, more nimble network. This threat of poaching has subsided, largely due to the regular activities of the PSPC and other stakeholders (*The Star*, 2021).

The pivotal cultural-cognitive factor in the environmental governance of Royal Belum is the PSPC's belief in the Orang Asli's involvement. According to Mohamed Shah Redza Hussein, a director of the PSPC, '[n]obody is going to take care of somebody else's house, somebody else's garden. ... But the indigenous community will look after and defend their own forest.' (Koshy, 2021). Hence, the PSPC established 'Menraq', a community-based wildlife protection patrol unit that consisted of Orang Asli of the Jahai tribe. Menraq means 'people' in the Jahai language. Enlisting the indigenous people in wildlife protection constitutes a recognition of their traditional knowledge and accords them a role in the stewardship of their environment (Lauber *et al.*, 2012). This cultural-cognitive factor in turn reinforces the positive regulative and normative factors in the governance framework.

Tourism governance

Sustainability is at the heart of the PSPC's tourism governance strategies. To begin with regulative factors, the PSPC instituted a multi-level visitor control mechanism. First, it established a licensing system for nature guides. An applicant must be accredited by MOTAC and must belong to the PPPT. The former ensures that an applicant is suitably qualified while the latter pegs an applicant to a private 'overseer'. Entry to Royal Belum must be accompanied by a PSPC-licensed nature guide. This measure seeks to prevent unlawful entry to Royal Belum. Second, an entry permit is required for tourist visitation. Fees chargeable go towards the PSPC's conservation fund. More importantly, the permit sets out the conditions of entry which, among other things, prohibit hunting and limit the catch size that can be removed when sport fishing. Failure to ensure visitors' compliance may result in the suspension or cancellation of the nature guide licence. Moreover, the licensed guides are co-opted into the waste management of the campsites and trekking trails. They are required to caution visitors against littering and advise on proper waste disposal, and as a last resort, remove waste from these sites. Third, dual checkpoints (manned by the armed forces and the PSPC rangers, respectively) monitor entry to and exit from Royal Belum. Since boat transportation is the only means to reach Royal Belum, the checkpoints are highly effective in deterring/detecting violations.

These regulative factors are reinforced by the norm of self-regulation among the licensed guides (Gan *et al.*, 2019). In other words, compliance with the regulations of the state park is largely voluntary, among a close-knit alliance of PPPT members. Thus, unauthorized visitor entry and violation of permit conditions are rare. Where a violation occurs, caution and monitoring by the PPPT invariably suffice. The need to impose draconian punishment hardly arises. Moreover, the PPPT members contribute to the maintenance of campsites and volunteer their own equipment in the event of campsite shortage.

The collaborative norm characterizes the PSPC's relationship with other stakeholders. Hence, the NCIA was open to the PSPC's input on the design of the public jetty, although funding was exclusively under the Northern Corridor Economic Region plan. The management of Royal Belum's tourism activities is also a collaborative effort. In particular, the state park is highly dependent on Gerik Municipal Council for waste management. More importantly, the stakeholders' consultative norm enables a systematic check on tourism growth. From the formal inception of tourism in Royal Belum in 2006, the number of tourist visits has systematically increased. Figure 7.2 shows the number of visitors to Royal Belum between 2006 and 2018, which indicates an approximately tenfold increase within two decades. Although the PSPC and other stakeholders consider the tourist number to be reasonable, steps are underway to manage the limits of tourism growth.

Mindful of the danger of over-supply of tourism services (Buhalis, 1999; Hernandez and Leon, 2013), the PSPC sought to pre-empt this problem at the Banding Island gateway. Excessive tourist accommodation is not a concern, since Banding Island is privately owned by a business entity (EMKAY Group) that operates the only resort there. However, the over-supply of houseboats plying Lake Temengor could undermine the sustainability of the segment. In the view of the regulator-stakeholders, over-supply of houseboats would lead to fierce competition and fewer bookings per operator, which might prompt the operators to pursue volume to make up for lower profitability. In 2016, there were 20 tourist houseboats with a collaborative culture among operators who engage with one another in friendly competition. In 2019, Gerik Municipal Council (upon consultation with the PSPC and other stakeholders) imposed a licence moratorium on houseboats. Essentially, upon the issuance of the licence for the 35th houseboat in Lake Temengor, no licence would be issued until further studies into the sustainability of ecotourism there. At the time, the number of applications had not exceeded the capped number. Unlike most tourism moratoria that are imposed as a response to over-supply of tourism services (Inchausti-Sintes and Voltes-Dorta, 2020), the houseboat licence moratorium was pre-emptive rather than reactive. Further, Gerik Municipal Council gave advance notice (in early 2018) and explained the rationale of the moratorium, namely, to ensure that the operators derive sustainable livelihood from ecotourism. This led some licence applicants to explore joint ventures for the

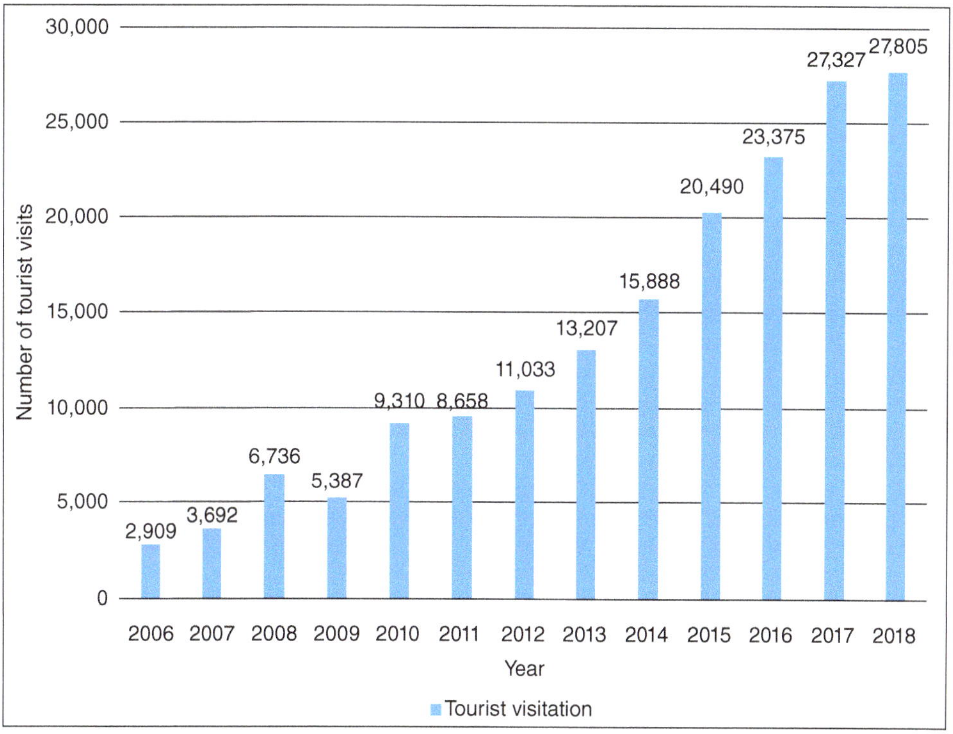

Fig. 7.2. Royal Belum State Park: tourist visitation. Author's own figure based on data supplied by Perak State Parks Corporation (PSPC), 2020.

bidding of the final tranche of licences. Overall, it can be seen that the regulative factors and the normative factors of Royal Belum's tourism governance framework promote sustainability.

Indigenous engagement and tourism

The ecotourism products in Royal Belum include visits to Orang Asli villages, although the mainstay of the segment is nature-oriented. Tourist visits to indigenous villages are conducted with the PSPC and JAKOA's permission. The latter's involvement is in the form of indirect monitoring, insofar that tourism may affect the welfare of the Orang Asli. Tourism villages in Royal Belum developed organically – a consequence of strategic charting of tour routes. The PSPC was not actively involved at the inception stage. Consequently, a community-based tourism model with equitable revenue sharing was not established. This meant that the Orang Asli derived limited benefits from tourism, apart from tourists' donation of food provisions and money (Gan, 2020).

This regulatory vacuum is partly caused by the PSPC's cultural-cognitive assumption about its role. Arguably, the PSPC's vision and mission warrant the prioritization of biodiversity conservation. However, it should be noted that the preservation of the biocultural heritage of the indigenous communities is central to the environmental protection of a protected area (Watson *et al.*, 2016). A holistic view of biodiversity conservation should encompass indigenous welfare, and where applicable, the promotion of tourism as a source of livelihood. Unfortunately, this approach was not adopted by the PSPC at the time. By contrast, the Orang Asli were co-opted in anti-poaching operations, arguably because the PSPC perceived this threat to fall directly within its sphere of concerns.

By 2012, a village called *'Ta-eng'* (which means 'river' in the Jahai language) became a popular destination. The PSPC renamed the village as 'Kampung Aman Damai' (which means 'village of peace' in Bahasa Malaysia) and designated the locale as the tourism village in Royal Belum (although visits to other settlements were not prohibited). However, the inhabitants gained little from tourism, as there was no agreed mechanism for the distribution of benefits. Today, Kampung Aman Damai no longer exists. Internal disagreements among the villagers over the distribution of tourists' donations and traditional lifestyle preferences led to the complete abandonment of the village. Over time, the forces of nature had whittled away the traces of its existence. Some of the villagers moved further into the interior while a sizeable number established a village called Kampung Kleweng that received visitors (Gan, 2020). Arguably, a more nuanced appreciation of the Orang Asli's expectations and the implementation of a community-based tourism model might have led to a different outcome.

The team's observations

Dahlia's team identifies various governance constructs from the analysis of the PSPC's role. In terms of regulative factors, statutory legitimacy, strong lead institution, clear vision and mission, government funding, the procedural operationalization of policies and the assumption of responsibility contribute to Royal Belum's good governance. These constructs are supported by important normative factors, namely, the multi-stakeholder interdependence traits of collaboration, consultation and participation. Additionally, the cultural-cognitive recognition that indigenous stewardship is central to environmental protection promotes stakeholder buy-in among the Orang Asli. Collectively, these factors strengthen the environmental governance of Royal Belum.

Tourism governance in Royal Belum also benefitted from the constructs identified above. The licensing of nature guides, entry control, clear communication of entry conditions, authorized accompanied entry and enforcement checkpoints are regulative factors that enhance tourism governance. This is reinforced by the normative factor of self-regulation among the PPPT members. Similarly, the norms of collaboration, consultation and participation promotes sustainability, as evidenced by the consensus approach towards the tourism moratoria. Dahlia's team uses Table 7.2 to map the governance constructs of Royal Belum in terms of categories and exemplifications.

In sum, an effective governance framework in Royal Belum ensures the twin goals of environmental integrity and sustainable tourism. However, as seen in Table 7.2, the dissonance in the tourism governance of Royal Belum is arguably the result of the PSPC's cultural-cognitive assumption about its role. A narrow interpretation of its mandate means that the PSPC does not intervene assiduously in the tourism development of indigenous villages. Consequently, tourism does not lead to a significant improvement in the Orang Asli's livelihood. A holistic view of biodiversity conservation that recognizes the indigenous people's role in the bio-heritage of the protected areas can broaden the PSPC's assumption of its vision and mission. A cultural-cognitive realignment is necessary to enhance indigenous engagement, which, in the Royal Belum context, is the key to convert 'good' to 'excellent' governance. The next challenge for Dahlia's team is in identifying the improvement strategies to realize this goal.

Discussion Questions

1. Contrast the Royal Belum case with a poor ecotourism governance framework. What are the decisive characteristics (regulative, normative, cultural-cognitive) that differentiate between the two governance frameworks?
2. Compare Royal Belum's tourist visitation with another state park. Critically comment on the acceptable growth for Royal Belum.
3. Evaluate the impact of Gerik Municipal Council's licence moratorium. Recommend a period for review of the moratorium and justify your recommendation.
4. Identify further measures that the Perak State Parks Corporation (PSPC) can adopt to enhance indigenous engagement in Royal Belum.

Table 7.2. Constructs and outcomes of Royal Belum governance framework. Author's own table.

Constructs	Factors	Outcomes for	
		Environmental governance	Tourism governance[a]
Regulative	• Statutory legitimacy • Strong lead institution • Clear vision and mission • Government funding • Operationalized policies • Assumption of responsibility	• Multi-stakeholder environmental protection taskforce: BT-JETF, 1M-BEON, Menraq	• Licensing of nature guides • Entry control • Permit conditions • Accompanied entry only • Enforcement checkpoints • Tourism moratoria
Normative	• Collaboration • Consultation • Participation • Self-regulation (nature guides)	• Managerial flexibility • Nimble stakeholder alliances	• Effective planning and design of tourism infrastructure • Effective waste management • Stakeholders buy-in for tourism moratoria
Cultural-cognitive	• Conducive: indigenous environmental stewardship • Dissonant: narrow perception of biodiversity conservation	• Indigenous anti-poaching task force	• Absence of community-based tourism model • Unequitable distribution of tourism benefits

[a]Shaded cell indicates weaknesses in indigenous engagement where tourism governance is concerned.

Teaching Notes

The Royal Belum case presents the learners with an example of a good tourism governance model. Too often, instances of poor tourism governance eclipse the pivotal role of tourism in the conservation of protected areas. Conceptualized using the institutionalist theory, the case systematically charts the governance attributes and outcomes. Using the positive constructs unveiled in the case, the learners can critically benchmark tourism governance frameworks of comparable tourist destinations. In this regard, it is recommended that the facilitator adopts the three-pronged approach of the institutional theory to analyse a governance framework in terms of: (i) regulatory effectiveness; (ii) normative influences; and (iii) cultural-cognitive influences.

Target audience

The case is suitable for students pursuing bachelor's degree programmes in hospitality and tourism, preferably in their second-year and above. It is also beneficial for graduate students undertaking research for master's degree-level dissertations. A background in conservation and protected areas is helpful but not a must.

Teaching methods and equipment

The Socratic approach of respectful argumentative dialogue between students is best suited to this case study. In a workshop setting, the instructor may consider providing contrasting case studies of tourism governance frameworks as reference points. A session can be supported by online access to various state park websites for benchmarking purposes.

Teaching instructions

In view that the Socratic approach is recommended, pre-class dissemination of the case study is imperative for critical discussion during a workshop. To facilitate active learning, students can be divided into distinct groups of three to five members each. A brief presentation of ideas by

Teaching instruction plan

Activities	Duration (min)	Teaching strategies
Case introduction	10	The instructor can begin by positioning protected areas as tourist destinations. To encourage participation, the instructor may ask the learners about their experience with ecotourism destinations. In particular, individual opinions about the environmental sustainability of such destinations serve as a backdrop for further discussion on the case.
Contextualization of the institutional theory to the case	10	The instructor can explain the meaning of 'institutions'. Some learners may well equate 'institutions' with establishments, hence this misconception should be clarified. As the instructor explains the three-pronged analysis of regulatory, normative and cultural-cognitive factors, the intangible elements of 'institutions' should become apparent.
Case summary	10	The instructor should briefly discuss the governance constructs of the case, as analysed through the institutional theory. This can be followed by a reminder on the ground rules of the group presentations.
Group presentations	10 (per group)	Each presenter group can be expected to address the discussion questions, as well as present a SWOT analysis with a contrasting case study that highlights poor tourism governance.
Q&A with the instructor and other students	5 (per group)	Where this activity is summative, the instructor should pose questions to the presenter groups to assess the breadth and depth of their learning on this topic. The instructor should also allow the presenter group to take questions from the floor, if any. Overall, the Q&A assesses the presenter group's preparedness and their ability to think on their feet.
Session takeaway	10	The session can be used to explain the rationale of marks allocation, where this activity is summative. The instructor can conclude the session with a summary of the various positions adopted by the presenter groups, the instructor's own observations, and recommendations for further learning.

each group can be followed by a broader discussion. The instructor can prescribe the discussion questions as presentation pointers. Further, where contrasting case studies have been provided, or where the instructor has prescribed further research on other governance frameworks, a SWOT (strengths, weaknesses, opportunities and threats) analysis of the different case studies should constitute part of the presentation. The following table sets out a teaching plan, on the premise that students attend the session with prior knowledge from pre-class preparation.

Duration

Approximately 120 min (for a class size of five presenter groups)

Analysis

1. Contrast the Royal Belum case with a poor tourism governance framework. What are the decisive characteristics (regulative, normative, cultural-cognitive) that differentiate between the two governance frameworks?

A certain level of subjectivity can be expected in the student's comparator choice of a 'poor tourism governance framework'. In reality, the three factors – regulative, normative and cultural-cognitive – overlap significantly, such that it is difficult to identify the critical determinant. A good discussion should highlight the interplay between these factors and how they influence a park authority's perception of its role, its collaborators and the ambit of its power.

2. Compare Royal Belum's tourist visitation with another state park. Critically comment on the acceptable growth for Royal Belum.

There are various methods to assess the sustainability of a tourist destination, from the concept of carrying capacity, limits of acceptable change, tourism optimization management model, visitor impact management, to visitor experience and resource management. Regardless of the impact appraisal method adopted, the baseline for analysis and the assumptions pertaining to growth ceiling or optimal use intensity are important. Students can analyse the PSPC's tourist number through any of these methods.

3. Evaluate the impact of Gerik Municipal Council's licence moratorium. Recommend a period for review of the moratorium and justify your recommendation.

Moratoria on tourism development are often belated responses to over-tourism. But in Royal Belum's case, the stakeholders do not consider the state park to suffer from over-tourism. A good discussion should explore the following issues: Is the moratorium unduly harsh to the licence applicants? Might one argue that the moratorium stifles potential tourism growth in Royal Belum? On the other hand, a belated imposition of a licence moratorium is disastrous to the ecosystem. Was the right balance struck? If so, how often should the moratorium be reviewed? Annually, biannually? Is state-level input necessary for this municipal-level decision?

4. Identify further measures that the Perak State Parks Corporation (PSPC) can adopt to enhance indigenous engagement in Royal Belum.

The literature on indigenous participation in tourism is abundant. Meaningful consultation, revenue sharing, community ownership, capacity building, and pre-emption of elite capture are among the solutions for successful indigenous engagement. A good answer must be positioned in the context of Royal Belum. Collaboration between the PSPC and the Orang Asli community is present in an anti-poaching operation.

References

Battaglia, M., Daddi, T. and Rizzi, F. (2012) Sustainable tourism planning and consultation: evidence from the Project INTER.ECO.TUR. *European Planning Studies* 20(2), 193–211.

Buhalis, D. (1999) Tourism on the Greek Islands: issues of peripherality, competitiveness and development. *International Journal of Tourism Research* 1(5), 341–358.

Gan, J.E. (2020) Uncovering the environmental and social conflicts behind residents' perception of CBT: a case of Perak, Malaysia. *Tourism Planning and Development* 17(6), 674–692. DOI: 10.1080/21568316.2020/1749121.

Gan, J.E., Nair, V. and Hamzah, A. (2019) The critical role of a lead institution in ecotourism management: a case of dual governance in Belum-Temengor, Malaysia. *Journal of Policy Research in Tourism, Leisure and Events* 11(2), 257–275. DOI: 10.1080/19407963.2018.1516076.

Hernandez, J. and Leon, C. (2013) Welfare and environmental degradation in a tourism-based economy. *Tourism Economics* 19(1), 5–35.

Improving Connectivity in the Central Forest Spine (IC-CFS) (nd) Hutanwatch – The dawn and rebirth of the Central Forest Spine. Available at: https://www.hutanwatch.com/single-post/2018/05/31/the-second-dawn-of-the-central-forest-spine (accessed 14 August 2022).

Inchausti-Sintes, F. and Voltes-Dorta, A. (2020) The economic impact of the tourism moratoria in the Canary Islands 2003–2017. *Journal of Sustainable Tourism* 28(3), 394–413.

Koshy, E. (2021) Orang Asli wildlife patrollers are the new heroes of the Royal Belum State Park. *New Straits Times*, 7 March. Available at: https://www.nst.com.my/lifestyle/sunday-vibes/2021/03/671647/orang-asli-wildlife-patrollers-are-new-heroes-royal-belum (accessed 29 July 2021).

Lapeyre, R. (2011) Governance structures and the distribution of tourism income in Namibian communal lands: a new institutional framework. *Tijdschrift voor Economische en Sociale Geografie* 102(3), 302–315.

Lauber, T.B., Decker, D.J., Leong, K.M., Chase, L.C. and Schusler, T.M. (2012) Stakeholder engagement in wildlife management. In: Decker, D.J., Riley, S.J and Siemer, W.F. (eds) *Human Dimensions of Wildlife Management*. The John Hopkins University Press, Baltimore, Maryland, pp. 139–156.

NCER (Northern Corridor Implementation Authority) (2021) *About NCER*. NCER, Malaysia. Available at: https://www.ncer.com.my/about-ncer/ (accessed 22 July 2021).

Pellis, A., Lamers, M. and Van der Duim, R. (2015) Conservation tourism and landscape governance in Kenya: the interdependency of three conservation NGOs. *Journal of Ecotourism* 14(2/3), 130–144.

PSPC (Perak State Parks Corporation) (2020) Vision and mission. Available at: https://www.royalbelum.my/eng/ (accessed 22 July 2021).

Scott, W.R. (2008) *Institutions and Organizations: Ideas and Interests*, 3rd edn. Sage Publications, Thousand Oaks, California.

Shields, B.P., Moore, S.A. and Eagles, P.F.J. (2016) Indicators for assessing good governance of protected areas: insights from park managers in Western Australia. *PARKS* 22(1), 37–50.

Stoffelen, A., Adiyia, B., Vanneste, D. and Kotze, N. (2020) Post-apartheid local sustainable development through tourism: an analysis of policy perceptions among 'responsible' tourism stakeholders around Pilanesberg National Park, South Africa. *Journal of Sustainable Tourism* 28(3), 414–432.

TenDam, H.W. (1987) Managerial flexibility: a strategic asset. *Leadership & Organization Development Journal* 8(2), 11–16. https://doi.org/10.1108/eb053612

The Star (2021) Poaching cases remain low in Belum-Temenggor forest. *The Star*, 30 May. Available at: https://www.thestar.com.my/news/focus/2021/05/30/poaching-cases-remain-low-in-belum-temenggor-forest (accessed 29 July 2021).

The Sun Daily (2016) National biodiversity resources under threat. *The Sun Daily*, 10 August. Available at: https://www.thesundaily.my/archive/1931911-MSARCH386734 (accessed 22 July 2021).

Toulmin, C. (2009) Securing land and property rights in sub-Saharan Africa: the role of local institutions. *Land Use Policy* 26, 10–19.

Watson, T.K., Burlando, C., Seagle, C., Crawhall, N. and Noshirwani, M.M. (2016) Introduction. In: From Solutions to Resolutions: Defining a New Social Contract for Just and Effective Conservation. *IUCN Commission on Environmental, Economic and Social Policy – Policy Matters* Issue 20, September. Available at: https://portals.iucn.org/library/sites/library/files/documents/Policy%20Matters-Issue%2020.pdf (accessed 22 July 2021).

Wearing, S., McDonald, M., Schweinsberg, S., Chatterton, P. and Bainbridge, T. (2020) Exploring tripartite praxis for the REDD+ forest climate change initiative through community based tourism. *Journal of Sustainable Tourism* 28(3), 377–393.

Wonderful Malaysia (2019) Royal Belum State Park. Available at: https://www.wonderfulmalaysia.com/royal-belum-state-park-malaysia.htm (accessed 29 July 2021).

WWF-Malaysia (2007) WWF applauds gazettement of Royal Belum State Park. Available at: https://www.wwf.org.my/?4160/ (accessed 23 May 2022).

Further Reading

Leung, Y.F., Spenceley, A., Hvenegaard, G. and Buckley, R. (2018) *Tourism and Visitor Management in Protected Areas: Guidelines for Sustainability*. International Union for Conservation of Nature (IUCN). Gland, Switzerland. Available at: https://portals.iucn.org/library/sites/library/files/documents/PAG-027-En.pdf (accessed 9 August 2022).

Nair, V., Hamzah, A. and Musa, G. (2020) *Responsible Rural Tourism in Asia*. Channel View, Bristol, UK.

8 COVID-19 a Disaster for Heritage Business Models – Victoria & Albert Museum Case Study

Colin Seymour,* Kalliopi Fouseki and Ilias Krystallis
University College London (UCL), London, UK

Abstract

The Victoria and Albert Museum (V&A), founded in 1852 and stemming from the Great Exhibition of 1851, had a mission to educate the public, designers, manufacturers and the public about art and design. Organizations like the V&A were intended to be initially funded from the proceeds of the Great Exhibition. But how did this funding model change with time? This teaching case study about the V&A makes a comparison between its pre- and post-coronavirus disease 2019 (COVID-19) business model (BM), aiming to map historic V&A value creation and value capture for visitors and stakeholders, while examining the underlying causes and effects of BM changes.

Subjects: business models (BMs), sustainable business models (SBMs), business model canvas (BMC), triple-layered business model canvas (TLBMC), Victoria and Albert Museum (V&A), benchmarking, COVID-19, PESTLE business tool (political, economic, social, technological, legal, environmental factors)

Learning Objectives

At the end of the case study, a learner will be able to:

1. Explain the business model (BM) concept.
2. Apply the BM concept to a real-world case study like the Victoria and Albert Museum (V&A).
3. Show the impact of political, economic, social, technological, legal and environmental (PESTLE) and sustainability factors, to an organization's BM.

Case

This case study about the Victoria and Albert Museum (V&A) (Fig. 8.1) business model canvas (BMC) is set in the context of the coronavirus disease 2019 (COVID-19) pandemic. The pandemic has decimated the heritage sector, with (for example) estimates that up to 30% of US museums will close permanently (Atkinson, 2020). The rationale for this case study stems from the enormous importance of heritage to

*c.seymour@ucl.ac.uk

 Teaching Cases in Tourism, Hospitality and Events (eds. S.K. Dixit and S. Piramanayagam)
DOI: 10.1079/9781800621022.0008

Fig. 8.1. Courtyard at the Victoria and Albert Museum (V&A). Photo courtesy of heatheronhertravels, available at: 'Courtyard at the V & A museum' licensed under CC BY 2.0 (accessed 22 November 2021).

the UK in terms of the economy (jobs, income, both locally and nationally, etc.) as well as health benefits. For example, heritage added an estimated £14 billion to the UK's gross domestic product (GDP) in 2018 (which included £5 billion for London) and sustained 198,000 UK heritage sector jobs (Centre for Economic Business Research, 2019). Moreover, there were more than 6 million visitors to English Heritage sites in 2018/19 (English Heritage, 2019), historic houses alone (independently owned houses) generate 34,600 full-time equivalent (FTE) jobs and contribute £510 million/year to the economy (Historic Houses, 2020). Furthermore, heritage institutions, while enduring reduced funding from central government sources, are having to diversify and adapt their business models (BMs), as well as justify their continued existence. This, coupled with the devastating impact of the COVID-19 temporary closures of heritage sites and the corresponding loss of visitor revenues, provides an excellent opportunity to take stock of the pre-COVID-19 BMs, as benchmarks for future BM innovations. The V&A was selected as a case study because its funding model is in transition. Hence, in some ways it represents heritage sites which are government funded and socially orientated, as well as characterizing those sites which are more commercial in nature.

Introduction to the problem

In 2020, Historic England highlighted the fact that the COVID-19 pandemic has severely impacted all heritage organizations, and that the very survival of many heritage businesses is at risk and some of them will fail (Historic England, 2020). Furthermore, they advised that resilience and capacity in the sector needed to be strengthened. Finally, they suggested that new BMs would improve sustainability in the sector and would also help it to avoid the risks of an approaching disaster.

The dilemma/research opportunity

This concern about sustainable business models (SBMs) in the heritage sector has also been reflected by Tristram Hunt, Director of the V&A:

> behind the growth [generated from tourism] is a business model [BM] which Covid-19 has crippled. As governments have systematically cut public funding to national museums, we have become ever more dependent upon paid-for exhibitions, membership schemes and commercial income – which will be hard to continue in an era of social distancing.
>
> (Hunt, 2020, p. 12)

Despite the growing urgent need to develop SBMs for the heritage sector, the area of SBMs, in the sector in particular, is not well researched.

The business model canvas (BMC)

BMs can be defined as how organizations create and capture value (Osterwalder and Pigneur, 2010). For Osterwalder and Pigneur, value creation ('value propositions') can have attributes which are quantitative ('e.g. price, speed of service') and qualitative ('e.g. design, customer experience') (Osterwalder and Pigneur, 2010, p. 23). Whereas (for these authors) value capture ('revenue streams') is the economic remuneration (cash generated) which an organization gains from 'customer segments' in exchange for access to, or maintenance of, the value proposition (Osterwalder and Pigneur, 2010, p. 30).

However, an older and broader definition of BMs, notable for the inclusion of 'benefit' and 'enhancing public good' is once again being utilized by researchers. The BMC (Osterwalder and Pigneur, 2010) has significantly advanced BM practice. It is a conceptual framework composed of nine themes, which allows an organization's BM to be mapped. Generally, the themes can be grouped under:

- the 'organization' (i.e. key activities, key partners, key resources, value proposition (e.g. the product or service)); and
- the 'market' (i.e. distribution/communication channels, revenue streams, visitor segments, visitor relationships).

The BMC also permits a degree of flexibility regarding the interpretation(s) and application(s) of the themes/components/building blocks.

The BMC: some strengths and limitations

While the BMC provides a very useful, simple, visual overview of how an organization creates and captures value, there are two limitations. First, it is constrained by the nine building blocks/themes themselves. Secondly, it omits external factors like political, economic, social, technological, legal and environmental (PESTLE) events. Furthermore, the BMC only provides a static snapshot of an organization (i.e. it is not temporal or longitudinal). The first and second objections can be overturned by simply expanding the framework to include these factors (see, for example, Fig. 8.2 the 'triple-layered business model canvas' (TLBMC) (Joyce and Paquin, 2016)). The BMC as a series of temporal snapshots of an organization provides a more nuanced view of its BM through time (potentially overcoming the third objection).

Policy development theory

A significant political element was identified during the course of producing this case study. Hence, a policy framework (Fig. 8.3) was necessary to contextualize the relevant information.

Overview of the organization

Formerly known as the South Kensington Museum, the V&A was renamed after Queen Victoria and her husband Prince Albert, in celebration of his role in its establishment. Profits from the Great Exhibition of 1851 were intended to be used to develop art and science education, cultural museums and colleges in South Kensington. The V&A, being the first of these institutions, moved to Exhibition Road in 1857. Over the course of time, the initial funding model changed, but how are great institutions like the V&A sustained now? In answering this question, this chapter takes an exploratory case study approach.

Fig. 8.2. Triple-layered business model canvas (TLBMC). From Joyce and Paquin, 2016, modified by Seymour, 2020.

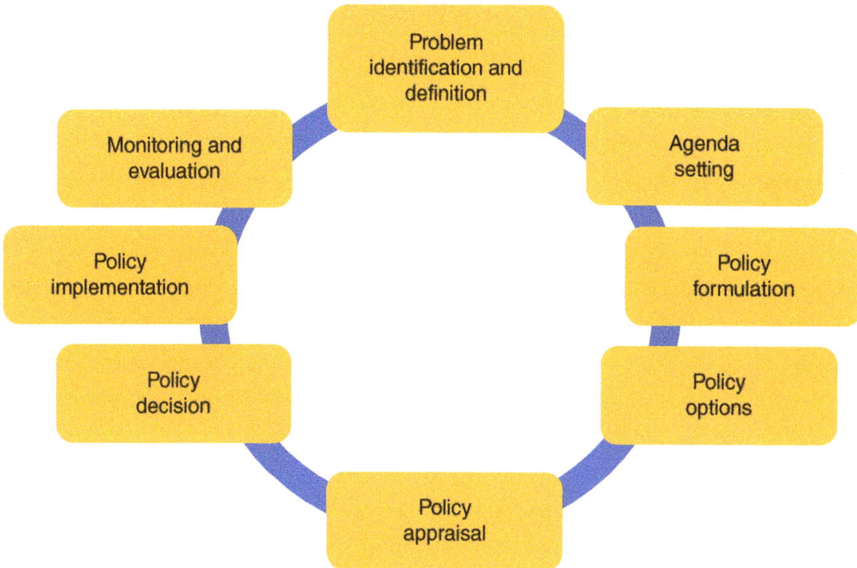

Fig. 8.3. The policy process, based on Her Majesty's Treasury (2011). From LOpen Government Licence v1.0. https://www.nationalarchives.gov.uk/doc/open-government-licence/version/1/open-government-licence.htm (accessed 24 May 2022).

The pre-COVID-19 V&A BM

Historically, the V&A gave face-to-face lectures and demonstrations to art and design students. For a long time, pedagogy was the only way that the V&A interacted with visitors. This lack of diversification potentially hints at a flaw in the initial V&A BM. The second significant finding concerned the phenomenally successful exhibition The Destruction of the Country House, held at the V&A in the autumn of 1974. This exhibition has now achieved 'legendary' status within the heritage sector. At a time when the UK government was considering implementing a wealth tax, The Destruction of the Country House exhibition revealed to the general public that large numbers of UK country houses were being demolished because owners could no longer afford the death duties. Heritage, including the contents of these great houses, was being lost on a vast scale.

So much public interest was generated by the exhibition that eventually the organizers were summoned to Parliament, to answer questions about the issues of the proposed tax and the destruction of heritage (House of Commons, 1975). This evidence (in conjunction with the 1975 European Architectural Heritage Year, promoted by the Council of Europe), led to a shift in government policy which had seismic repercussions for UK heritage and the V&A. However, it should be noted that while The Destruction of the Country House exhibition was a 'huge' success for the V&A and UK heritage, the visitor attendance figures for 1974 were only 1,060,311, which included 135,279 visitors at the Bethnal Green Museum (Table 8.1). This was fewer visitors than in 1895–1899, but was also admittedly in the context of the 1973/74 world oil crisis. From this point onwards, the V&A broadens its mission to also include the preservation of heritage. Parliament also makes provisions for a V&A Conservation Centre. While The Destruction of the Country House exhibition greatly impacted the political and public spheres external to the V&A, the final significant finding indicated that it was an external event which would most impact the V&A. This event was the National Heritage Act of 1983.

The National Heritage Act of 1983 was implemented in the wake of growing public

Table 8.1. V&A visitor attendance data.[a]

Year	Visitor attendance figures	Actual change in visitor numbers from previous period	Change in visitor numbers from previous period (%)	Comment/context	Reference
1895	1,395,876	N/A	N/A	Early V&A attendance figures generally show visitor numbers falling year on year.	Board of Education (1900)
1974	1,060,311	–335,565	–24	The huge success of The Destruction of the Country House exhibition at the V&A led to a change in government policy. Economic disruption caused by the 1973/74 oil crisis.	House of Commons Hansard (1975)
1982	2,058,480	998,169	94	Pre-National Heritage Act.	House of Commons Hansard (1982)
1983	2,221,000	162,520	8	National Heritage Act of 1983 and associated government heritage marketing campaign. English Heritage founded. V&A Board of Trustees established. Scope/ mission of the V&A expanded.	House of Commons (1989)
1984	2,079,000	–142,000	–6		House of Commons (1989)
1985	2,067,000	–12,000	–1		House of Commons (1989)
1986	1,431,000	–636,000	–31	Introduction of admission charges due to reduced government funding. New disease BSE (mad cow disease) caused genuine fear of the unknown which may have impacted tourism and visitor numbers.	House of Commons (1989)
1987	1,392,000	–39,000	–3		House of Commons (1989)
1988	997,000	–395,000	–28	V&A explores a market research strategy to identify visitor needs.	House of Commons (1989)
1995	1,490,000	493,000	49	Pre-introduction of admission charges due to reduced government funding.	House of Commons Hansard (1997)
1996	1,600,000	110,000	7	Introduction of admission charges due to reduced government funding.	House of Commons Hansard (1997)

Continued

Table 8.1. Continued.

Year	Visitor attendance figures	Actual change in visitor numbers from previous period	Change in visitor numbers from previous period (%)	Comment/context	Reference
1999	1,288,396	–311,604	–19	V&A was encouraged by the Department of Culture Media and Sport (DCMS) to be more 'inclusive' and 'attract new audiences' (from under-represented groups, i.e. families, young people, cultural minorities and those with disabilities) via targets aligned to the Funding Agreement (DCMS, 2000). Government policy to implement free entry to museums begins. The V&A begins a programme of audience development. V&A website gets 20,000–40,000 hits/day.	V&A Museum (2001)
2000	1,342,079	53,683	4	V&A continues to be encouraged by the Department of Culture Media and Sport (DCMS) to be more 'inclusive' and 'attract new audiences' (from under-represented groups, i.e. families, young people, cultural minorities and those with disabilities) via targets aligned to the Funding Agreement (DCMS, 2000). Government policy to implement free entry to museums continues. The V&A continues a programme of audience development.	V&A Museum (2001)
2007	3,760,000	2,417,921	180	Rapid expansion in visitor numbers as a result of government policy tied to the funding regime as well as free entry to museums. Financial crisis (credit crunch).	V&A Museum (2008)
2008	5,221,400	1,461,400	39	Rapid expansion in visitor numbers as a result of government policy tied to the funding regime as well as free entry to museums. V&A website had 20,000,000 visits in a year and parts of the collection were digitized. Financial crisis (credit crunch).	V&A Museum (2009)
2017	4,396,557	–824,843	–16	Pre-COVID-19	V&A Museum (2018)

Continued

Table 8.1. Continued.

Year	Visitor attendance figures	Actual change in visitor numbers from previous period	Change in visitor numbers from previous period (%)	Comment/context	Reference
2018	4,293,188	−103,369	−2	Pre-COVID-19	V&A Museum (2020)
2019	4,080,696	−212,492	−5	Pre-COVID-19	V&A Museum (2020)
2020	874,897	−3,205,799	−79	Impact of COVID-19 and lockdowns on V&A visitor numbers (first year).	Sharpe and da Silva (2021)
2021	857,742	−17,155	−2	Impact of COVID-19 on V&A visitor numbers (second year). Sustainability Plan (Net Zero).	Sharpe and da Silva (2021)

[a]BSE, bovine spongiform encephalopathy; COVID-19, coronavirus disease 2019; N/A, not applicable.

enthusiasm, as well as concern, for UK heritage. From this Act, the English Heritage organization was founded. Regarding the effect of the Act on the V&A, three elements emerge: (i) a board of trustees was established; (ii) income-earning activities were developed via V&A Enterprises Limited (a new company which was 100% owned by the V&A); and (iii) the scope/mission of the V&A was also expanded. Dealing with these three elements in turn, although the V&A received a grant from the Office of Arts and Libraries, it seems that the new arrangements meant that this grant would be gradually reduced over time. The trustees were granted control of the V&A by the Act of 1983, and it seems that the trustees were responsible for ensuring that the shortfall (caused by the reduction in aid from the Office of Arts and Libraries), was met through income-earning activities. At this time, the V&A was exploring a 'market research strategy' to identify visitor needs. This strategy appears to have been highly successful, as V&A visitor numbers (including the Bethnal Green Museum) increased from 1,600,000 in 1996/97, to 5,221,400 in 2008/09.

V&A visitor research

Between 1986 and 1996, V&A visitor research indicated that the typical (i.e. the majority of) V&A visitors were white, British, female, general studies or arts graduates, aged 55+, in social groups AB or C1 (i.e. middle class), living in Greater London (or the South East of England), working full-time and who were visiting the V&A for the first time. Significantly, the V&A attracted proportionally twice as many AB and C1 visitors than those reflected in the UK population. The numbers of V&A visitors aged under 24 declined during this period and the museum began to attract more visitors aged over 45. It should also be noted that at this time, visitors to the Bethnal Green Museum, were mainly women aged 25–44.

However, between 1999 and 2000, the V&A was encouraged by the Department of Culture Media and Sport (DCMS) to be more 'inclusive' and 'attract new audiences' (from under-represented groups, i.e. families, young people, cultural minorities and those with disabilities) via targets aligned to the Funding Agreement. Through attempting to understand these groups using market research, trying to meet the targets and a campaign to change perceptions of the V&A ('Contemporary V&A'), the percentage of first-time visitors to the South Kensington site increased by 8% (although overall visitor numbers across all sites fell by 13%). DCMS subsequently published a document which outlined how museums could become 'agents of social change' through inclusive-access 'audience development' and the adoption of a six-point, strategy-based plan.

V&A VISITOR SEGMENTS

V&A visitors form six segments:

- families;
- schools;
- students;
- professionals in the creative industries;
- groups; and
- individual adults not in other categories.

V&A VISITOR NEEDS. Each identified V&A visitor segment has specific visitor needs. For families these needs include: (i) level access for baby buggies; (ii) activities for children; (iii) direct experience and play for children; (iv) accessible displays at suitable heights for children; (v) easy text for adults to scan and answer questions from their children; and (vi) questions and topics of discussion for adults.

For schools (pupils aged 5–18), visits are made in line with the National Curriculum or post-16 public examination courses. School visitors specifically need: (i) 'objects as evidence'; (ii) objects as sources of information (both reliable or unreliable); (iii) insights into 'collection bias'; (iv) objects which are open to interpretation; (v) examples of contemporary art and design; (vi) the opportunity to sketch objects; (vii) space for large groups (of 30–35 or more); (viii) additional visual and written resources for coursework; and (ix) displays at suitable heights.

Likewise, students aged over 16 who are studying at college or university need to study individual objects close up and in depth. They need additional facilities and space for group work, briefings, orientations, guided lectures, as well as seating areas. Also they need: (i) greater access to objects, curators and experts; (ii) information about objects (as well as bibliographies and key sources); (iii) museum resources (i.e. printing, archives, photo library, databases and websites, as well as V&A publications); and (iv) dedicated events and activities for students.

Professionals in the creative industries have similar needs to students, in addition to needing access to information on peer groups as well as opportunities for networking, extended research, access to collections plus advice and support for educators.

For groups, including 'commercial tours', 'adult education', 'clubs and associations', they need quick access to 'key objects', space, seating and 'pleasurable and entertaining' activities. Additionally, foreign language groups need: (i) simple text on signage and panels (and/or translations of the same); (ii) cultural awareness (i.e. of cultures other than English/UK) on behalf of the museum; (iii) information about British culture; (iv) subtitles for films and videos; and (v) gallery programmes (literature) in a variety of other languages.

The individual adults not in other categories don't participate in institutionalized learning (i.e. they are aged 16+ and are not connected with schools, colleges or universities). They are 'motivated by curiosity' and 'manage their own learning'. They have their own social/professional networks and utilize 'word-of-mouth sources' to hear about events and exhibitions. They need multimodal information (which can cope with 'different learning styles'), as well as facilities for independent learning (e.g. books, computers/laptops, desks with good light).

Income-earning activities

Regarding the income-earning activities element, while the V&A had always generated income from admissions fees for special exhibitions, catalogues and photographs, etc., the 1983 National Heritage Act now enabled it to do so on a grand scale. Self-generated income increased from £112,000 in 1974/75 to £63,036,000 in 2018/19. This 2018/19 figure was £22,360,000 more than the grant received from the DCMS for that financial year.

The final element is perhaps the most significant, in that the 1983 Act redefined the remit of the V&A most dramatically in the following area: 'to generally promote the public's enjoyment and understanding of art, craft and design by means of the Museum's collections and by such other means as is considered appropriate' (Office of Arts and Libraries, 1988, p. 34).

In comparison to the original V&A mission 'to educate designers, manufacturers and the public in art and design', this change is perhaps the most striking. Although study and research are also mentioned in the Act, the main emphasis is no longer on education (and perhaps knowledge), but also enjoyment and understanding. Instead of the specific aims of looking to enhance UK manufacturing and design the target is now more diffuse, more ethereal.

During the recovery of COVID-19: V & A Business Model (BM)

While it might be premature to talk about the post-COVID-19 V&A BM, especially in the light of the Omicron variant of the virus, some general observations can be made.

Visitor numbers

Post-lockdown COVID-19 V&A visitor numbers from 17 May to 13 June 2021 were 67,697. This represents an approximate fall of 78% compared with the corresponding period's (pre-COVID) 2019 V&A visitor numbers. However, annual 2020/21 visitor numbers were down almost 79% against 2019/20 numbers.

Explore the Collections

The V&A's Explore the Collections was launched in January 2021. It gives online access to 1.2 million objects in the V&A collection, and spans 5000 years of history. This is a major BM innovation. However, this online collections' platform was originally conceived in 2019 (i.e. pre-COVID-19).

V&A Academy Online

V&A Academy Online, the V&A's adult learning programmes (in art history, sewing, drawing and professional development, etc.) were launched online in financial year 2020/21.

Retail sales

With a new worldwide audience, the V&A expanded online e-book and retail sales. Year 2020/21 online sales were up 112% in comparison with 2019/20. Although in 2020/21 physical V&A shops achieved only 5% of the annual turnover that was reached in 2019/20.

Infrastructure

The V&A is investing heavily in the renovation and rebranding of the Museum of Childhood (to be known as Young V&A), as well as the construction of a new site at Queen Elizabeth Olympic Park (V&A East).

Social media

The museum has made significant gains with its social media presence, now having 3.7 million followers across Facebook, Twitter and Instagram (V&A Museum, 2021).

Human resources

In 2020/21, staffing changes saw reductions from an initial headcount of 1093 at the beginning of 2020, of almost 11%.

Finance

COVID-19 has shattered the V&A's financial situation. However, the museum was very fortunate in that it received an almost £26 million increase in its 2020/21 government grant in aid compared with 2019/20, which (along with £7 million furlough income and an £18.3 million reduction in 2020/21 V&A expenditure) helped significantly to balance the books.

Conclusion

The V&A has a long and treasured history. However, it seems that this case study has highlighted some flaws and gems related to its BM. On the one hand, concerning visitor attendance figures, the original V&A BM was unsustainable. On the other hand, it generated history-making and impactful exhibitions. Changing, or perhaps enhancing the BM in 1983 may have repaired the original flaws, but possibly also created others. For example, museums which have grown increasingly dependent on visitor numbers and self-generated income, faced a very difficult time due to COVID-19.

It may be useful to further explore the drivers behind the V&A visitor expansion.

For example, apart from the DCMS policy identified above, there could be other reasons why this expansion happened at the time that it did (i.e. additional external political, economic, social, technological, legal and environmental (PESTLE) reasons).

This case study has also highlighted how initially the V&A influenced government policy in 1975 through its 1974 exhibition The Destruction of the Country House and how subsequently (in 1983 and 2000) government policy impacted the V&A, to the point where such centrally funded museums have now become vehicles for government-driven social engineering. Understanding how and why this shift occurred could help to formulate strategies to produce more economically, culturally and environmentally sustainable museums.

More pre- and post-COVID-19 research into BM innovation in the heritage sector could be very interesting. A selection of case studies from across a spectrum of heritage sites (if appropriately benchmarked) might achieve some very useful insights into how these institutions adapt. Using data from a range of sources, may provide a wider evidence base upon which to advise interested parties on how best to 'future proof' these important organizations.

Discussion Questions

1. The original V&A business model (BM) was unsustainable, due to falling visitor numbers. UK government policy changing museum BMs in 1983 repaired the original flaws, but also created other weaknesses. Can you explain what BM weaknesses COVID-19 exposed?

2. Using the political, economic, social, technological, legal, environmental (PESTLE) business tool, review Table 8.1 V&A visitor attendance data. How do the various PESTLE factors impact the V&A BM?

3. Applying Fig. 8.2, the triple-layered business model canvas (TLBMC), try to map a heritage organization's BM. Your group needs to use the pre-lesson task resources.

- **(a)** Present findings to the class for 10 mins. What problems did you encounter?
- **(b)** How can the problems be resolved?

Teaching Notes

The case intends to introduce students to the business model canvas (BMC), within the setting of a real-world example. The primary aim is to equip postgraduate, master's level students with the ability and skills to analyse a heritage site or organization's BM. They will gain this knowledge via the three learning objectives. First, through explaining the BM, students will encounter the notions of value creation and value capture. At a deeper level, they will also come to understand the different meanings of value in the context of sustainable business models (SBMs). For example, through incorporating economic, social, environmental and cultural aspects of sustainability, the tensions that exist within and between these elements will readily become apparent. Secondly, the process of mapping an organization's BM encourages systems thinking, promoting the understanding of strategy, design creativity and visual languages (Walls, 2019). Finally, in applying the BM concept to a case study while incorporating the internal and external PESTLE factors, students will gain valuable skills in analysis.

Target audience

The target students are on postgraduate, master's level heritage/tourism/built environment-related courses. Some of these students will have questions that are connected to sustainable heritage/resource management. This case study and the BMC will give students a conceptual framework with which to answer some of their questions.

Teaching methods and equipment

The teaching methods include desk research, group work/collaboration and presentations.

The main teaching resources include: (i) Ondruš and Lyytinen's (2011) simplified business model canvas (Fig. 8.4); (ii) the 'triple-layered business model canvas' (Joyce and Paquin, 2016); (iii) paper, pens and Post-It Notes®; (iv) desktop PCs/laptops for

Infrastructure aspects	Value proposition	Market aspects
Financial aspects		

Fig. 8.4. Simplified business model canvas by Ondruš and Lyytinen (2011). Reproduced with permission of IEEE, from Ondruš and Lyytinen, 2011.

students; and (v) an interactive whiteboard. If the lesson is being conducted online/ blended, a MURAL board (see Boye and Machet, 2021) can replace pens, paper and Post-It Notes®.

Teaching instructions

The case can be taught in 1 day (6 hours excluding breaks and pre-lesson tasks). It is assumed that students will be given a pre-lesson task of selecting a heritage site or organization and obtaining sources such as:

- the history of the heritage site or organization;
- annual accounts;
- annual reports;
- plans;
- policies;
- mission/vision statements;
- news items;
- visitor numbers/visitor research;
- visitor reviews;
- events;
- stakeholder/community information;
- employee information;
- management information;
- environmental information; and
- technology information.

In completing the pre-lesson task, the students should be separated into groups and allocated specific types of these sources to bring to the lesson.

It is advisable to provide students with an explanation of the four aspects of the simplified business model canvas (infrastructure, market, financial and value proposition) at an early stage in the session.

It is suggested that the assessment is formative 75% group work/presentation and summative 25% individual assignment, with perhaps a report of 1500 words. The report could evaluate the efficacy of using the TLBMC to understand heritage organizations/sites and heritage/resource management.

At the start of the session, and especially before the presentations begin, it is important that the instructor manages student expectations/behaviour and sets ground rules.

Duration

360 min

Analysis

1. The original V&A BM was unsustainable, due to falling visitor numbers. UK government policy changing museum BMs in 1983 repaired the original flaws, but also created other weaknesses. Can you explain what BM weaknesses COVID-19 exposed?

Using Table 8.1 V&A visitor attendance data, it can be seen that there is a 24% fall in V&A visitor numbers between 1895 and 1974. Additionally, in the 3 years pre-COVID (2017–2019), V&A visitor numbers fell year on year. The case study indicated that the reasons for the original flaws in the V&A BM were because at the time:

- The V&A was catering to a very narrow audience (e.g. white, middle-class females, aged 55+, living in Greater London, or the South East of England).
- The V&A had a very narrow focus in terms of its remit (education).
- The V&A had very specific value propositions related to education, but these were being negatively impacted because of

Teaching instruction plan

Discussion point	Time required (min)	Teaching strategies
Introduction to the business model concept and the simplified business model canvas	10	The instructor begins the session by asking if anyone in the class has heard of the business model (BM) concept. If they have heard of the BM, the instructor can then ask the class if they can define what a BM is. The instructor can provide the Osterwalder and Pigneur (2010) definition of a BM in terms of how an organization creates and captures value. Using the interactive whiteboard, the instructor can then provide an explanation of the four aspects of the simplified business model canvas (infrastructure, market, financial and value proposition).
Overview of the triple-layered business model canvas	30	Using the interactive whiteboard, the instructor provides an overview of the triple-layered business model canvas (TLBMC) commencing with the economic layer, secondly with the environmental layer, and finally with the social layer. Figure 8.2 is useful in this regard, because it utilizes lots of questions which can be asked of the students with reference to each of the individual building blocks of the TLBMC. The instructor can ask the students: 'What might be missing from the TLBMC?' Potentially, the answer is things like intellectual property, product alternatives, governance, etc. (see Leschke, 2013) and more cultural elements (see Kimbell, 2019a, b; Rex *et al.*, 2019).
Introduction to the PESTLE business analysis tool	10	The instructor introduces the PESTLE business analysis tool, with reference to each of the various elements (political, economic, social, technological, legal and environmental). The instructor can ask the students: 'How do you think the PESTLE business analysis tool might be useful in conjunction with the TLBMC?'
Summary of the case study	10	The instructor briefly provides an overview of the case study, in addition to informing the students about the learning aims, objectives and learning outcomes. The intent is to gradually build on what has been discussed earlier in the session. The instructor will also inform the students about the assessments, as well as the discussion questions.
Discussion question 1	60	**The original V&A BM was unsustainable, due to falling visitor numbers. UK government policy changing museum BMs in 1983 repaired the original flaws, but also created other weaknesses. Can you explain what BM weaknesses COVID-19 exposed?** The instructor examines the case study in detail with the students. The students should find (at least) that the V&A BM was flawed in that its original mission, as well as visitor segments were too restricted and its value proposition was too limited. Additionally, students should understand that COVID-19 exposed BM flaws caused by government policies (see Fig. 8.3 The policy process, based on Her Majesty's Treasury (2011)). Namely, the government encouraged museums like the V&A to be more entrepreneurial, to depend more on self-generated income through high visitor numbers and be less reliant on government grants. The impact of COVID-19 and lockdowns was no visitors and hence no self-generated income. The V&A (and the entire UK heritage sector) needed significant government support to survive.

Continued

Teaching instruction plan Continued.

Discussion point	Time required (min)	Teaching strategies
Discussion question 2	60	**Using the political, economic, social, technological, legal, environmental (PESTLE) business tool, review Table 8.1 V&A visitor attendance data. How do the various PESTLE factors impact the V&A BM?** With reference to the PESTLE factors and Table 8.1, the instructor facilitates examination of the case study by the students. Students should be able to identify, at least, that the economic, political, social and legal factors led to major changes in the V&A BM, especially related to visitor numbers and the reliance on entrepreneurial income-generating activities.
Discussion question 3	160	**Applying Fig. 8.2, the triple-layered business model canvas (TLBMC), try to map a heritage organization's BM. Your group needs to use the pre-lesson task resources.** **(a) Present findings to the class for 10 mins. What problems did you encounter?** **(b) How can the problems be resolved?** The students will get together in their groups. They will map their chosen heritage organization's BM using the TLBMC, as well as the resources from the pre-lesson task. The instructor will move between the groups and facilitate the discussion and the work by asking questions. Each group of students will also prepare and present a 10 min presentation on their chosen heritage organization to the rest of the class. At the end of each presentation, the rest of the class are encouraged to ask the presenting group questions about their presentation and BM. It is important that the instructor effectively manages student expectations and sets ground rules before the presentations begin.
Closing remarks and discussion	20	The instructor concludes the presentations by giving feedback to each group. The instructor will also recap on the session, by providing a synopsis of what was discussed in the lesson.

demographic changes (i.e. reductions in the number of visitors aged under 24 and increases in the number of visitors aged 45+).

The 1983 change in UK government policy, incentivized museums like the V&A, to:

- be more inclusive in terms of attracting a wider audience;
- expand their remits to include:
  - income-generating activities; and
  - a focus on promoting the public's enjoyment.

Coupled with a policy of free entrance to museums from 2007, these changes led to a rapid expansion in visitor numbers. Table 8.1 shows a 180% increase in V&A visitor numbers between 2000 and 2007. However, with the onset of COVID-19 and lockdowns in 2020, the decline is almost equally dramatic. Table 8.1 shows a 79% decrease in V&A visitor numbers between 2019 and 2020. With its ability to self-generate income being highly constrained, the V&A essentially needed a £26 million government funded bailout (excluding the amount received from the government's Furlough Scheme).

2. Using the political, economic, social, technological, legal, environmental (PESTLE) business tool, review Table 8.1 V&A visitor attendance data. How do the various PESTLE factors impact the V&A BM?

An example of the answer to this question is as follows.

Political

- The National Heritage Act of 1983 and associated government heritage marketing campaign led to an expansion of the V&A's visitor segments and also impacted the V&A BM visitor relationships and visitor channels.
- When English Heritage was founded this led to a change in the V&A's value proposition.
- When the V&A Board of Trustees was established the scope/mission of the V&A expanded. This led to a change in the V&A's: (i) value proposition; (ii) key activities; (iii) key partners; (iv) key resources; (v) revenue streams; and (vi) cost structure.

Economic

- Economic disruption caused by the 1973/74 oil crisis might have negatively impacted the V&A's: (i) revenue streams; and (ii) cost structure.
- Economic disruption caused by the 2007/08 financial crisis (credit crunch) might have negatively impacted the V&A's: (i) revenue streams; (ii) cost structure; (iii) key partners (some partners might have suffered financially); and (iv) key resources.
- The revenue losses and cost structure impacts might have been mitigated by self-generated income from increased visitor numbers.

Social

- The changing demographic (reduction in the number of visitors aged under 24 and an increase in visitors aged over 45) led to a change in the V&A's visitor segments.
- The government's policy of 'social inclusion' led to a change of the V&A's visitor segments.

Technological

- The rise of the Internet (1980s) had an impact on the V&A, for example the V&A website had 20,000–40,000 hits/day in 1999 which led to an expansion of the V&A's visitor segments. This also impacted the V&A BM visitor relationships and visitor channels.
- The V&A website had 55,000 hits/day in 2008 (V&A Museum, 2009) which led to an expansion of the V&A's visitor segments and impacted the V&A BM visitor relationships and visitor channels.
- Digitization of parts of the V&A collection led to a change in the V&A's: (i) value proposition; (ii) key activities; (iii) key partners; (iv) key resources; (v) revenue streams; and (vi) cost structure.

Legal

- The National Heritage Act of 1983 had implications for the V&A BM (see 'Political', above).
- During the COVID-19 pandemic, lockdowns were imposed as a legal way to reduce the spread of the virus in 2020/21. This had implications for the V&A BM and led to a rapid reduction in the V&A's visitor segments, and impacted the V&A BM visitor relationships and visitor channel.
- Now the V&A uses social media and online technologies to reach its audience. This has led to changes in: (i) the value proposition; (ii) key activities; (iii) key partners; (iv) key resources; (v) revenue streams; and (vi) cost structure.

Environmental

In 2021 the V&A implemented its Sustainability Plan (Net Zero). This impacts:

- the V&A BM visitor segments and value propositions; and
- changes the following in the V&A BM: (i) key activities; (ii) key partners; (iii) key resources; and (iv) revenue streams.

3. Applying Fig. 8.2, the triple-layered business model canvas (TLBMC), try to map a heritage organization's BM. Your group needs to use the pre-lesson task resources.
   (a) Present findings to the class for 10 mins. What problems did you encounter?
   (b) How can the problems be resolved?

For example, mapping the BM of the V&A using the TLBMC 'economic layer' might give the V&A business model canvas (BMC) shown in Fig. 8.5.

For example, mapping the BM of the V&A using the TLBMC 'social layer' might give the V&A BMC shown in Fig. 8.6.

For example, mapping the BM of the V&A using the TLBMC 'environmental layer' might the V&A BMC shown in Fig. 8.7.

(a) Present findings to the class for 10 min. What problems did you encounter?

- Mapping a heritage organization's BM using the TLBMC produces a static picture of how the organization creates and captures value.
- Some very relevant points relating to the V&A were omitted in the

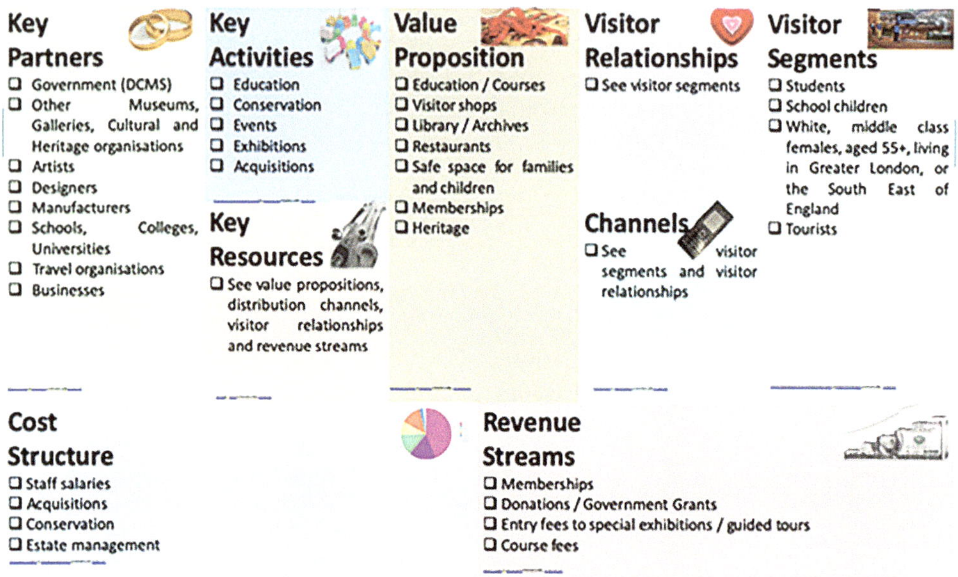

Fig. 8.5. V&A business model canvas (BMC) (economic layer). Authors' own diagram.

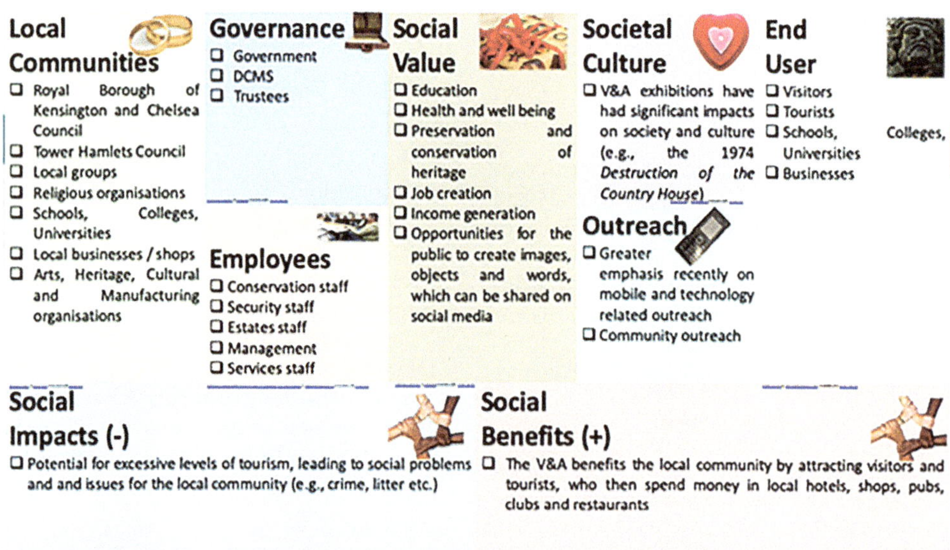

Fig. 8.6. V&A BMC (social layer). Authors' own diagram.

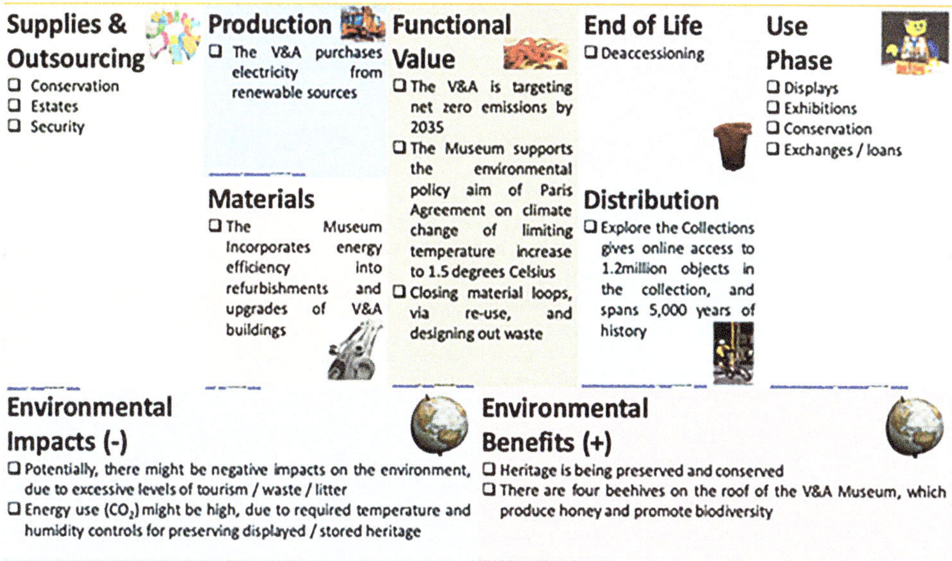

Fig. 8.7. V&A BMC (environmental layer). Authors' own diagram.

mapping. For example, the TLBMC does not incorporate political factors (e.g. the introduction of the 1983 National Heritage Act).

- The process of mapping BMs can be very laborious and time consuming, especially if there are lots of data.

(b) How can the problems be resolved?

- It might be better to try to incorporate some of the temporal (i.e. longitudinal) elements into the BM mapping process.
- Combining the TLBMC with business tools, like PESTLE, allows the inclusion of elements such as politics and government policy when mapping BMs.
- Dividing the work among the group helps to improve the speed and efficiency of the BM mapping process. For example, if there are six people in a group, two can focus on each layer of the TLBMC (i.e. two on the economic layer, two on the social layer and two on the environmental layer).

References

Atkinson, R. (2020) One third of American museums at risk of closure. *Museums Association*, 23 November. Available at: https://www.museumsassociation.org/museums-journal/news/2020/11/one-third-of-american-museums-at-risk-of-closure/ (accessed 31 March 2021).

Board of Education (1900) *Report of the Board of Education 1899–1900 Volume II Appendix to the Report (Secondary Education)*. HMSO, London.

Boye, T. and Machet, T. (2021) *Mixed-mode Teaching: Emerging from COVID-19 to Future Practice*. SEFI (European Society for Engineering Education), Brussels.

Centre for Economic Business Research (2019) The heritage sector in England and its impact on the economy. Available at: https://historicengland.org.uk/content/heritage-counts/pub/2019/the-heritage-sector-in-england-and-its-impact-on-the-economy-pdf/ (accessed 3 February 2021).

DCMS (Department of Culture Media and Sport) (2000) Centres for Social Change: Museums, Galleries and Archives for all: Policy Guidance on Social Inclusion for DCMS funded and local authority museums, galleries and archives in England. Available at: https://ia902500.us.archive.org/31/items/b32225386/b32225386.pdf (accessed 17 August 2022).

English Heritage (2019) *English Heritage Annual Report 2018/2019.* Available at: https://www.english-heritage.org.uk/siteassets/home/about-us/eh-annual-report-2018_19_full.pdf (accessed 2 March 2020).
Her Majesty's Treasury (2011) The Green Book: appraisal and evaluation of central government. Available at: https://www.gov.uk/government/publications/the-green-book-appraisal-and-evaluation-in-central-government (accessed 23 May 2022).
Historic England (2020) *Annual Report and Accounts 2019–2020*. Historic England, London.
Historic Houses (2020) Historic Houses Key Stats Infographic. Available at: https://www.historichouses.org/resources/all-resources/key-stats-infographic.html (accessed 7 May 2020).
House of Commons (1975) *Select Committee on a Wealth Tax 1974–1975*, Volume III. HMSO, London.
House of Commons (1989) *Should Museums Charge? Some Case Studies*. House of Commons, Parliament Session 1989–1990. HMSO, London.
House of Commons Hansard (1975) *Written Answers (Commons) of Monday 19th May 1975*. House of Commons, Parliament 1974–1975. HMSO, London.
House of Commons Hansard (1982) *Written Answers (Commons) of Monday 17th January 1983*. House of Commons, Parliament 1982–1983. HMSO, London.
House of Commons Hansard (1997) *Written Answers (Commons) of Thursday 30th March 1997*. House of Commons, Parliament 1996–1997. HMSO, London.
Hunt, T. (2020) With a renewed mission, museums can play a vital role in our recovery. *Evening Standard*, 24 April. Available at: https://www.standard.co.uk/comment/comment/with-a-renewed-mission-museums-can-play-a-vital-role-in-our-recovery-a4423291.html/ (accessed 25 May 2022).
Joyce, A. and Paquin, R.L. (2016) The triple layered business model canvas: a tool to design more sustainable business models. *Journal of Cleaner Production* 135(1), 1474–1486.
Kimbell, L. (2019a) Insights from management literature to understand business models in arts and cultural organisations. (*Creative Lenses Working Paper* No. 2.) Stronger Arts and Cultural Organisations for a Greater Social Impact. Available at: https://ualresearchonline.arts.ac.uk/id/eprint/14859// (accessed 25 May 2022).
Kimbell, L. (2019b) Modelling shared value and mediating values: describing business models in performing arts organisations and cultural venues. (*Creative Lenses Working Paper* No. 3.) Available at: https://ualresearchonline.arts.ac.uk/id/eprint/14858/ (accessed 25 May 2022).
Leschke, J. (2013) Business model mapping: a new tool to encourage entrepreneurial activity and accelerate new venture creation. *Journal of Marketing Development and Competitiveness* 7(1), 18–26.
Office of Arts and Libraries (1988) *Museums & Galleries: Accounts for Year Ending 31 March 1987*. Office of Arts and Libraries, London.
Ondruš, J. and Lyytinen, K. (2011) Mobile payments market: towards another clash of the Titans? In: *Proceedings of the 2011 10th International Conference on Mobile Business*. IEEE Computer Society, Washington, DC, pp. 166–172. doi: 10.1109/ICMB.2011.41.
Osterwalder, A. and Pigneur, Y. (2010) *Business Model Generation: a Handbook for Visionaries, Game Changers, and Challengers*. Wiley, London.
Rex, B., Kaszynska, P. and Kimbell, L. (2019) *Business models for arts and cultural organisations: research findings from Creative Lenses*. Project Report. Trans Europe Halles, Lund, Sweden. Available at: https://ualresearchonline.arts.ac.uk/id/eprint/14392/ (accessed 25 May 2022).
Seymour, C. (2020) Heritage Business Models. Presented at Doctoral Skills Research Poster Competition 2020. University College London (UCL), London.
Sharpe, E. and da Silva, J. (2021) Visitor figures 2020: top 100 art museums revealed as attendance drops by 77% worldwide. *The Art Newspaper.* Available at: https://www.theartnewspaper.com/2021/03/30/visitor-figures-2020-top-100-art-museums-revealed-as-attendance-drops-by-77percent-worldwide (accessed 26 April 2022).
V&A Museum (2001) *Annual Report and Accounts 2000–2001*. HMSO, London.
V&A Museum (2008) *Annual Report and Accounts 2007–2008*. HMSO, London.
V&A Museum (2009) *Annual Report and Accounts 2008–2009*. HMSO, London.
V&A Museum (2018) *Annual Report and Accounts 2017–2018*. Williams Lea Group on behalf of the Controller of Her Majesty's Stationery Office, London.
V&A Museum (2020) *Annual Report and Accounts 2019–2020*. APS Group on behalf of the Controller of Her Majesty's Stationery Office (HMSO), London.
V&A Museum (2021) *Annual Report and Accounts 2020–2021*. APS Group on behalf of the Controller of Her Majesty's Stationery Office (HMSO), London.
Walls, A.J. (2019) Mapping a business model in 5 steps. Available at: https://medium.com/swlh/a-new-methodology-for-business-model-mapping-6a122f140579 (accessed 18 February 2022).

how destructive such natural catastrophes can be. Thousands of locals and visitors were killed and badly injured in such incidents, and thousands of citizens were impacted. Aside from mortality, two primary dangers that catastrophic event regions may confront are displacement and financial damage. Both man-made and natural disasters can affect a tourist destination. Disasters caused by natural calamities include floods, tsunami, landslides, cloudbursts, famines, cyclones and earthquakes, while man-made disasters include terrorist attacks, wars and various anti-social activities – both types of disaster may affect the life of not only the tourist community but also the host community. Tourists desire a place to relax, so the existence of a threat of a natural calamity or an exigency situation in a touristic location may encourage tourists to seek out other recreational destinations to enhance the value they may obtain from the leisure engagement. As a consequence, economic gains gained by a destination drop substantially in parallel with the plummeting tourist numbers, compounding the damage caused by a natural catastrophe. Furthermore, even if no real natural catastrophe has happened and such events are simply potential, the effect of a natural hazard persists. To look at it another way, vacation choices are made by assessing the risks. These natural disasters tend to lower total foreign visitor flows to any tourist site.

Odisha is very much vulnerable to disaster because of its geographic and socio-economic conditions. This is the world's sixth most cyclone-prone area with the east coast having the highest risk of cyclone landfall. Out of the 1019 cyclonic storms in the Indian subcontinent in the last century, 890 occurred along the eastern coast and 260 of these made landfalls on the Odisha coast. For beach tourism, Puri and Ganjam districts are the most popular destinations in the state of Odisha. Because of the growing importance and growth in number of tourists visiting the beaches of Puri and Ganjam, these districts require much more attention from the government for their promotion and strategic development, to protect these destinations from potential problems from such disasters and to create a positive image in the mind of tourists forever. Many cyclones that have made landfall in and around the Puri and Ganjam coastal belt have badly affected the tourism industry. Being famous beach tourism destinations and revenue generators for the state, in the absence of a robust disaster management strategy, these destinations may lose their importance resulting in a decrease in tourist arrivals for years to come.

Destination 1 – Puri District

Located in the eastern part of Odisha, Puri District is a coastal district. It is the abode of Lord Vishnu, popularly known as Lord Jagannath of Odisha. The district is named after the heritage city of Puri, one of the four pilgrimage sites of India. The district is filled with gracious tourist attractions. Puri Beach, on Odisha's tropical coast, is one of the most beautiful tourist spots, with sun-kissed beaches that wet visitors' feet and a soft breeze that lifts them from the earthy perfumes to the clouds. The *Casuarina*-fringed Balighai Beach is one of the most well-maintained and serene beaches and it is an enchanting sight to watch the waves crash on the low-lying rocks below and admire the golden rays of the setting sun from the many viewpoints. Sun Temple, a proclaimed heritage hotspot in Konark, residing by the sand-filled beaches, is a masterpiece to reckon with and is accentuated by the week-long dance festivities that showcase artists from various corners. The Chandrabhaga Beach of Konark and the International Sand Art Festival offer a breath-taking experience for visitors.

Destination 2 – Ganjam District

Another most prominent district in Odisha is Ganjam District. The district has derived its name from '*Gan-i-aam*', which means 'food grain warehouse'. Ganjam is located on the Bay of Bengal's eastern shore, and as a result it is known for its various unique beaches, which contribute to the city's year-round tourism. This magnificent location is rich in historic ruins and is adorned with verdant greenery, accentuated by huge mountains and magical rivers. Ganjam is an area dotted with holy and ancient temples where hundreds of worshippers' flock to seek miracles from the Almighty. Some of its most beautiful sandy beaches, such as Aryapalli and Huma Kantiagada, are unrivalled.

Cyclones in Odisha

Cyclones have never been unfamiliar to Odisha. Being a cyclone-prone state, it is also known as the national disaster capital of India. The state has experienced everything from minor to major disasters. It has overcome the difficulties caused by the big tides, flourishing again after the waters wiped away the state's splendour and left it in ruins. Many cyclones spin and twirl in the Bay of Bengal, including those that make landfall in Odisha. Furthermore, the part of Odisha from Puri to Bhadrak extends further out towards the sea than the rest of the coast, making them more prone to cyclonic activity.

Table 9.1 lists the different cyclones that have affected the state in the last few years since 1999 and indicates their intensity.

Disaster management approach

For effective disaster management at the national, state, district and local levels, the Disaster Management Act 2005 lays down the institutional and coordination mechanisms in India. The Government of India developed a multi-tiered institutional structure comprising: (i) the National Disaster Management Authority (NDMA) headed by the Prime Minister; (ii) the State Disaster Management Authorities (SDMAs) headed by the respective chief ministers; and (iii) the District Disaster Management Authorities (DDMAs) headed by the district collectors. Together these authorities were tasked with formulating the National Disaster Management Plan (NDMP), as per United Nations Environment Programme (UNEP) guidelines, to provide a framework and direction to national disaster response. This dynamic document, the NDMP, is constantly being improved, keeping up with the latest developments in knowledge about disasters and the emerging global best practices in disaster management. This document is designed to be ready to respond to an emergency at any time for any part of the county. It is developed in such a way that it may be applied as needed in all phases of disaster management in a flexible and scalable manner such as prevention, mitigation, preparedness, response, recovery and reconstruction (see Fig. 9.1). The NDMP is aligned with United Nations (UN)-supported practices across the world, particularly the Sendai Framework for Disaster Risk Reduction 2015–2030. The Sendai Framework intends to reduce the risk of disaster and loss of lives, livelihoods and health, as well as the physical, economic, social, environmental and cultural assets of individuals, enterprises, communities and nations, by 2030. India has committed to take efforts to help the world in achieving its goals by continuously advancing the whole disaster management

Table 9.1. List of cyclones that have affected Odisha (1999–2019). Compiled by authors.

Name of cyclone	Year	Winds (km/h)	Landfall area	Affected areas
Super cyclone	1999	276	Jagatsingpur	Fourteen coastal districts and 28 coastal towns
Phailin	2013	215	Ganjam	Odisha coastlines
Hudhud	2014	185	Near to Ganjam (Vizag)	Ganjam, Gajapati, Koraput and Puri
Titli	2018	110	Ganjam	Ganjam, Puri, Gajapati, Bhadrak, Balasore and Khurda
Fani	2019	250	Puri	Puri, Ganjam, Cuttack, Kendrapara and Jagatsingpur

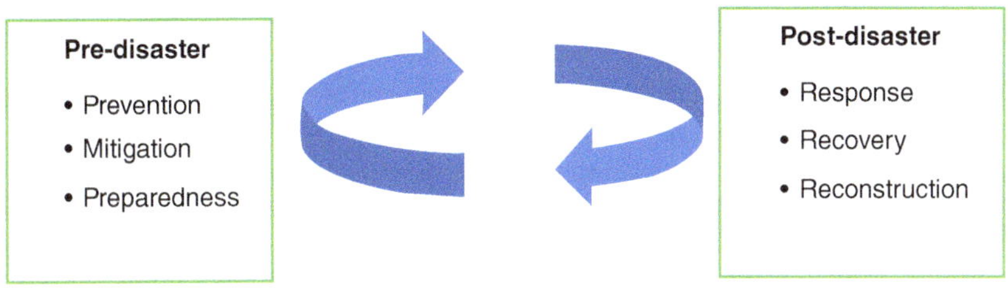

Fig. 9.1. The disaster management cycle. Compiled by the authors.

cycle by following the Sendai Framework's recommendations and adopting globally recognized best practices.

The following four actions are emphasized under the Sendai Framework:

- recognizing the risk associated with disaster;
- establishing a robust risk governance system for disaster risk management;
- developing resilience by investing in disaster risk reduction; and
- managing disaster preparedness for 'Build Back Better'.

The NDMP is a framework that covers all areas of the disaster management cycle such as risk reduction and mitigation, preparedness, response, recovery, rehabilitation, etc. It advocates that effective disaster management in an area requires an all-inclusive framework encompassing multiple hazards. It uses an integrated strategy to ensure that government agencies, the private sector, a variety of other relevant organizations, and local communities are all involved. It proposes the structure shown in Fig. 9.1 as a disaster management cycle for carrying out disaster management operations in disaster-affected areas in various phases, depending on the kind and magnitude of the catastrophe.

Impact of cyclones on coastal tourism destinations

Odisha is a rich amalgamation of spiritual, historical, cultural, religious and nature-based tourism resources. Cyclones substantially affect Odisha's tourism sector. Electricity, telephone infrastructure, transportation services and other business operations are badly impacted by a cyclone. Major roads get obstructed owing to uprooted trees, broken road signs and damaged culverts and there are power outages in many parts of the state for about a week. Residential houses are also severely damaged by high wind speeds, leaving many people homeless. Agricultural activities, fisheries, livestock, etc. all suffer significant losses. Coastal tourism destinations such as Puri, Chillika Lake, Konark and Gopalpur located in the coastal areas of the state are severely affected (Mohapatra, 2020).

The disasters shake up the tourism industry in such destinations. A large number of tourists, especially international tourists, cancel their bookings generating losses for many local tourism suppliers. The uprooting of trees on a large scale damages the landscape including gardens and the walls that surround historical monuments, leaving the scenic beauty attractions unrecognizable. The beach's strip stands eerily empty. Hotels, restaurants and shops lay vacant. The hotel industry is the worst sufferer of a cyclone hit. Cultural and promotional events get cancelled. Many of the performing artists and craftsmen face distress due to loss of livelihood because of the cancellation of cultural events and damage to the art-and-craft-making centre. The annual tourist traffic to the state is affected, causing a huge loss for the tourism industry. The cyclone's aftermath, on the other hand, provides an opportunity for all stakeholders to review the existing situation and progress towards more robust alternatives. It gives government officials, civic society, communities, the corporate sector and concerned individuals an opportunity to work together.

Implementation of plans for the disaster management cycle

The disaster management cycle comprises two stages. The first stage, the pre-disaster action plan comprises *prevention, mitigation and preparedness* focusing on reduction of human and property losses caused as a result of the probable threats. The second stage consists of post-disaster initiatives such as *response, recovery and reconstruction* following a disaster, to accomplish early rehabilitation and repair of affected victims such as tourists and the local community of coastal destinations.

Disaster prevention and mitigation in coastal destinations refers to a variety of efforts that are thought to prevent or lessen the detrimental consequences of a disaster in the short and long term. Thus, mitigation and risk reduction are connected with the established legal and organizational framework for: (i) risk reduction; (ii) assessing the intensity of hazards and communicating this to the local community; (iii) early warnings; (iv) making use of existing resources; and (v) sharing the necessary technology, etc. to prevent, reduce and mitigate disaster. All actions and activities that provide the community with the ability for an effective and structured reaction are included in the 'preparedness' or pre-impact phase. The UN suggested the APELL (Awareness and preparedness for emergencies at the local level) process for managing preparedness in coastal destinations. Over the last two decades, the APELL programme worked to bring the APELL mechanism to global attention, to help local partners foster long-term stakeholder partnerships, and conduct capacity-building projects, thereby boosting preparedness and response systems in several industries, including tourism and communities in both developed and developing nations. APELL's approach consists of implementation of the ten steps shown in Fig. 9.2.

This mechanism is very much pertinent to raising awareness and capabilities in communities that haven't been included in national contingency plans or early warning systems before. This assists local inhabitants in the development of information and decision-making mechanisms for coping with the disaster that affects them. During the post-disaster phase, some activities are undertaken in the immediate aftermath of a disaster to achieve early recovery and restoration of the affected area and its local community. These are referred to as actions of response and recovery. Response activities include efforts to reduce the risks posed by a calamity which includes search, rescue and providing some disaster aids.

The local community plays an important role in undertaking such activities as rescuing the local people and the tourists. In the immediate hours and days after the disaster, providing food and non-food items, sanitation and hygienic facilities, health and shelter interventions are very much vital. The immediate purpose of the recovery stage is to restore some degree of normalcy to the devastated areas of such coastal locations. The activities of recovery include providing temporary or transitional shelters, food and drinking water, and enabling people to resume their daily lives. This also includes: (i) arranging temporary accommodation; (ii) transportation for the visitors/tourists to return to their hometown; and (iii) assisting tourism and other businesses to operate normally.

In contrast to the recovery stage, reconstruction refers to long-term development support that can aid the people in affected areas in rebuilding their lives and meeting their current and future requirements. It takes into consideration future disaster risk reduction. Damaged structures and resources may not be restored to their original shape or location as a result of rehabilitation. However, it will replace the temporary arrangements and upgrade the infrastructure. Thus, reconstruction must be thoroughly incorporated into long-term development plans, taking into consideration future disaster risks and the potential for mitigating them via suitable measures. During reconstruction, the property's location or construction materials are taken into account.

To carry out the rescue, recovery and reconstruction task, optimum involvement of the local community, local tourism stakeholders, environmentalists, police administration

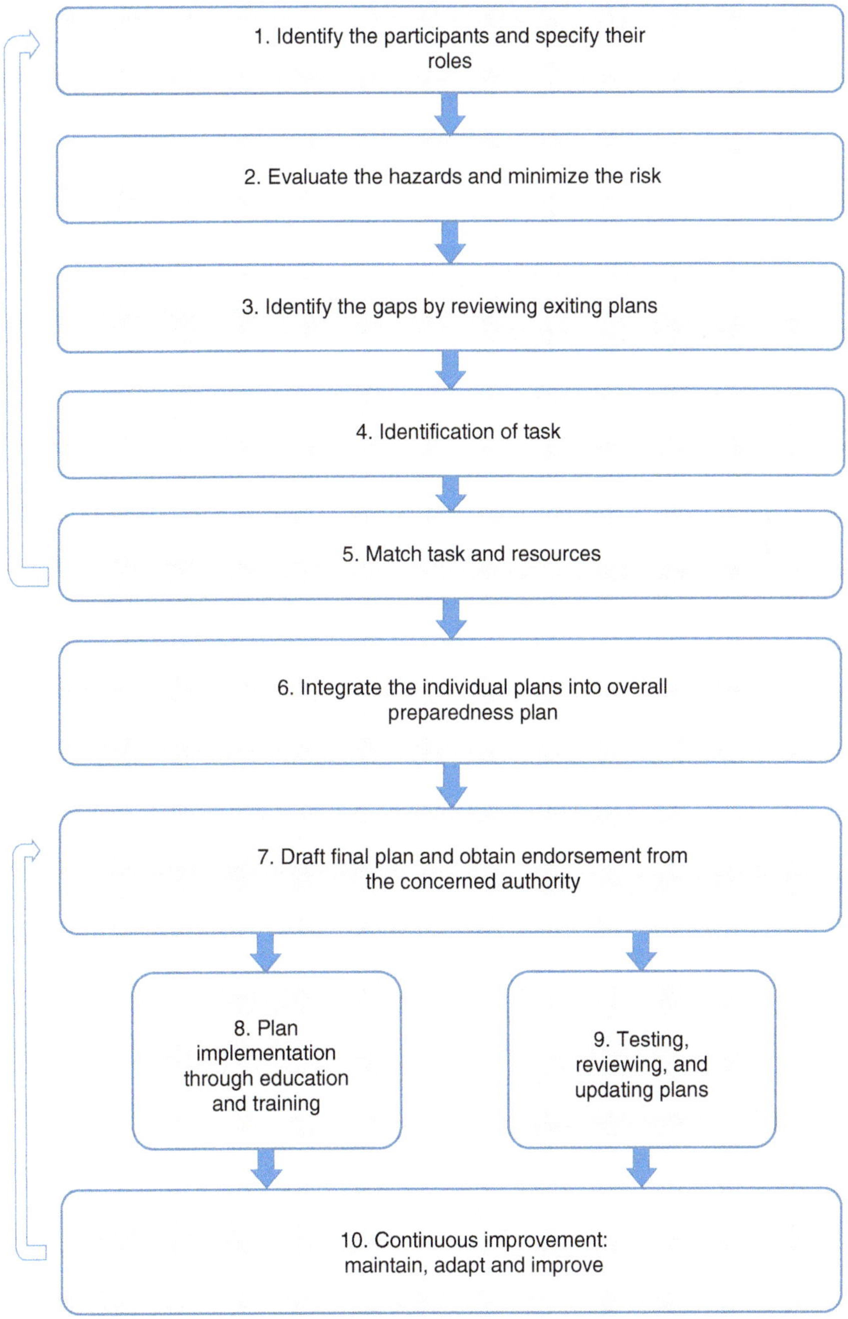

Fig. 9.2. APELL's ten-step process. APELL, Awareness and preparedness for emergencies at the local level. Compiled by the authors.

and various volunteer groups should be taken into consideration. Government aid and facilitation also must be important resources to undertake the response, recovery and reconstruction activities and to evacuate people from anticipated affected areas.

Several tourists visiting the coastal destinations can be utilized as volunteer tourists to help carry out various recovery procedures for cyclone-affected areas. The volunteer tourists are a set of individuals who are willingly ready to volunteer themselves to offer their help to support the victims in affected areas. The tourists from within the country and various parts of the world may help by providing a wide range of services such as: (i) rebuilding homes; (ii) serving as medical interns in hospitals; and (iii) volunteering for non-governmental organizations (NGOs) to re-establish sustainable development of the local community and overcome post-disaster effects. The volunteer tourist community in post-cyclonic disaster areas may use their expertise, skills, knowledge and support with their available resources to help the local governing body to implement their rescue action plan. Volunteering tourists can play a vital role as members of the recovery task force constituted by the local government or any non-governmental body to undertake the evacuation process and recovery task.

Pre-disaster and post-disaster activities by Odisha

In October 1999, a super cyclone slammed down the entire Odisha coast. Over 10,000 people were killed and millions were displaced due to its severe effect. This disaster is forever engraved in minds of the people as a cataclysmic failure of the government, as the huge winds wreaked devastation, revealing people's terrible susceptibility to nature's whims. Another cyclone named Phailin destroyed the state in 2013, impacting over 1.1 million people and incurring Rs. 4240 crores in losses. This made the government be alert and careful in planning and preparing to face future such disasters. In 2019, Fani, a 'rare' summer storm, might have caused havoc in the districts of Puri and Khurda if it hadn't been for the state government's vast pre-disaster preparedness. This helped the government in saving many lives and reducing damage to its natural and heritage resources. However, this did not happen just a day before the cyclone; the state had made a concerted effort to build capacity throughout time, especially at the community level. As part of the National Cyclone Risk Mitigation Project, it has: (i) issued community-level alerts; (ii) constructed multi-purpose cyclone shelters; and (iii) established an early warning dissemination system with last-mile connectivity. As a result, the capacity of communities to deal with natural disasters has increased dramatically.

Tourism is dependent on a healthy ecosystem as well as institutional mechanisms that can respond to community needs. As a result, disaster preparedness and management should be a key component of any destination's integrated management strategy. Disasters have always existed and will continue to exist. Coastal areas like Puri and Ganjam districts will become increasingly vulnerable to sea-level rise and weather-related calamities due to global warming, which needs to be managed properly with well-designed mitigation, preparedness, recovery and reconstruction activities. However, the lesson from the super cyclone of 1999 is to have an effective disaster management system with more emphasis on mitigation preparedness. In such a scenario, the implementation of the UN's APELL process for improving disaster preparedness and capabilities of communities is very much vital for managing the disasters.

Discussion Questions

1. Discuss how natural disasters affect the tourism industry using suitable examples.
2. Describe the pre-disaster and post-disaster activities of a disaster management cycle.
3. Discuss the APELL mechanism for disaster preparedness and risk reduction in coastal destinations of the Puri and Ganjam districts.
4. Identify the measures taken by the Government of Odisha for managing disasters in coastal destinations since the super cyclone of 1999.
5. What are the current challenges faced by the Government of Odisha in implementing the action plans for disaster management in coastal destinations?

Teaching Notes

This case study has been developed to familiarize students with the concept of the disaster management cycle in managing coastal tourism destinations affected by cyclones. The case aims to find a resilience disaster management approach for successfully managing coastal tourism destinations by minimizing the risk of cyclonic turmoil and taking resilience measures that adversely affect Puri and Ganjam district's coastal belt. The case study will enable the learners to apply the concept of disaster management with special reference to the coastal tourism destinations, which also will help them to deal with similar situations in their professional careers. The learners will be given an opportunity to analyse and understand each activity of the disaster management cycle and recommend how such activities could be designed and implemented. The instructor must be able to correct the learners in case of any inappropriate decision and lead them in the right direction, justifying the reasons and relevance.

Students can work in small groups of four to five individuals for this case study.

Target audience

The case study will be used for bachelor's and master's programmes in tourism management, general management and disaster management.

Teaching method and equipment

In the beginning, the case will be initiated as a situation analysis and which can take the shape of a group discussion. The students will be required to collect data relating to how many cyclones had affected Odisha in the last 5 years. A detailed analysis will be made by the students on the impact of such cyclones on the tourism industry of the state. The situation analysis will be then narrowed down to analyse the impact of such cyclones on coastal tourism destinations in the Puri and Ganjam districts of Odisha. The students are then required to access the Internet and search for literature relating to the subject. The instructor may help the students to review similar kinds of literature from online sources to have a better understanding of the scenario. The students need to note down the relevant takeaway messages based on which appropriate measures could be recommended by them. To facilitate the group discussion and towards the obtainment of more concrete ideas, audio-visual equipment such as computers and projectors can be used.

Teaching instructions

The case can be discussed and taught in a classroom setting for a time of 2 hours, as per the teaching instruction plan shown below. The teaching plan will enable the learners to become familiar with disaster management techniques for 'building back better' destinations affected by cyclones.

First, the instructor will make the learners familiar with various types of natural disaster and how it affects a tourist destination, substantially reducing tourist inflows, by citing the various examples (provided earlier in the case study). A thorough discussion will begin with the students knowing destinations visited by them that have been affected by such natural disasters. This will enable them to visualize the scenario and practically understand the devastating impact of such disasters and damage to the tourism industry from various perspectives. Also the learners will be exposed to the role of the various disaster management authorities (NDMA, SDMA and DDMA) and the NDMP framework.

The second hour will obtain the learner's ideas on how to manage such natural disasters in a tourist destination through a brainstorming session. First, students will be made to understand the destination disaster management cycle shown in the Fig. 9.1.

Teaching instruction plan

Teaching point	Time required (min)	Teaching strategies
Introduction	10	The instructor will initiate the case by discussing natural and man-made disasters that affect the tourism industry, asking the students if they have been subjected to such kind of experience before. The instructor may ask them to share their experiences based on their recent visits to any of such disaster-affected destinations.
Briefing about the disaster management cycle	20	The instructor will make the students aware of the disaster management authorities (NDMA, SDMA and DDMA) and National Disaster Management Plan (NDMP) providing a framework of disaster management cycles consisting of different phases and actions that have to be undertaken in each phase to manage a tourist destination affected by a natural disaster.
Overview of the case study	20	The case study focuses on the state of Odisha and how due to its peculiar geographical location it is very prone to cyclonic storms. Puri and Ganjam districts are the districts that are most affected by cyclones and are included under the scope of the case study. The case study utilizes the disaster management approach recommended by the United Nations Environment Programme (UNEP) in managing and mitigating the impact of cyclones in the coastal tourism destinations of Puri and Ganjam districts. Students can work in small groups of four or five individuals to answer the discussion questions.
Assignment question 1	20	**Discuss how natural disasters affect the tourism industry using suitable examples.** The background of the case will be discussed and the impact of cyclones on the tourism industry affecting different components of tourism in the coastal tourism destinations will be emphasized.
Assignment question 2	10	**Describe the pre-disaster and post-disaster activities of a disaster management cycle.** Having analysed how a cyclone affects the coastal destinations, the students will be required to respond to each phase of the pre-disaster and post-disaster management cycle by contributing their ideas that will serve as measures for each such phase. Their answers should be prepared and presented in the form of a PowerPoint presentation justifying the rationale behind choosing the particular course of action.
Assignment question 3	10	**Discuss the APELL mechanism for disaster preparedness and risk reduction in coastal destinations of the Puri and Ganjam districts.** The UN's APELL process (Awareness and preparedness for emergencies at the local level) is a significant mechanism to create awareness about natural hazards and minimize hazard risk reduction. The learners will be required to find out and recommend the roles and responsibilities of all community members in probable emergency readiness and response measures concerning the coastal destinations of the Puri and Ganjam districts.

Continued

Teaching instruction plan Continued.

Teaching point	Time required (min)	Teaching strategies
Assignment question 4	10	**Identify the measures taken by the Government of Odisha for managing disasters in coastal destinations since the super cyclone of 1999.** The students will be familiarized with the lesson learned by the state from the super cyclone of 1999 and its preparedness to manage and mitigate the future impacts of the disaster on coastal tourist destinations of Puri and Ganjam districts. The state has now established a robust disaster management plan with the constitution of the Odisha State Disaster Management Authority, an early warning dissemination system, multi-purpose cyclone shelters, and capacity building in the local community to deal with cyclones that could save many lives, and natural and physical resources of coastal destinations in Puri and Ganjam districts.
Assignment question 5	10	**What are the current challenges faced by the Government of Odisha in implementing the action plans for disaster management in coastal destinations?** The learners will be made aware of the current challenges faced by the Government of Odisha which need to be looked after for the successful implementation of the disaster management plan. Currently, there is a lack of coordination among the authorities at different levels. Various disaster funds provided for the purpose are not utilized in a proper way. Further, there is a delay in providing relief to the local people and suppliers of tourism services. Many local people are also not educated to respond to the government's disaster management action plan.
Closing remarks and discussion	10	The instructor will conclude the case analysis with remarks to answer further queries from the learners about the disaster management cycle. Also the instructor will share additional optional Internet resources with the students for their reference and for further enhancement of the students' knowledge on the topic.

Duration

120 min

Analysis

1. Discuss how natural disasters affect the tourism industry using suitable examples.

The effect of natural hazards is harmful to the tourist industry even though it affects the other economic sectors (Ritchie, 2004). Not only does it put a stop to the activities of tourism enterprises (Nguyen *et al.*, 2017), but also its cascading effect damages the key infrastructure of a tourist destination (Pescaroli and Alexander, 2016). The broad and delicate supply networks of tourism are being disrupted by the hazards (Becken and Hughey, 2013). Most significantly, it disrupts the tourism demand by reducing the tourist inflows to a destination and creating a negative image of the destination among the tourists. For example, cyclone Gaja hit South India in 2018, destroyed the towns and rural areas, killed 45 people, and displaced more than 250,000 people. This cyclone impacted all tourist destinations in Kerala, Karnataka and Tamilnadu. Being affected by the flood, Kerala, one of the most beautiful destinations in India, lost its economy due to the cancellation of many bookings and the closure of tourism industries for months.

2. Describe the pre-disaster and post-disaster activities of a disaster management cycle.

For a proactive and efficient response, a systematic approach to disaster management is essential. Following recent natural and man-made disasters that have wreaked havoc on the worldwide tourism sector, tourism destinations should place a stronger emphasis on disaster preparedness planning. Thus, the disaster management cycle as recommended by UNEP is an effective approach for managing disasters at tourist destinations (UNEP, 2008). The disaster management cycle is also referred to as the emergency management cycle; it involves two stages for managing the crises. The first stage, the pre-disaster action plan comprising *prevention, mitigation and preparedness* emphasizes reduction of risk to the lives of local people and visitors and losses of tourism resources caused due to probable threats. The second stage consists of *response, recovery and reconstruction* activities that include: (i) rescue operations; (ii) provision of food and drinking water; (iii) medical treatment; (iv) bringing back the regular activities of local people and tourism activities into normalcy; and (v) long-term development support for the local people and the tourists (Henderson, 2004).

3. Discuss the APELL mechanism for disaster preparedness and risk reduction in coastal destinations of the Puri and Ganjam districts.

Disasters invariably result in crises in any tourist destination. The degree to which the local communities are prepared for the consequences of the disasters determines whether the communities will be susceptible or resilient. Coastal areas will become increasingly vulnerable to sea-level rise and weather-related calamities due to global warming. So, minimizing the disaster risk, in conjunction with measures to adapt to climate change and improve living circumstances is unquestionably today's key global challenge. Thus, UNEP has developed and recommended the application of the APELL mechanism in coastal tourism destinations in Asia. This mechanism has proven successful in three tsunami-affected tourist destinations such as Kanyakumari in Tamil Nadu, Patong in Phuket, and Pi-Pi Island in Krabi, Thailand. Hence it will assist to augment the disaster preparedness capacities of the local communities (residents, authorities and the tourism industry) (UNEP, 2007).

The first task in successfully implementing the APELL mechanism in the coastal destinations of Puri and Ganjam is to form a coordinating group at the local community level. Such a group will include: (i) representatives from the government; (ii) local people; and (iii) other stakeholders such as hotel associations, tour operator representatives, auto rickshaw associations, tourist guide associations, restaurant representatives and NGOs. This will help in improving risk communications about disaster concerns in the area between local authorities, the private sector and exposed residents and tourists, allowing for a more effective and efficient response in crises situations, ultimately increasing tourist trust and confidence in visiting coastal destinations.

4. Identify the measures taken by the Government of Odisha for managing disasters in coastal destinations since the super cyclone of 1999.

The Government of Odisha is now successful in minimizing the loss of life; this was no minor feat and needed a great deal of work. Further, the state can sustain significant infrastructure losses, notably in the areas of electrical infrastructure, housing, public buildings, and transportation infrastructure from the impact of cyclones. When a cyclone is forecast, the Indian Meteorological Department (IMD) is now able to actively monitor the cyclone's path, and reveal where exactly the storm would cross and when it would hit land. Based on these forecasts, the Government of Odisha has demonstrated a high level of readiness and can evacuate people well in time. In April 2018, Odisha was the first Indian state to put into action an early warning dissemination system (EWDS) for natural disasters that may affect its residents living along its 482-km-long coast. The EWDS now covers 1205 villages spread across 22 blocks in the coastal districts of Puri, Jagatsinghpur, Ganjam, Balasore and Bhadrak. The state has now formed the Odisha Disaster Rapid Action Force (ODRAF) consisting of a fully skilled workforce capable of dealing with several types of calamities. The emergency management centre is outfitted with 66 different types of modern equipment to cope with disasters (Mohanty, 2021).

5. What are the current challenges faced by the Government of Odisha in implementing the action plans for disaster management in coastal destinations?

On the coast of Odisha, a cyclone occurs every 2 years or a big cyclone every 4 years, according to the statistics. As a result, coastal residential and industrial buildings in Odisha are vulnerable to cyclones and severe rains. Similarly, the electricity infrastructure is entirely above ground, making it particularly vulnerable to the occurrence of a natural disaster. When the electrical system is disrupted, it has a cascade effect on health systems, water supply, communication and transportation, among other things, all of which become even more important during natural disasters to ensure that relief reaches the most vulnerable. The state Government of Odisha should seriously consider investing in safer homes, particularly in coastal areas, as well as underground electrical lines. Not only would this reduce the need for evacuation, but it will also save money and other resources.

References

Becken, S. and Hughey, K.F.D. (2013) Linking tourism into emergency management structures to enhance disaster risk reduction. *Tourism Management* 36, 77–85.

Henderson, L.J. (2004) Emergency and disaster: pervasive risk and public bureaucracy in developing nations. *Public Organization Review* 4(2), 103–119.

Mohanty, D. (2021) How Odisha's model of disaster preparedness came into being. Available at: https://www.hindustantimes.com/cities/cyclone-yaas-how-odisha-s-model-of-disaster-preparedness-came-into-being-101621969683964.html (accessed 12 March 2022).

Mohapatra, G. (2020) Impact of cyclones on Odisha economy. Available at: https://www.odisha.plus/2020/05/impact-of-cyclones-on-odisha-economy/ (accessed 12 March 2022).

National Disaster Management Authority (2016) *National Disaster Management Plan 2016*. National Disaster Management Authority, Government of India, New Delhi.

Nguyen, D.N., Imamura, F. and Luchi, K. (2017) Public–private collaboration for disaster risk management: a case study of hotels in Matsushima, Japan. *Tourism Management* 61, 129–140.

Pescaroli, G. and Alexander, D. (2016) Critical infrastructure, panarchies and the vulnerability paths of cascading disasters. *Natural Hazards* 82(1), 175–192.

Ritchie, B.W. (2004) Chaos, crises, and disasters: a strategic approach to crisis management in the tourism industry. *Tourism Management* 25(6), 669–683.

UNEP (United Nations Environment Programme) (2007) *Disaster Reduction through Awareness, Preparedness, and Prevention Mechanisms in Coastal Settlements in Asia: Demonstration in Tourism Destinations, Disaster Risk Reduction in Tourism Destinations*. UNEP, Paris.

UNEP (United Nations Environment Programme) (2008) *Disaster Risk Management for Coastal Tourism Destinations Responding to Climate Change: a Practical Guide for Decision Makers*. UNEP, Paris.

Further Reading

Granville, F., Mehta, A. and Pike, S. (2016) Destinations, disasters, and public relations: stakeholder engagement in multi-phase disaster management. *Journal of Hospitality and Tourism Management* 28, 73–79.

The World Bank (2019) What states can learn from Odisha in disaster preparedness and mitigation. The World Bank, 14 June. Available at: https://www.worldbank.org/en/news/speech/2019/06/14/odisha-fani-disaster-preparedness (accessed 12 March 2022).

10 Start from Scratch: Managing the Life Cycle and Sustaining Tourist Destinations

Senthilkumaran Piramanayagam,[1*] Saurabh Kumar Dixit[2] and Alan Vijay[3]
[1]*Manipal Academy of Higher Education, Manipal, India;* [2]*North-Eastern Hill University, Shillong, India;* [3]*Jamal Mohamed College, Tiruchirappalli, India*

Abstract
Tourist attractions are often fragile and need efficient management to sustain the positive benefits while reducing the negative impact of tourism on the local community. The attraction could be a destination such as a city or a museum. The Tourism Area Life Cycle (TALC) model propounded by Richard W. Butler in the 1980s is widely used by scholars. The TALC model has retained its relevance even today after four decades. The current case study narrates the life cycle of six tourism destinations in India's newly formed state province. The case study helps learners understand the life stages of a destination and the implication of the destination life cycle on tourism planning and marketing.

Subjects: destination marketing, strategic management, tourism planning, research methods in tourism

Learning Objectives

At the end of the case study, a learner will be able to:

1. Explain the concepts of the Tourism Area Life Cycle (TALC) model.
2. Identify the life stage of various destinations by adopting the TALC model.
3. Explain the relationship between the destination life stage and tourism planning and marketing of the destination.

Case

Mr Venkatesh, the Chairman of the Special Task Force on Tourism Development of the newly formed state, is on his way to the secretariat of the state, recently promulgated by the Government of India, for an important discussion on tourism planning and development, which later needs to be presented to the Council of Ministers. After six decades of struggle between different leaders and communities in the state concerning how the state should be reorganized, the State Reorganization Commission of India has divided the state into two different states after a prolonged legislative discourse. As a result of the reorganization, many existing tourism destinations that attract numerous domestic and international tourists have, fortunately, been equally distributed between the two states. Now it is the responsibility of the individual states to identify,

*senthil.kumaranp@manipal.edu

 Teaching Cases in Tourism, Hospitality and Events (eds. S.K. Dixit and S. Piramanayagam)
DOI: 10.1079/9781800621022.0010

develop and promote tourism that is specific to the region. The new Cabinet, headed by the eminent Chief Minister, unanimously wants to capitalize on the potential benefits of tourism in generating employment and mobilizing new investment and regional growth.

Many Cabinet Ministers opined that instead of creating new tourism attractions, development and regulation of tourism in existing destinations in the region could be a viable option as they have very limited resources. Mr Venkatesh, a civil service officer, serving as a deputy secretary of tourism in the federal government, has been invited to take the position of Chairman of the Special Task Force on Tourism Development. With his long-standing experience in tourism policy formulation, destination planning and development, Mr Venkatesh has accepted the role as he was very familiar with the current Cabinet team in the state. The Cabinet Ministers also requested Mr Venkatesh to start from scratch on tourism development to sustain the positive benefits while simultaneously reducing the negative impact of tourism. Being a son of the soil, he is very familiar with each region's geographical, socio-demographic and cultural uniqueness in the state. As the tourism resource inventory and visitor statistics are the basis for any tourism policy and planning, Mr Venkatesh has invited four tourism management experts to be a part of the special task force.

The expert team and the task

Mr Acharya, the Deputy Secretary of Tourism in the state, has been requested to compile the visitor statistics for the state for the last 10 years. Dr Ramana, a member of the expert team who is currently serving as a professor at the central university, has been requested to identify suitable destinations in the regions that need immediate attention and strategies to develop, promote, manage and control tourism activities. Dr Rithika D'Souza, a professor of marketing in a well-known business school in the capital of the newly proposed state, has been allocated the task of formulating the marketing strategies for the tourist destinations identified by Dr Ramana and his doctoral scholars. Mr Anwar, an expert in tourism and strategist and a former destination marketing officer in the neighbouring state, is invited to collaborate and support other strategic planning experts. In his previous assignment, Mr Anwar successfully turned the state into the most favoured destination among domestic and international travellers.

The main task of the team is as follows:

- collection of domestic and international tourism arrival data related to six existing tourism destinations in the state;
- collection of data on visitor expenditure and the average length of stay by both domestic and international tourists in the above-identified tourism destinations;
- to identify the current state of tourism attractions and tourism resources in terms of their life stage; and
- formulating marketing strategies to promote and sustain the benefits while reducing the negative impacts of tourism in each destination with different life stages.

Description of the destinations identified

Dr Ramana and his scholars have identified six very diverse tourism destinations. The description of each destination is given below:

Destination 1

This is a popular religious centre in the state gloriously narrated in the Hindu scriptures. The temple is one of the most pivotal landmarks in Hindu mythology, and pilgrims used to visit for prayer to fulfil their wishes. It currently attracts a maximum number of visitors in the state. It is located on a hill encircled by a major river in the state. There are a lot of tourism attractions in the vicinity of the religious centre. Recently, it has been declared as a national park as it is the home of some primates, elephants and red sandalwood. In the depth of the jungle, one can witness an ancient cave, which is believed to be sacred, located next to a waterfall.

Destination 2

This is a beautiful hill station located in a rural district in the south-west part of the state. The economy of this area mainly depends on agriculture, plantations, forest products and tourism. The region is popular for its coffee plantations. A visitor can find several beautiful waterfalls, wildlife, mountain ranges, savannahs and lovely valleys. Climatic characteristics in this destination remain pleasant throughout the year, especially in summer. The average temperature during the summer between March and June is about 26°C. The average annual rainfall in the destination is about 260 cm. During the winter, between December and February, it is quite cold and the average temperature is around 18°C. In winter, in some parts of the upper hilly region, the temperature may go below 0°C. However, the majority of residents were against tourism development and tourists because overtourism has become a major problem, being blamed for causing deforestation, pollution, increased traffic and crime activities.

Destination 3

This is a port city and well-known industrial centre in the state. The place is known for various beaches with different natural environments. The destination is surrounded by a lot of lakes, hill regions and a valley. The location boasts many tourist attractions that reflect the state's culture, food, ways of life and civilization. Historically it is understood that the port city was a famous urban trade centre of a famous kingdom. The city is also known for its ancient temples and century-old mosque and church. Malls, theatres, an art gallery and state-operated museums are part of this destination.

Destination 4

As the state hosts at least ten heritage sites and structures, the state has great potential to be a heritage tourism destination. Dr Ramana and his team identified this group of monuments as having the potential to make it onto the United Nations Educational, Scientific and Cultural Organization's (UNESCO) World Heritage List, if they were well conserved. Apart from the historical background and the structures themselves, which are known for their grandeur, traditions, cultural values, beauty and architecture, each site has its unique biological, geological and heritage significance and tourism values. A tourism circuit covering Buddhist remains, a sprawling fort spread across 48 acres, and a very ancient temple will be an attractive tourism resource that could attract domestic and international visitors.

Destination 5

This is another urban centre located on the east coast, famed for its long sandy beaches, inland canals and water sports facilities. The place has long been acknowledged as the state's most prominent tourist destination, fortunate with natural and built infrastructure. The destination's appeal as a tourism destination first began to gain momentum during the mid-1990s. However, recent data on visitor arrivals indicates that visitor numbers are decreasing.

Destination 6

This is an area of dense forests known for camping, where the team identified a privately operated community-based tourism village, which was declared as an ecotourism resort by the past government. The area has plenty of fauna and flora and is blessed with several waterfalls and rivers. Also the village offers a panoramic view of the Eastern Ghats, a range of mountains along India's eastern coast. Many international visitors indulge in various adventure activities here such as rock climbing, hiking, trekking, birdwatching, going on a wildlife safari and river cruising. There are various options for accommodation including a treehouse, log huts and wild camping and the destination attracts a significant number of Generation Z. Well-trained naturalists, mostly from the tribal community in the village, is another supportive element to encourage ecotourism at this destination and there is considerable participation from the local community, including the opportunity for visitors to learn 'the language of the jungle'. Now slowly this destination is attracting researchers and experts from around the world for its biodiversity.

Visitor data

After considerable difficulty, Mr Acharya collated the visitor data specific to the six destinations as suggested by Dr Ramana. The data shown in Tables 10.1 and 10.2 was provided to the expert team for formulating marketing strategies based on the tourist destination life cycle.

Destination life cycle – what is it?

Even though Mr Venaktesh is very familiar with tourism development and planning, he wanted to understand the concept of the destination life cycle, as every expert in the task force emphasized the life cycle for destination planning, marketing and management. After reading a classic tourism book suggested by Dr Raman, the text revealed that although tourism destinations might look similar, they are unique in nature, as it is more complex than any other product or service. Tourism destinations are dynamic and are continuously evolving. A tourism destination as a product is an amalgam of five components: (i) access; (ii) accommodation; (iii) attractions; (iv) activities; and (v) amenities. This is simply called the '5 A's of tourism'. In other words, tourism products are not a hotel bed or a cover in a restaurant but an amalgam of many components and experiences.

The '5 A's of tourism' are as follows.

- The *access* element of a destination describes how a visitor can reach a particular tourism destination. However, 'access' is not limited to reach, but also refers to destination infrastructure, facilities that cater for the tourist with a disability, and easy access to tourism attractions.
- The *accommodation* element describes the option for visitors to stay. A destination with a wider choice of accommodation attracts more travellers. The accommodation may vary from a luxury hotel to a simple hostel.
- The *attraction* element indicates any element of a destination that attracts the visitors. It could be a beach, an amusement park, a world heritage site, an art gallery, an agricultural field, wildlife, an ethnic group or even local food.
- As a destination's effort to involve and engage the visitors, *activities* are key elements that convert a visitor's experience from ordinary to memorable and extraordinary. The activities could be hiking, scuba-diving, paragliding, wine tasting at the winery, interacting with tribes or experiencing rituals in a festival.
- The '*amenity*' component of a destination describes various facilities and services provided at the destination that help the destination meet the needs and expectations of visitors. The facilities may include local transport, a communication network, medical facilities, currency exchange, shopping malls, theatres, guide services and dining facilities. However, the availability of these elements varies between destinations, which is one way to make a destination unique.

Among the destination elements, managing 'attractions' that are often fragile or deteriorate by excessive use is critical for sustaining the benefits and making sense of the other four A's of tourism.

The 'destination life cycle' is similar to the product life cycle (PLC); it is a very significant element that reflects the progress of a tourist destination. The destination life cycle as a concept describes the development of destinations over successive stages of a time period. Tracking and understanding the destination life stage is essential to develop appropriate and timely strategies and action plans to avoid the destination's decline. In the past, several destinations have declined as they were not managed well due to a lack of expertise and a mechanism to track the progress of the destination. A decline in the destination may cause unemployment, adversely impact the local economy, lead to a

Table 10.1. Tourist arrivals (in thousands) to selected destinations. Authors' own table.

Period	Destinations											
	1		2		3		4		5		6	
	Total arrivals	International	Total arrivals	International	Total arrivals	International	Total arrivals	International	Total arrivals	International	Total arrivals	International
2010	100	8	240	11	328	4	21	4	135	2	166	14
2011	128	10	300	14	350	4	35	4	128	2	156	15
2012	149	11	333	16	378	5	48	5	140	3	148	14
2013	161	13	328	16	220	3	56	3	132	2	152	16
2014	189	14	355	17	344	4	59	4	136	2	154	15
2015	214	15	326	16	320	4	67	4	143	2	146	15
2016	245	17	287	10	214	2	78	2	140	1	155	14
2017	217	19	280	7	367	2	89	2	190	2	150	15
2018	256	20	212	5	199	3	102	3	243	3	157	16
2019	278	22	156	3	201	3	134	3	280	3	160	17

Table 10.2. Average length of stay (number of days) by domestic and international tourists. Authors' own table.

Period	Destinations											
	1		2		3		4		5		6	
	Domestic	International	Domestic	International	Domestic	International	Domestic	International	Domestic	International	Domestic	International
2010	2.0	3.0	2.8	3.2	1.5	2.6	2.8	1.5	1.0	1.0	1.5	2.0
2011	2.3	3.2	3.0	3.4	2	3.0	3.0	2.2	1.2	1.1	1.3	2.4
2012	2.1	3.2	3.0	3.4	2.1	3.2	2.6	1.2	1.0	1.2	1.2	2.6
2013	2.4	3.0	3.0	3.4	2.3	3.4	2.9	1.3	1.1	1.2	1.3	2.4
2014	2.0	2.4	3.0	4.0	2.2	3.6	3.0	1.1	1.0	1.2	1.3	2.3
2015	2.1	2.4	2.0	2.8	2.1	3.4	3.0	1.2	1.0	1.1	1.3	2.3
2016	1.8	2.3	2.0	1.8	2.1	3.5	2.7	1.9	1.2	1.3	1.2	2.2
2017	2.3	3.0	1.6	1.2	2.4	3.6	3.0	1.8	1.1	1.2	1.3	1.9
2018	1.9	3.0	1.5	2.2	2.5	3.8	2.8	1.6	1.1	1.1	1.2	1.9
2019	1.5	2.9	1.4	1.8	2.5	4.2	2.7	1.8	1.3	1.2	1.2	1.5

rise in social problems, and have an adverse impact on the environment. To sustain the benefits of tourism and avoid any unfavourable outcomes of tourism in the region, destination marketing organizations (DMOs) adopt destination life cycle models to develop timely marketing intervention strategies based on the life stage of the destination.

One of the most prominent destination life cycle models, known as the Tourism Area Life Cycle (TALC) model was proposed by Richard W. Butler, a geographer and professor of tourism (Butler, 2006). The TALC model describes how a tourist attraction changes over time. Butler narrated that an attraction goes through multiple life stages over a period of time, varying from explorations to decline. He also emphasized that an attraction could face decline or rejuvenation, depending on the DMO's marketing strategies and interventions. Also he explained that if an attraction is overdependent on a long holiday market, witnessing short duration of tourist stays, low visitor spending, being highly seasonal, acute opposition for tourism by locals, and a limited appeal to overseas visitors, these are indications of decline of a visitor attraction.

The team's main task is to create a tourism plan based on the data provided and formulate strategies for tourism development and marketing, emphasizing sustainability and reducing the negative impact of tourism on the community.

Discussion Questions

1. Identify the life stage of various destinations selected by Dr Ramana and his team, referring to the TALC model.
2. Identify two destinations of your choice from the selected destinations that need to be developed as an international tourism attraction. Justify your choice using the visitor statistics.
3. Identify the destination(s) that reached its stagnation in the selected list of destinations and list various indicators of maturation.
4. In your opinion, is it right or wrong to promote the existing destinations instead of identifying new tourism as decided by the Cabinet Ministers? Justify your answer with rationale.
5. As it finds itself struggling with overtourism, is it possible for destination 2 to reinvent its glory again? Discuss the strategies using evidence from the research articles published in the scientific journals.

Teaching Notes

The intention of the case study is to familiarize learners with the concept of destination life cycle using a seven-stage TALC model developed by Richard W. Butler. The main objective is to enable postgraduate tourism students and research scholars to identify and appraise the life stage of a tourist destination and subsequently apply a higher level of knowledge and concepts learned through tourism marketing and strategic management. In higher education, the execution of this case study in the classroom, after lecture sessions on tourism destination, destination life cycle and the TALC model, will help learners understand and apply the concepts in their professional careers. Analysing the trend in visitor statistics by calculating a simple year-to-year growth and compounded annual growth rate (CAGR) would result in students thinking beyond the classroom discussions to real-time application of concepts that could help in making informed decisions. Using different visitor statistics such as annual tourist arrival, visitor spending at the destination, and the average length of stay in a destination will help students understand the significance of each statistic in strategy formulation. The calculation of the annual growth rate and CAGR using a spreadsheet will enable students to adopt statistical packages in analysing the trends and patterns of tourism flow, giving a strong thrust to research based on secondary data sets. However, students are expected to learn from their mistakes. The facilitator or tutor is recommended to discuss the nature of destination, alternative destination life cycle models, and their implication in tourism planning and marketing destinations. The case has been executed among postgraduate students of tourism management in actual classroom settings and has proved to be an effective training tool to teach the destination life cycle.

The case is designed for a group of four students, where each student takes the position of an expert in the task force described in the case. The number could be extended to eight, where a pair of students will work on each assigned task or question given for discussion. The instructor of the case has been expected to take the role of the Chairman of the Special Task Force on Tourism Development. If there are more groups of students in the class, the tutor may instruct the student team to identify the life stage of the destination using different visitor statistics (i.e. annual tourist arrival, visitor spending at the destination and average length of stay).

Target audience

This case can be used in postgraduate-level courses on introduction to tourism, tourism planning and tourism marketing. Also it is appropriate for corporate practitioners in executive management programmes.

Teaching methods and equipment

The case can be activated through scenario analysis and group discussions. Year-to-year growth analysis and CAGR need to be done first if the facilitator would like to adapt the group discussion. The decision and answer to each discussion question should be based on simple year-to-year growth analysis and CAGR calculated through computer spreadsheets or Microsoft Excel. The rationale for any strategies should be supported by the scientific research literature published in scientific journals. The instructor may ask the students to read specific research articles and provide evidence to justify their answer to the fifth question before the case analysis. Chart papers and graph sheets need to be provided to each team to answer the discussion questions. Although there is no need for any other equipment, the use of computers, projectors and flipcharts can be encouraged before their presentation to the Chairman of the Special Task Force.

Teaching instructions

The case can be taught in a 2-hour class using the following teaching instruction plan. The suggested plan assumes that learners are familiar with the concept of the destination life cycle and TALC model. The instructor should evaluate the understanding of learners' familiarity with the concept by interacting with them. If the students are not very familiar with the concept, the tutor could begin by asking learners their experience of a recent visit to a tourist destination and their overall satisfaction with the trip. The discussion will help the tutor to ensure a common understanding of what the case discussion is about.

The case may be discussed in two phases so that after the initial phase of the case study (outlined here in the teaching instruction plan), it may be extended to a second session to discuss the marketing strategies for destinations at different life stages using evidence that the students have found from reading the research articles on the TALC model published in scientific journals. The instructor may instruct the students to collect visitor data for any three destinations in the state or region of their interest. Students need to identify the life stage and investigate the actual strategies adopted by the DMOs to sustain the tourism arrival in the destination.

Duration

120 min

Analysis

1. Identify the life stage of various destinations selected by Dr Ramana and his team, referring to the TALC model.

Teaching instruction plan

Discussion point	Time required (min)	Teaching strategies
Introduction	10	The instructor begins the case analysis by asking questions about the learners' recent visit to a tourism destination and their experience. The instructor asks a few randomly selected students to narrate the good and bad elements that impacted their travel experience at the destination. If any students wish to volunteer a response, the instructor asks the students to raise their hands. The instructor writes the points on the board. The instructor asks a few questions to verify the students' understanding of reading materials on the Tourism Area Life Cycle (TALC) model supplied before the case study analysis.
Briefing on concepts related to the TALC model	10	The instructor introduces the concepts of: (i) the '5A's of tourism'; (ii) the destination life cycle; and (iii) the TALC model. The instructor asks randomly selected students to provide an example for each element of the 5A's and asks students to specify the name of a destination in the region or country to narrate the stages of the TALC model. If any volunteers intend to answer, the instructor tells them to raise their hands.
Summary of the case	10	The instructor briefly discusses the background information, the scenario, the protagonist, the dilemma or decision to be taken, and the discussion questions. The instructor also discusses the adaption of the spreadsheet to calculate annual growth and CAGR. After summarizing the case, the instructor creates multiple teams to answer the discussion questions.
Assignment question 1	25	**Identification of the life stage of various destinations** Now, the instructor allows the teams to identify the life stages of all six destinations using a spreadsheet to calculate the annual growth and CAGR. After the calculation, the instructor asks the teams to provide the result of the life stage analysis of destinations on a sheet of paper and asks the teams to plot the TALC model on chart paper and position each destination on the chart based on its life stage. The instructor must tell the teams not to disclose their results to other teams. After receiving the team's response, the instructor needs to write the answer given by each team on the board and display the chart created based on the TALC model. Now, instructors ask the team to justify the rationale behind their answers referring to the TALC model. If there is any variation in the answer found between teams, the specific teams need to further clarify their decision.
Assignment question 2	15	**Identification of two destinations to be developed as an international tourism attraction** The instructor now facilitates the teams to debate on their choice with their reasons. Each team must defend their choice using the data and rationale.
Assignment question 3	15	**Identify the destination(s) that reached its maturity** The instructor now asked the teams to list the destinations that reached the maturity stage of life. The instructor asks the teams to debate their choice with their rationale. Each team must defend their choice using the data and the TALC model.

Continued

Teaching instruction plan Continued.

Discussion point	Time required (min)	Teaching strategies
Assignment question 4	15	**In your opinion, is it right or wrong to promote the existing destinations instead of identifying new tourism as decided by the Cabinet Ministers?** **Yes** **No** The instructor asks the students to participate in the poll using any online app or a Google Form link. After finishing the poll, the instructor shows the output as a percentage. The instructor now allows each team to discuss the question and identify the pros and cons of promoting existing destinations or developing new destinations. The instructor writes all the teams' responses on the board.
Discussion question 5	10	**As it finds itself struggling with overtourism, is it possible for destination 2 to reinvent its glory again?** **Yes** **No** The instructor asks the students to participate in the poll using any online app or a Google Form link. After finishing the poll, the instructor shows the output as a percentage. The instructor now allows each team to discuss the possibilities of managing the impact of overtourism and measures to rejuvenate destinations based on the prior readings supplied to the students.
Closing remarks and discussion	10	The instructor concludes the case analysis session with his/her analysis and answers for the discussion questions using the TALC model and statistical analysis.

The result can be obtained by calculating the annual growth rate or CAGR using the domestic tourist arrival data of the selected destinations (Table 10.3 and Fig. 10.1).

2. Identify two destinations of your choice from the selected destinations that need to be developed as an international tourism attraction. Justify your choice using the visitor statistics.

There can be multiple answers to this discussion question. By applying the annual growth rate on the average length of stay by both domestic and international visitors to the selected destination, a destination that attracts most international tourists can be promoted as an international destination. As per the average length of stay, destination 3 has been the most favoured destination by international visitors (Table 10.2). Destination 4, with a rich cultural heritage and destination 6 that is rich in natural beauty, can be promoted among international tourists. Age and motivation-based segmentation data will help further understanding as the given data may not be sufficient to decide.

3. Identify the destination(s) that reached its stagnation in the selected list of destinations and list various indicators of maturation.

Destinations 2, 3 and 6 have reached the maturity stage as the annual growth of visitor arrivals data and average annual growth rate indicates that these destinations have reached saturation (destination 3), or are either stagnant or have started to decline (destinations 6 and 2; see Table 10.4). Destination management organizations and planners need to develop strategies to revive tourist inflow for these destinations.

4. In your opinion, is it right or wrong to promote the existing destinations instead of identifying new tourism as decided by the Cabinet Ministers? Justify your answer with rationale.

Table 10.3. Tourist arrivals between 2010 and 2019: annual growth rate. Authors' own table.

Period	Destination 1	Annual growth rate	Destination 2	Annual growth rate	Destination 3	Annual growth rate	Destination 4	Annual growth rate	Destination 5	Annual growth rate	Destination 6	Annual growth rate
2010	100		240		328		21		135		166	
2011	128	28.0%	300	25.0%	350	6.7%	35	66.7%	128	–5.19%	156	–6.0%
2012	149	16.4%	333	11.0%	378	8.0%	48	37.1%	140	9.38%	148	–5.1%
2013	161	8.0%	328	–1.5%	220	–41.8%	56	16.7%	132	–5.71%	152	2.7%
2014	189	17.4%	355	8.2%	344	56.4%	59	5.4%	136	3.03%	154	1.3%
2015	214	13.2%	326	–8.2%	320	–7.0%	67	13.6%	143	5.15%	146	–5.2%
2016	245	14.5%	287	–12.0%	214	–33.1%	78	16.4%	140	–2.10%	155	6.2%
2017	217	–11.4%	280	–2.4%	367	71.5%	89	14.1%	190	35.71%	150	–3.2%
2018	256	17.9%	212	–24.3%	199	–45.8%	102	14.6%	243	27.89%	157	4.7%
2019	278	8.9%	156	–26.4%	201	1.0%	134	31.4%	280	15.23%	160	1.9%

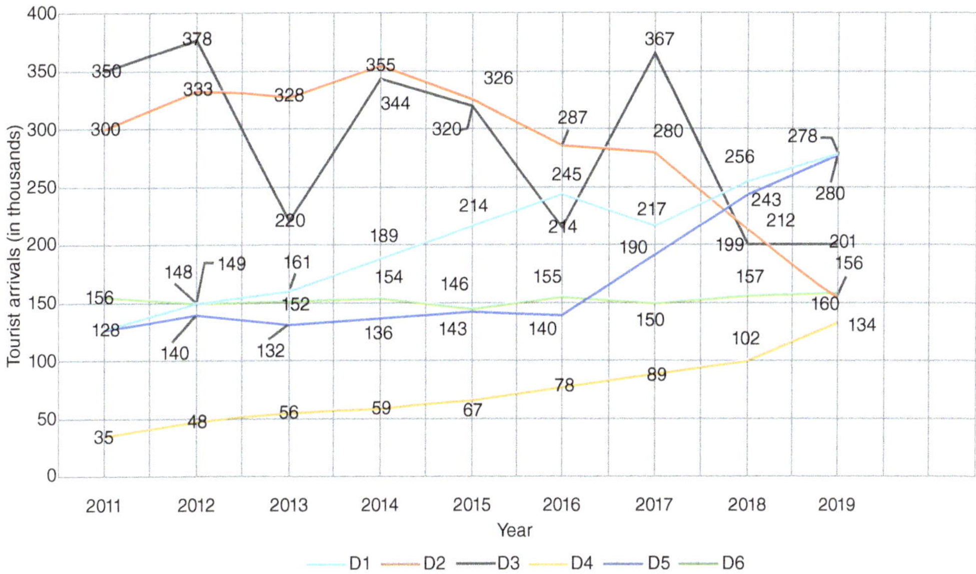

Fig. 10.1. Tourist arrivals (in thousands) at the six selected destinations (D1–D6) for each year between 2011 and 2019. Authors' own image.

Table 10.4. Life stage of destinations based on average annual growth rate. Authors' own table.

Destination	Life stage	Average annual growth rate (%)	Indications
Destination 1	Development	12.52	A steady inflow of tourists with an average annual growth over 12%
Destination 2	Decline	–3.39	Steady decline in flow of tourists with a negative average annual growth of 3.39%
Destination 3	Saturation	1.77	An average annual growth of 1.77%, a very slow growth in flow of tourists, indicates this destination is at saturation
Destination 4	Exploration	23.99	Steady inflow of tourists with an average annual growth of 23.99%. Compared with other destinations, the rise in tourists visiting this destination indicates that destination 4 is in the exploration stage
Destination 5	Development	9.27	Growth has started only after 2016, with an average annual growth rate of 9.27%
Destination 6	Stagnation/ decline	–0.31	Negative average annual growth rate indicates the stagnation or decline in flow of tourists

There can be alternative answers to this discussion question. However, stating the resources as a major constraint for the newly created state or region, it is right to promote the existing destinations. As per the TALC model, newly developed destinations need much time and effort to reach their tourism potential.

5. As it finds itself struggling with overtourism, is it possible for destination 2 to reinvent its glory again? Discuss the strategies using evidence from the research articles published in the scientific journals.

Generally, the falling number of tourists is considered to be a symptom of the decline of the destination. However, considering only the number of tourists has its own disadvantages as it does not consider price, quality, and other attributes that limit the number of visitors to a destination. Tourism development exceeding a destination's carrying capacity, overcrowding, congestion, deterioration of tourist attractions, pollution, increased cost of expenditure, increased crime rate, poorly perceived tourism experience, and conflict between tourists and residents are considered indicators and cause for the decline of a destination. Some of the strategies used to rejuvenate the destination are:

- regulating the tourist arrival;
- equal spatial distribution and promotion of new tourism destinations;
- sustainable tourism practices;
- matching supply and demand;
- reducing the queue by using timed ticketing and dynamic pricing;
- demarketing to deter particular market segments or reduce demand pressure;
- regulation to change stakeholders' behaviour;
- taxing tourists in the form of a development fee;
- developing supportive infrastructure to avoid crowding; and
- limiting the number of visitors.

Students should be able to demonstrate support for some of these strategies to avoid overtourism and rejuvenate a destination, using evidence from their research, having read articles published in relevant journals.

Reference

Butler, R.W. (ed.) (2006) *The Tourism Area Life Cycle, Vol.1: Applications and Modifications*. Channel View Publications, Bristol, UK.

Further Reading

Buhalis, D. (1999) Tourism on the Greek Islands: issues of peripherality, competitiveness and development. *International Journal of Tourism Research* 1, 341–358. https://doi.org/10.1002/(SICI)1522-1970(199909/10)1:5<341::AID-JTR201>3.0.CO;2-0

Chapman, A. and Speake, J. (2011) Regeneration in a mass-tourism resort: the changing fortunes of Bugibba, Malta. *Tourism Management* 32, 482–491. https://doi.org/10.1016/j.tourman.2010.03.016

Mihalič, T. (2000) Environmental management of a tourist destination: a factor of tourism competitiveness. *Tourism Management* 21, 65–78. https://doi.org/10.1016/S0261-5177(99)00096-5

11 Water Management in Hotels Using Tap Specs in Conjunction with the Ishikawa Approach

Naresh P. Nayak,* Julia N. Albrecht and Willem Coetzee
University of Otago, New Zealand

Abstract
The hotel and tourist industries rely heavily on average water consumption. The establishment and maintenance of a water conservation culture in hotels is the responsibility of industry professionals in this field. For determining the root cause of a problem and devising a workable solution, researchers often use the Ishikawa technique invented by Ishikawa in 1943, also called the fishbone diagram. This case study demonstrates how water tap specs can be used in conjunction with the Ishikawa approach to achieve water savings in hotels. As an added benefit, this case study illustrates how the outcomes of this approach can be assimilated in standard operating procedures (SOPs) in the accommodation business.

Subjects: Ishikawa approach, water conservation management, standard operating procedure (SOP), tap specs

Learning Objectives

At the end of the case study, a learner will be able to:

1. Identify water use in guest rooms and conduct an evaluation of existing equipment using tap specs at the accommodation business.
2. Explain the concepts of the Ishikawa approach.
3. Develop a standard operating procedure (SOP) for water conservation for this standalone business based on the Ishikawa approach.

Case

A standalone motel, run by David Tait is a modern '4 Star Plus', Qualmark hotel offering rooms, a restaurant, a café and a guest laundry. Qualmark is a New Zealand-specific rating system that assesses quality as well as environmental performance. To qualify as a Qualmark '1 Star' to Qualmark '5 Star' in New Zealand, an accommodation business must meet strict standards in terms of amenities, cleanliness and overall quality (Qualmark, 2021). There has been an increasing focus on

*nareshp.nayak@postgrad.otago.ac.nz

DOI: 10.1079/9781800621022.0011

environmental credibility since 2008, and Qualmark performs its own certification in this area. Enviro-accreditation involves: (i) hotel waste; (ii) consumption of consumables; and (iii) recycling programmes at the property. The motel has 26 rooms and a dining room with 60 seats, all on the same level. There is a refrigerator, microwave, tea/coffee maker, and a selection of drinks in each room's kitchenette. Recycling bins are accessible in every room, and luxury dispensers have been fitted to remove single-use plastics from the hotel's inventory. Although the hotel had a housekeeping document that defined their basic cleaning methods, David and his team in this motel find it challenging to implement water-saving measures in their properties, despite their stated commitment to environmental stewardship. This may be in part because travellers use more water when staying in hotels than they do at home (Gössling *et al.*, 2012).

Inspection team – guest room and bathroom

The inspection team decided to meet the motel manager and after discussion regarding the issue, the team was permitted to carry out a regular inspection of the guest rooms' furnishings and equipment with a focus on the toilets (WCs). David's motel is eco-conscious, and he and his staff care about the environment as well as his customers. As a result of their proactive approach to water conservation at the hotel, they have taken the necessary steps to reduce their water footprint and improve the health and well-being of their guests. After making detailed observations of the room cleaning procedures, spending a substantial amount of time doing fieldwork and interacting informally with the housekeeping staff, it was noted that the staff members were generously using a lot of water throughout their cleaning procedure, which contributes to the hotel's overall water use. Water consumption for 2021 is shown in Table 11.1.

The water meter readings are taken by the city council every 4 months and the overall water consumption from February 2021 to November 2021 suggests that there has been an overall water increase. In the month of December water consumption dropped by comparison to November as the motel was closed over Christmas and New Year.

Table 11.1. Overall water consumption. Authors' own table.

Month	Year	Overall water consumption (l)
February	2021	25,393
May	2021	26,555
August	2021	27,455
November	2021	28,314
December	2021	27,917

The following observations were made concerning water use during room cleaning:

- Washbasin taps ran for an average of 1 min during cleaning as some rooms had water goblets to be washed and some water was needed for the dusters as well.
- Water from the bath-tub tap or the showerhead was not used much, and the floor surface was wiped clean with the chemical and a duster.

Water use is heavily influenced by the habits of visitors. The 'pleasure factor', in which people enjoy taking hot showers or baths while they are at a hotel or other place of lodging, is one cause for the increased water usage (Eurostat, 2009). David wants his guests to be able to have a stress-free showering experience without having to worry about water pressure or flow. However, he was receptive to any interventions targeted at enhancing facility and procedure efficiency for water management and cost savings. So while he does not wish to reduce water flow while showering, to reduce water use the motel manager needed an innovative solution that would need minimal retrofitting and would have no adverse effect on guest satisfaction. Retrofitting means modifying or enhancing existing procedures, equipment or buildings as opposed to replacing them (Pyloudi *et al.*, 2015). This may involve adding, removing, rearranging or replacing components (Baldwin *et al.*, 2018). To achieve the criteria of a green building, it saves resources and energy while providing residents with a healthy, safe and pleasant living environment. Showers, taps (faucets) and toilets are the primary water-consumption sources in the rooms (Table 11.2). These locations in guest rooms are targeted for water-saving options in order to conserve water. While inspecting the bathroom we found the motel used 8 l of water for a full flush and 4 l for a reduced flush, making dual-flushing toilets a good choice for the bathroom (Fig. 11.1a). Eco-friendly showerheads (Fig. 11.1b) were installed that used less than 12 l of water/min, while the specifications for the bathroom washbasin tap (Fig. 11.1c) indicated a water flow of 8 l/min. Based on these values for the current eco-showerhead, it is hypothesized that 26 visitors using the shower for 14 min, would result in a daily water consumption of 4368 l if each guest had a shower twice a day. Moreover, according to Styles *et al.* (2015) regarding toilet flushing, it is reasonable to allow four flushes/day per guest and two flushes/day per occupied room required for cleaning. On this basis, and considering cleaning 26 rooms, using the existing specs of 4 l and 8 l for a dual-flush toilet, it was calculated that the total water consumed for David's motel was 624 l. Finally, the specs of tap in the washbasin were 8 l/min, with 26 guests using taps for 9 min each, which includes water use by housekeeping staff, the water consumption was around 1872 l.

The Ishikawa approach and its problem-solving potential

The Ishikawa diagram used in this case study makes it easier to understand the motel's excessive water usage and its repercussions. Kaoru Ishikawa created this type of diagram in 1943, and he used it to help track down process and service quality variations. This tool may be used to monitor process variation or the occurrence of an event, and the elements that contribute or interact with one another to produce the end result can be examined (Ilie and Ciocoiu, 2010). The Ishikawa or fishbone diagram, often known as a cause–effect diagram, is one of the most important tools in the field of problem solving (Shinde *et al.*, 2018). According to Lira *et al.* (2017), this diagram is used to show the sources of a problem within the four Ms: (i) method; (ii) labour (originally 'manpower' but we prefer 'labour'); (iii) materials; and (iv) machinery. It is often used to look at concerns inside an organization. Consequently, discovering and resolving errors is simpler, and recurrences are less frequent.

The fishbone diagram has been utilized in a few studies to analyse employees' perceptions of nepotism using qualitative research. In the tourism and hotel industry, nepotism, hiring, training and development were shown in this way. Corruption in the workplace, flexible working hours and unfair employee incentives influence employees' perceptions of nepotism (Elsayed and Daif, 2019). Similarly, hazard analysis and critical control point (HACCP) implementation challenges have been mapped using a fishbone diagram, and it was revealed that 'reducing the difficulties caused by personnel and system/documentation in introducing HACCP' is of critical importance for international tourist hotels (Tseng *et al.*, 2016, p. 67).

Table 11.2. Water consumption under current specification. Authors' own table.

Component	Tap specification	Quantity hypothesized through benchmark practice	Total water consumption (l/day)
Showerhead (eco-shower)	12 l/min	364 l	4368
Tap	8 l/min	234 l	1872
Dual-flush toilet[a]	4 l and 8 l per flush	156 flushes/day[b]	624/1248
Total daily water usage			6864

[a] Water usage for cleaning the dual-flush toilet is included in the calculations for water consumption.
[b] This 156 flushes/day is for all 26 rooms (i.e. the whole hotel).

Fig. 11.1. Bathroom fixtures and equipment specification: (a) dual-flush toilet; (b) showerhead for an eco-shower; and (c) tap (also known as a water faucet). Photos courtesy of the authors.

The fishbone approach

The Ishikawa diagram's design is clearly influenced by the skeleton of a fish (Fig. 11.2). Using vertical line segments that lean horizontally, it is possible to represent the many causes and sub-causes of a specific effect or problem. The fish's head or mouth is where the issue or consequence (i.e. the effect) is seen. On the smaller 'bones', possible contributing factors are indicated under several cause categories.

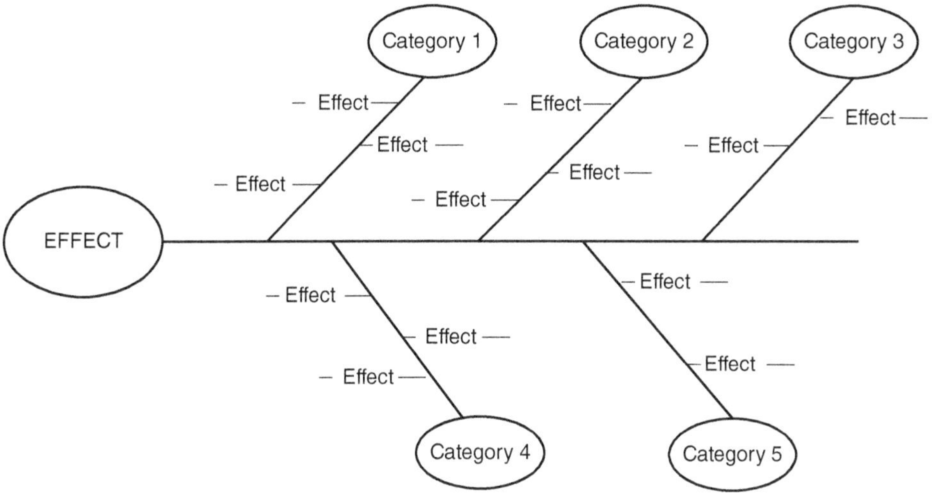

Fig. 11.2. Cause and effect – the fishbone diagram. Authors' own diagram.

Water management using the Ishikawa approach

Room cleaning processes were repeatedly inspected, with a focus on water used in guest bathrooms, and this was followed by informal discussions with the housekeeping crew regarding current practices in order to understand water use. The main aim of having this conversation with all staff members was because it allows the researcher to probe for any additional information to explore the reasons for the current water consumption. The findings were used to create a fishbone diagram by identifying the primary parameters that have the most significant impact on water conservation. Using the Ishikawa (fishbone) diagram helps as it breaks down the nature of the issue into subdivisions, offering a clear sense of the scope and nature of the problem, which in this case is to find ways of saving water in the hotel. The primary parameters were identified as the most important aspects of service management, which are: (i) processes; (ii) personnel (or workers); (iii) policy; (iv) measurements; and (v) the environment (Fig. 11.3). The Ishikawa diagram makes it possible to examine all the challenges and aspects that affect water conservation. Evaluating the impact of the different components will help achieve the desired outcome, which is to develop a standard operating procedure (SOP) for water conservation in the hotel that suits both the management and the staff. For each parameter, the following issues were identified as important in order to establish which processes require major modifications to achieve water savings.

Processes

In this motel, each room undergoes up to 18 unique cleaning phases, with one room needing to be cleaned within 30 min. The housekeeping team can make a huge difference in savings in water usage while cleaning the room. For example, the cleaning staff should begin to reduce their water usage by making modest alterations like flushing toilets just once or emptying separate waste bins into appropriate recycling receptacles. Although the existing showerheads, dual-flush toilets and tap aerators are considered to be water saving, a significant amount of water can be reduced without affecting the water pressure and jeopardizing the comfort of guests. Promotional events, brochures and persuasive messages that may be placed in the room can be used to encourage visitors to save water at the hotel. Over half of hotel visitors' water use is likely to come from their showers (Smith *et al.*, 2009). To keep visitors informed, they should be reminded to: (i) take

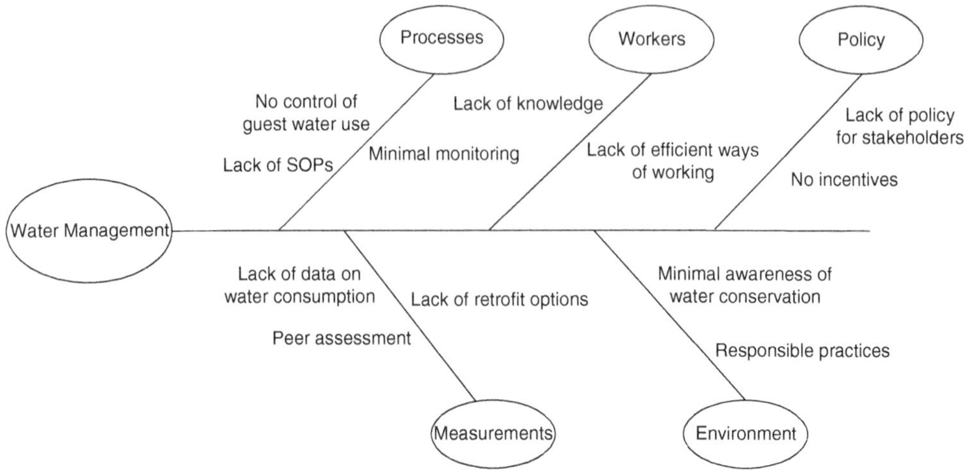

Fig. 11.3. Water management in the motel using the Ishikawa approach. The Ishikawa or fishbone diagram shows the causes or parameters (processes, workers, policy, measurements and environment) that contribute to water management (the effect). Evaluating the impact of the different components could help to develop a standard operating procedure (SOP) for water conservation in the hotel. Authors' own diagram.

shorter showers using digital or sand timers; (ii) turn off the water while brushing their teeth; and (iii) avoid flushing when it isn't necessary. Also it is essential to convey to them that water conservation is a shared issue because it is a common problem. Often guests are ready to pick the most environmentally friendly option when presented with a simple choice. Likewise, reducing water usage while cleaning guest rooms, as identified by Styles *et al.* (2015), should always be included in the SOPs.

Workers

There was a need to educate staff members on water conservation via initiatives and training programmes. Training in important operational responsibilities to improve water efficiency should be promoted and those duties should be frequently monitored. Management can focus on educational programmes for staff, and informative signs on how to save water addressing tourists (Gössling *et al.*, 2012).

Policy

The motel has established standards for the reuse of towels and linens, but there are no long-term SOPs in place for water management. Because of the environmental impact caused by the overuse of water, a sound water management plan is highly recommended for both the equipment and the physical processes that use water, as well as for visitor water consumption. In addition, department-level water reduction via programmes and incentives should be introduced for staff after evaluating their tasks and completing their water-saving targets. Efforts are clearly needed to improve incentives and increase awareness and knowledge about water-saving measures like habits and attitudes among hotel staff and guests (Torres-Bagur *et al.*, 2019).

Measurements

Based on the observation in this motel, it is suggested that regular water-use monitoring and standardization should be a part of the SOP. Metering and sub-metering water use in the guest rooms is necessary to keep tabs on guest-room water consumption. This will assist in determining the average water consumption per guest per night. Sub-metering guest rooms may uncover hidden problems and avoid further water wastage. As identified by Juvan and Dolnicar (2014) water metering in rooms may be promoted to create a nice hotel atmosphere and change visitor habits. On the other hand, housekeeping employees can influence visitors' behaviour and ensure that the equipment in the rooms they clean is working properly. This would offer sufficient data on water consumption from both the demand and the supply perspective. It is essential that water-saving activities be evaluated by colleagues to keep staff motivated. Low-cost solutions, such as altering the flush valve in toilets, may also help save water. Water conservation initiatives and incentives at the departmental level, as well as reducing water loss due to leaks in pipes, are additional potential solutions. The water efficiency of the fittings determines how much water may be saved via various activities.

Environment

A motel's environmental management plan should include water conservation and efficiency to improve productivity, cut operating costs, and lessen the impact on natural resources, all of which help to build customer loyalty and increase repeat business for the motel. The motel management team needs to keep track of actions in order that they know how close they are to industry standards or environmental requirements. The easiest way to do this is to give every single individual the responsibility for an issue that will result in a more effective water usage.

The inspection team's task

As explained above, using the information supplied and the observations made at David's motel, the team used the Ishikawa (fishbone) diagram to identify the root causes of poor water management practices, which in turn could help them to produce ideas for future improvement. The main task of the team is to create a SOP based on the observations and to design water management practices that emphasize sustainability and mitigate the negative impact of tourism on the environment.

Discussion Questions

1. Identify the various points mentioned on the fishbone diagram and draft a standard operating procedure (SOP) in reference to water management in a guest room.
2. Based on the hypothesized calculation (shown in Table 11.2), identify the water consumption values for the bathroom components in a guest room using tap specs.
3. 'Standardization' and 'quality control' are 'two wheels of the same cart'. In your opinion, is it right or wrong to have SOPs for water management in a hotel guest room? Discuss the strategies using evidence from the research articles published in scientific journals (see 'Further Reading' section).

Teaching Notes

The purpose of this case is to acquaint students with the notion of the Ishikawa method, which was established by Kaoru Ishikawa in 1943, and to use it to identify water management concerns in any accommodation (lodging) business. They will learn how to evaluate water consumption patterns in guest rooms using hypothesized computation of water use through taps, showers and toilets and then suggest how water consumption can be reduced using low-cost interventions and adjustments, as well as implementing precise standard operating procedures (SOPs).

Target audience

This case study is suitable for both undergraduate and postgraduate students studying hospitality and tourism management. Undergraduates may be studying subjects such as hotel housekeeping management and accommodation operations management, while postgraduate-level courses may include subjects like service quality management in hospitality, tourism and leisure, hotel operations management and total quality management in hospitality.

Teaching methods and equipment

A mixture of classroom sessions, field trips and experiential learning activities may be helpful in making the case more productive. Students and instructors work together to gather data in an unfamiliar setting, with open-ended and surprising learning outcomes for students dependent on the data collection.

In preparation for the case study the instructor should briefly discuss water usage in hotels, its impact on environmental sustainability, the nature of guest behaviour, and introduce the Ishikawa diagram to the students.

For the field trip, the instructor needs to find a hotel, or other business offering lodging/accommodation, that is willing to permit a visit from the instructor and his/her students. Any paperwork required giving specific approval for the trip needs to be completed prior to the visit. The hotel management must be consulted and agreement reached before establishing the monthly water consumption (if rooms are sub-metered) and tap specification for guest rooms. A semi-structured interview with the manager and a few members of the housekeeping staff must also be organized. Every environmentally certified lodging business will keep a water-use record, which the facilitator would want to access, and teachers should be flexible and prepared to traverse challenging and/or unproductive conclusions in order to encourage healthy dialogue. These encounters with real businesses offer great potential for students' learning as they demonstrate how to apply theoretical concepts to solve practical issues concerned with hotel management.

Back in the classroom, deciding and responding to each discussion question in this case study should be guided by the case study's objectives and is best done in groups or teams of two–four students. Scientific evidence published in peer-reviewed journals should support any strategy's rationale, so prior to the case analysis students may be asked to review specific research articles and evidence to support their answers. Chart paper must be provided to each group/team to answer the discussion questions. Although computers, projectors and PowerPoint presentations are not necessary, they may be recommended prior to their presentation to the hotel manager, if this is planned.

Teaching instructions

The case may be used in a 3-hour class using the teaching instruction plan outlined as follows. As mentioned above, it is assumed that students will be familiar with the Ishikawa technique and water management in hotel businesses, but the instructor can assess students' knowledge and understanding of these by asking them questions. Next, the students should be taught or reminded how to use the fishbone diagram technique which combines thinking and mind mapping to find root causes. To start drawing a fishbone diagram, the first step is to agree and write down details of the exact causes/concerns. The second step is, on a whiteboard, write the problem statement (the effect, in this case 'Water management') at the fish's head of the horizontal line representing the spine (as per the diagram in Fig. 11.2). Draw a line from the fish's spine to the cause you identify and label it. Step three starts with the brainstorming process as you find likely reasons for the problem associated to each element. The skeleton is now ready to be filled in with the students' justifications. A comprehensive fishbone diagram detailing all possible causes of the problem statement/effect should be prepared by now.

The teaching plan can be split into two parts: (i) the field visit; and (ii) work carried out in the classroom. In preparation for the field visit the instructor may discuss with the students how to have a successful conversation with the hotel management, gather data on tap specs and water usage statistics, conduct a focus group semi-structured interview with the housekeeping staff, and examine a few room-cleaning techniques. The students must identify and list any possible concerns with water conservation techniques at the hotel/motel using the fishbone diagram (they may take notes for this during the visit). Prior to the classroom session, the instructor will need to provide evidence from scientific journals or other comparable publications to support use of the Ishikawa method and water management solutions (e.g. see 'Further Reading' list).

Teaching instruction plan – field visit (duration 60 min)

Discussion point	Time required (min)	Teaching strategies
Meeting with the manager	20	After the instructor has obtained the appropriate approvals to visit the hotel/motel, the instructor should take the students to meet the hotel manager. The meeting should begin with informal talks with the hotel manager to establish, for example, the size of the property, occupancy, and any eco-friendly initiatives. The questions should be open-ended so that respondents can answer without the interviewer's prejudice or control. The interview questions should be mainly focused on topics found in the literature such as: (i) knowledge and awareness of water-saving practices; (ii) sustainable innovations; and (iii) the water management policy.
Focus group semi-structured interview	20	Following a meeting with the hotel manager, the housekeeping staff may be interviewed based on the following themes: • knowledge and awareness of water conservation practices; • any innovative practices followed in the hotel/motel to save water; and • discussion about existing standard operating procedure (SOP) in the hotel/motel on water saving if any.
Observe room cleaning	20	List the detailed steps in bathroom room cleaning and note down any specific points pertaining to misuse of water handling in the bathroom.

Teaching instruction plan – classroom (duration 120 min)

Discussion point	Time required (min)	Teaching strategies
Introduction	10	The instructor begins the case study by asking the students about their most recent hotel or motel visit and their experience. At random, a few students are asked to share their thoughts on hotel water management best practice. Students put up their hands if they want to answer. When the instructor has finished speaking, he or she writes down what the students have said on the board.
Briefing on concepts related to the Ishikawa model	10	The instructor discusses the fundamentals of the Ishikawa model and explains the fishbone diagram in depth – drawing it on the whiteboard in a three-step process as described above in the 'Teaching instructions'. The lecturer invites randomly chosen students to offer an example for each of the points (i.e. processes, workers (personnel), policies, measurements, and the environment) as well as any difficulties they anticipate in a hotel.
Summary of the case study	10	Once the projected values for showerheads, taps and toilets have been calculated, the instructor demonstrates how to calculate water usage and recommend SOPs. Afterwards, the instructor splits the students into groups/teams of two–four individuals and asks them to respond to the discussion questions.
Assignment question 1	50	**Identify the various points mentioned on the fishbone diagram and draft a standard operating procedure (SOP) in reference to water management in a guest room.**

Continued

Teaching instruction plan – classroom Continued.

Discussion point	Time required (min)	Teaching strategies
		Using a copy of the basic fishbone diagram drawn earlier with the specific headings (i.e. processes, workers (personnel), policies, measurements, and the environment) drawn onto chart paper, each student group begins by addressing each of the most significant factors or causes in turn. For example, starting with 'Processes', they should use key points from their meeting with the hotel manager, interviews with housekeeping staff and observations made during the field visit, so that each member of the group comes up with a different reason or sub-cause for the cluster. These sub-causes/reasons should be added to the fishbone diagram. No comments or evaluations are made on the chart, as is customary in a brainstorming session. When all the points are gathered for 'Processes', then the process is repeated for the next cause (e.g. 'Workers'). Each of the remaining 'bones' is treated in the same manner until the fishbone diagram is complete. The instructor now encourages the groups/teams to discuss their fishbone diagrams with the rest of the class and justify their decisions using evidence and reasoning gathered during the field visit. Still working in their groups, now the students should brainstorm potential solutions that may help by reducing water consumption and use these ideas to draft an SOP in reference to water management in a guest room. Once the students have had time to do this, the instructor should lead a discussion about this by asking each group in turn to contribute one item for the SOP and writing the contributions on the whiteboard, and continuing until all the ideas have been discussed and written on the board.
Assignment question 2	20	**Based on the hypothesized calculation (shown in Table 11.2), identify the water consumption values for the bathroom components in a guest room using tap specs.** The instructor now explains to the groups/teams on how to calculate total water usage from showers, taps and toilets using predicted values and tap specifications. The hypothesized value for taps, toilets and showers are 234 l, 156 flushes and 364 l, respectively (see Table 11.2, which includes the tap specs). These values are taken keeping in mind 26 guests using the shower for 14 min, twice a day; regarding flushing, four flushes/day per guest and two flushes/day per occupied room for cleaning 26 rooms; and with 26 guests using taps for 9 min which includes water use by housekeeping staff. Following the calculation, request that each group/team identify key areas where water consumption could be reduced and brainstorm potential solutions as to how this might be done without compromising the guests' comfort and hence satisfaction with their stay.

Continued

Teaching instruction plan – classroom Continued.

Discussion point	Time required (min)	Teaching strategies
Assignment question 3	10	**'Standardization' and 'quality control' are 'two wheels of the same cart'. In your opinion, is it right or wrong to have SOPs for water management in a hotel guest room? Discuss the strategies using evidence from the research articles published in scientific journals (see 'Further Reading' section).** **Yes** **No** The instructor should ask students to participate in a poll to answer this question using an online app, or a Google Form link. Once students have cast their votes, the instructor displays the results as a percentage for 'yes' and 'no' and asks the students the reasons for their answers. The instructor now leads a discussion with the students, asking them to describe how the housekeeping department's water use in guest rooms may be detected and how the environmental impact can be avoided in light of their findings from their reading of research articles previously provided.
Closing remarks and discussion	10	The instructor closes the case analysis session with his/her analysis of the discussion questions utilizing the Ishikawa model and assumptive computations.

Overall duration

Three hours (60 min field work plus 120 min class work)

Analysis

1. Identify the various points mentioned on the fishbone diagram and draft a standard operation procedure (SOP) in reference to water management in a guest room

There can be various answers to this discussion question as problems at each hotel/motel will be different. However, water conservation or efficiency is part of an overall environmental management plan for improving efficiency. As a result the SOP has to be drafted keeping in mind water management solutions of that particular lodging. The SOP may include the following points:

- check and report water leaks immediately;
- in order to create attainable goals, install sub-meters to assess water consumption in particular areas, such as guest bathrooms;
- install toilets with a dual flush as these can save water by enabling guests to select a full or half flush. Alternatively install a cistern volume-adjusting device as these can save water too;
- flow restrictors or aerators may be added to manual valve taps;
- depending on the property's specifications, low-flow showerheads, such as those that mix air into the stream so that pressure seems strong, might cut water use;
- set up a system for tracking and analysing outcomes, and keep an eye on it all the time;
- retrofit bathroom fixtures for lower water consumption 'water efficient fittings'; and
- encourage guests to take shorter showers using persuasive messages and suggest guests do not leave the tap running when brushing their teeth.

2. Based on the hypothesized calculation (shown in Table 11.2), identify the water consumption values for the bathroom components in a guest room using tap specs.

The retrofit tap specifications for each lodging establishment and the hypothetical

Table 11.3. Total water consumption using retrofit specs. Authors' own table.

Component	Quantity hypothesized through benchmark practice	Retrofit device	Total water consumption (l/day)
Showerhead (eco-shower)	364 l	9 l/min	3276
Tap	234 l	6 l/min	1404
Dual-flush toilet[a]	156 flushes/day[b]	3 l and 6 l per flush	468
Total daily water usage			5148

[a] Water usage for cleaning the dual-flush toilet is included in the calculations for water consumption.
[b] This 156 flushes/day is for all 26 rooms (i.e. the whole hotel).

values specified in Table 11.2 of the case study are used to create the solution to this problem.

Regarding David's motel described in the case study, eco-bath fixtures are clearly present in the bathroom, but flow rates can be reduced by installing aerators or flow restrictors to limit water flow. A significant amount of water may be saved with tap upgrades, including: (i) cutting the average water flow of showerheads to 9 l/min and that of taps to 6 l/min; and (ii) opting for 3 l and 6 l flush options in the dual-flush toilets. Retrofitting taps and showers has been shown to reduce water consumption. Therefore, the total water use is lowered from 6864 l (see Table 11.2) to 5148 l (see Table 11.3). The success of a hotel depends on the quality of the visitor experience, and this should be the motel's main goal. Guests are clearly interested in supporting ecologically friendly hotels, as seen by their selection of an eco-friendly motel. Low-flow taps and showerheads conserved water, but the entire visitor experience was unaffected. Hotels must be designed and operated in a way that has a minimal impact on the environment.

3. 'Standardization' and 'quality control' were 'two wheels of the same cart'. In your opinion, is it right or wrong to have SOPs for water management in a hotel guest room? Discuss the strategies using evidence from the research articles published in scientific journals (see 'Further Reading' section).

To put it another way, standardization is the act of creating and putting into effect norms. A standardization effort results in standards, which in the context of a quality system include quality papers or documents connected to quality management systems. For your hotel, however, a detailed SOP manual is critical to your business and is an essential aspect of maintaining quality control. SOPs are process documents that outline in detail how an operator should complete a certain task. The studies listed in the 'Further Reading' section aim to uncover different approaches used to implement SOPs in hotels.

References

Baldwin, A.N., Loveday, D.L., Li, B., Murray, M. and Yu, W. (2018) A research agenda for the retrofitting of residential buildings in China – a case study. *Energy Policy* 113, 41–51.

Elsayed, K. and Daif, R. (2019) Use of the Ishikawa diagram to understand the employees' perceptions towards nepotism in tourism and hospitality industry. *Journal of Tourism and Hospitality Management* 7.

Eurostat (2009) *Europe in Figures: Eurostat Yearbook 2009*. Office for Official Publications of the European Communities, Luxembourg.

Gössling, S., Peeters, P., Hall, C.M., Ceron, J.P., Dubois, G., *et al.* (2012) Tourism and water use: supply, demand, and security – an international review. *Tourism Management* 33, 1–15.

Ilie, G. and Ciocoiu, C.N. (2010) Application of fishbone diagram to determine the risk of an event with multiple causes. *Management Research and Practice* 2, 1.

Juvan, E. and Dolnicar, S. (2014) The attitude–behaviour gap in sustainable tourism. *Annals of Tourism Research* 48, 76–95.

Lira, L.H., Hirai, F.E., Oliveira, M., Portelinha, W. and Nakano, E.M. (2017) Use of the Ishikawa diagram in a case-control analysis to assess the causes of a diffuse lamellar keratitis outbreak. *Arquivos brasileiros de oftalmologia* 80, 281–284.

Pyloudi, E., Papantonui, S. and Kolokotsa, D. (2015) Retrofitting an office building towards a net zero energy building. *Advances in Building Energy Research* 9, 20–33.

Qualmark (2021) What does Qualmark mean for you? Qualmark, New Zealand, January. Available at: https://www.qualmark.co.nz/ (accessed 20 December 2021).

Shinde, D.D., Ahirrao, S. and Prasad, R. (2018) Fishbone diagram: application to identify the root causes of student–staff problems in technical education. *Wireless Personal Communications* 100, 653–664.

Smith, M., Hargroves, K.C., Desha, C. and Stasinopoulos, P. (2009) Water transformed: sustainable water solutions for climate change adaptation. The Natural Edge Project (TNEP), Griffith University, and Australian National University, Australia. Available at: https://eprints.qut.edu.au/85206/ (accessed 20 December 2021).

Styles, D., Schoenberger, H. and Galvez-Martos, J.L. (2015) Water management in the European hospitality sector: best practice, performance benchmarks and improvement potential. *Tourism Management* 46, 187–202.

Torres-Bagur, M., Ribas, A. and Vila-Subiros, J. (2019) Incentives and barriers to water-saving measures in hotels in the Mediterranean: a case study of the Muga river basin (Girona, Spain). *Sustainability* 11, 3583.

Tseng, S.Y., Tseng, S.W. and Lin, B.S (2016) How to improve the HACCP system in an international tourist hotel group? Cause and effect analysis and PDCA application. *International Journal of Tourism and Hospitality Reviews* 3, 60–68.

Further Reading

Englart, S. and Jedliskowski, A. (2019) The influence of different water efficiency ratings of taps and mixers on energy and water consumption in buildings. *SN Applied Sciences* 1.

Essex, S., Kent, M. and Newnham, R. (2004) Tourism development in Mallorca: is water supply a constraint? *Journal of Sustainable Tourism* 12, 4–28.

Kim, Y.J., Kim, W.G., Choi, H.M. and Phetvaroon, K. (2019) The effect of green human resource management on hotel employees' eco-friendly behavior and environmental performance. *International Journal of Hospitality Management* 76, 83–93.

12 Paradise Splendour

Snigdharani Mishra, Vasantha Lakshmi and Preeti Narendra*
Institute for Technology and Management (ITM), Mumbai, India

Abstract
The current case study examines the work environment in the hospitality industry including customer expectations while also touching on the sensitive aspects of human behaviour. This describes the journey of an employee in his new job and the challenges faced by him due to the attitudes and behaviour of his colleagues. It clearly depicts that the job can be well managed, but to manage employee perceptions is an effort that is necessary to succeed in organizations. The case study helps learners understand the various facets of human behaviour such as perception, individual decision making, inclusion, and the importance of healthy relationships in the workplace. This case study discussion in class will help young managers gain knowledge of perception and its impact in the work environment thereby leading to their ability to manage themselves, people and the organization in a more balanced way.

Subjects: hospitality industry, employee workplace relationship, perception, attitude, behaviour

Learning Objectives

At the end of the case study, the learner will be able to:

1. Understand the concept of perception and its impact on human behaviour and performance.
2. Help young managers understand the variables and impact of the power of perception.
3. Understand the role of attribution theory in the perception process and its impact on self and others in professional and personal life.

Case

Nestled in dense forests that surround the valley, Paradise Splendour is a five-star facility for tourists of taste who are visiting the Northern Hill State of Uttarakhand. Formed in November 2000 as the 27th Indian state, Uttarakhand was carved out from Uttar Pradesh. A hilly state, Uttarakhand is a poet's delight, an artist's fantasy, a spiritual seeker's dreamland and a traveller's destination. This is primarily because the foothills of the Himalayan and Siwalik range of mountains grace a large part of the state. The state is

*preetin@itm.edu

DOI: 10.1079/9781800621022.0012

famous for its rivers, lush and dense forests, snow-adorned mountain peaks and glaciers, besides being host to famous Hindu religious sites such as Badrinath, Kedarnath, Gangotri and Yamunotri. Nothing more, perhaps, would have been needed to make this state an ideal investment location for the hospitality industry.

Among the many hotels and resorts that thrived and prospered, Paradise Splendour stood out as a rare, unique and a-tale-to-tell five-star property – in short, one of its kind. The surrealistic location, its exquisite grandeur, the splash of rock carvings, the liberal doses of Indo-British decor, its mesmerizing surroundings, and a plethora of rich gentry's pastime options were some of the top-notch attractions of Paradise Splendour that lured the rich and the famous to stay. Attractions include a golf course, live classical music renditions, an in-built museum, an internationally recognized and accredited Ayurvedic spa, a beautifully designed swimming pool, and on-site live villages that house people from three different, prominent local communities of Kumauns, Gahrwalis and Gujjars.

A 25-year-old Palash Patil was a front office supervisor at Paradise Splendour. Just about 9 months into his second job, which he had applied for after reading an advertisement on the Naukri.com portal, Palash had not quite settled into his new milieu. Academically, he had a postgraduate diploma in hotel management from a not-so-famous hotel management institute 'down south'. His professional career began when he was picked from the campus by a casino in Goa. As his stint as a management trainee ended, he was absorbed into a permanent customer support role at the casino. PP, as he was lovingly addressed by his friends and colleagues, had higher aspirations though. His search for better opportunities yielded results when he was selected for the role of front office supervisor at Paradise Splendour. Figure 12.1 depicts the organization chart of the front office and the reporting relationships of PP.

These 9 months at Paradise Splendour had been quite eventful for PP. Having been catapulted from a casino set-up to a five-star environment was no mean achievement for him. At the same time, getting adjusted to a different milieu was no cakewalk either. Most of his peers had pedigree academic backgrounds accompanied by work experience and PP could quickly make out that getting accepted among his peers would be a definite struggle for him. For instance, during lunch breaks, he was not invited by his peers to join in. A couple of times when he tried to join them, he felt ignored since no one gave him the context of the topic under discussion. Hence, he could neither contribute to the ongoing lunch-table discussion nor be part of the jokes and laughter. Whenever he tried to change the topic, he found that no one was sharing their perspective on the same, yet politely hearing him out, before going back to continue their discussion on the earlier topic.

It was hurting. PP was feeling low, and all his excitement about a new job was lost in the first few weeks of his tenure. It was then that he remembered an uncle of his, who had been a good mentor and guide for him throughout his childhood and college days. He called him up and shared his situation. His uncle said:

> Why are you surprised? We live in a perceptive world. Perception is the prime driver of whatever we think, know and believe. It drives our hopes and dreams, our sense of dressing, walking, talking, the professions we choose and to that matter, drives most of our day-to-day actions and behaviours. It appears that your new colleagues have a perception of you without knowing you. Give them some more time to know you better. Perhaps, things will change. In the meantime, try to stay away from perceiving that their behaviour is because of your humble pedigree. That will come in the way of your work delivery and your chances of building a healthy relationship with them.
>
> (PP's uncle)

Convinced, as sound counsel, from someone he trusted, PP went about his job chores and work-related conversation with the very same colleagues who were ignoring him, as if nothing had happened. He was friendly and helpful when engaging with

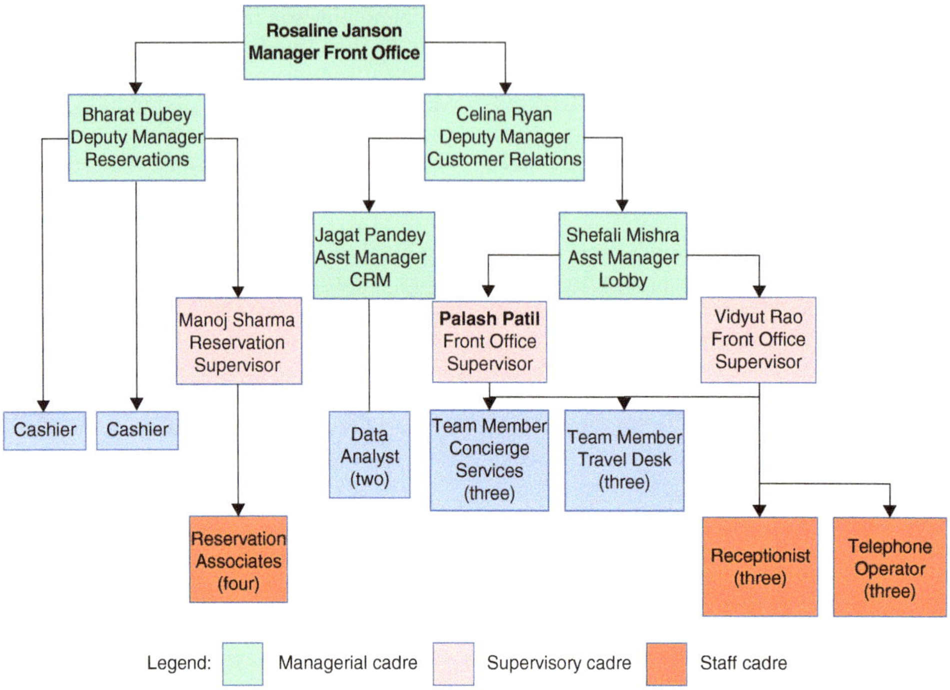

Fig. 12.1. Organization of the front office at Paradise Splendour showing Palash Patil's reporting relationships. CRM, Customer relationship management. Authors' own diagram.

them in work-related conversations while continuing to join them during lunch breaks. For instance, once, Manoj Sharma, the Reservation Supervisor was completely off-colour while on duty. PP could notice that not all was well with Manoj. He walked up to him and asked, 'Hope you are okay Manoj? I notice, you look quite pale and exhausted'. Manoj was quiet for a few seconds and then replied:

> My wife's parents have tested COVID positive, and she has gone to her hometown to attend and assist them. For the last 3 days, I have been struggling hard to manage my child, other household chores, as well as the demands of my job. ...
>
> (Manoj)

Before, Manoj could continue, PP said, 'Can I help you for a few days, by doubling up my job and yours too? I have some experience of the reservation's role during my days at the casino.' Manoj's face lit up and instantly he hugged PP. After Manoj came back from his leave, PP observed that there was a marked change in the demeanour and behaviour of Manoj, Vidyut, Shefali and Jagat. PP was finally getting to feel a sense of acceptance by his colleagues.

As PP's time with Paradise Splendour increased his work was getting gradually appreciated and talked about. Shefali, his manager, was very much impressed with the way PP dealt with the guests at Paradise Splendour. On one occasion there was a celebrity check-in at the resort. Bharat Dubey (Deputy Manager Reservations) had been assigned to do the reservation for the celebrity and his family. As the guest arrived at the lobby, Bharat, Celina (Deputy Manager Customer Relations) and Manoj (the Reservation Supervisor) were there to welcome the guests and usher them to their suite. However, upon arriving at the suite, the celebrity was visibly annoyed. He turned to Bharat and said, 'I presume this is not a Presidential Suite'. To this Bharat said:

> Yes, Sir. This is not. I had clarified to you, during our conversation, that all four of our Presidential Suites were booked and that

> I would only offer you a normal suite for the time being. I had also promised that subject to availability, during your stay with us, I would upgrade you to a Presidential Suite.
> (Bharat)

The celebrity was furious. Raising his voice he said, 'This is the problem with you guys. You promise something and offer something else. I distinctly remember you having said that the upgrade will happen upon my arrival.' Visibly cowed down, Bharat said, 'Sorry Sir, how could I make such a promise when I know that the Presidential Suites are booked ...'. Before Bharat could offer further clarification, the celebrity started throwing tantrums and his face turned red with anger. He turned to Shefali and said, 'My dear lady, could you please do something about this situation. I would not like to talk to this person. Make my upgrade to a Presidential Suite right away or just connect me to the owner of this resort.'

As this commotion was happening, PP got wind of it from one of the butlers. He immediately informed Rosaline (the Front Office Manager) and rushed to the top floor of the facility, where all suites were located. Soon Rosaline also joined the rest. However, by that time, the celebrity was livid and was threatening to walk away from the resort. Rosaline was trying to convince the guest to stay. She reasoned that in 2 days one of the Presidential Suites would be vacated and the guest could be upgraded for the rest of his stay. Shefali, Bharat, Manoj and PP were hurdled up at a distance and Shefali was visibly worried about the dent to the brand, should the celebrity post about this situation he and his family were going through on social media. Bharat was angry since he thought that he was being falsely targeted by the guest for no fault of his. Manoj was grim and glum and was trying to pacify Bharat by sharing his opinion about celebrity tantrums. It was at this point that PP suggested to Shefali:

> What if the celebrity would have heard Bharat differently or just forgotten about the candid conversation that Bharat had with him on the phone? Or is it just possible that he promised his family the luxury of a Presidential Suite and, now since it is not available, is feeling embarrassed before his family?
> (PP)

Shefali, Bharat and Manoj had surprised looks on their face. The thoughts running through their minds were – 'Is PP taking sides?', 'Is he thinking that Bharat had not clearly communicated the non-availability of the Presidential Suite?' Putting aside speculation, Shefali asked, 'What are you trying to suggest PP?' To this, PP replied:

> Shefali, I am just looking at this situation from a different lens. Probably, our guest has perceived the service at Paradise Splendour very differently, because of something that has put him on a spot, rather than because of a communication error from our end. Perceived from the lens of the guest, our effort should be to pull him out of his predicament and support him in standing tall before his family.
> (PP)

'Isn't that what we have been trying to do all this time PP?' said Shefali. She continued, 'We can't build a Presidential Suite for him out of thin air by the wave of a magic wand. He needs to wait for at least 2 days to get upgraded, which he is not ready to understand'. PP said:

> That is precisely the point. It is about thinking about how we can create the ambience of a Presidential Suite for him with our existing suites so that he perceives that we have gone out of our way to do a service recovery and his family perceives the power of his celebrity status.
> (PP)

Shefali's face lit up immediately. All the anxiety lines that were showing on her face were gone. Smiling, she whispered, 'That is a brilliant idea PP. Come let us intervene and suggest this to the guest. I am sure Rosaline will be okay if we barge into her conversation, if your idea clicks.' Saying this, she almost dragged PP by his hand to join the conversation that Rosaline was having with the celebrity. 'May I suggest something Sir', said Shefali as soon as she could find a pause in the conversation between Rosaline and the celebrity. The irate guest turned his head to face Shefali with his eyes, red as if they were spitting out fire. 'Yes, Mam, please go ahead', said he, showing no remorse for his anger or behaviour. Ignoring his demeanour, Shefali calmly said, 'Palash, would you please share your plan with our esteemed guest?' PP walked a few steps forward and said:

> Sir, I need just 30 min to have two of our adjacent normal suites, which have a connecting door between them, to be provided with all the luxuries that come with the Presidential Suite. You will not feel the difference between a Presidential Suite and a normal suite Sir, with the arrangements I have in mind, especially for you and your family. This is a promise, Sir. Further, we will not charge you the rates of a Presidential Suite for these two normal suites, even though you will enjoy all the facilities of a Presidential Suite. From our side this waive off will be a token of our appreciation for your flexibility to agree to our suggestion and forget whatever has happened just now.
> (PP)

The last sentence was blurted out by PP in a manner that conveyed that Paradise Splendour was making up for a serious mistake that had been pointed out by the guest.

A sudden calm descended. The temperatures cooled down, the face of the celebrity looked more welcoming and accommodating, the perspiration on Rosaline's forehead started to dry up. After a thoughtful pause the celebrity replied, 'Well, that is an idea worth considering. However, if I and my family miss the ambience of a Presidential Suite, despite all your effort, I will expect an immediate upgrade as soon as one of your Presidential Suites becomes vacant.' 'Without question, Sir', chipped in Rosaline with a large smile on her face.

Guess what, PP was the star of the day. That night Rosaline gathered all her team members for an impromptu party to celebrate and raise a toast to the health of PP for having averted an imminent service disaster that could have had far-reaching negative brand publicity for Paradise Splendour.

Discussion Questions

1. Discuss in your group the difference between perception, feelings and attitude. Thereafter, identify the case facts where these three important concepts have been demonstrated by the pertinent characters of the case.

2. What factors influenced the process of perception formation in the mind of Palash? Discuss in your group with reference to the factors cited by Robbins and Judge in their book, *Organizational Behaviour*, 18th edition, Chapter 5, 'Perception and Individual Decision Making'. Cite case data to justify your answers.

Teaching Notes

Summary

Marcel Proust, the legendary French novelist and critic, famous for his seven-volume novel *In Search of Lost Time*, had once famously said, 'The only true voyage of discovery [would be] to possess other eyes, to behold the universe through the eyes of another' (Proust, 1999, originally published in 1923). Interpersonal perception is a field of study in psychology that has far reaching interdisciplinary impact. Be it in sociology, political science, economics, business, anthropology, or any other domain which involves the study of human beings, the ramification of interpersonal perception has far-reaching consequences. The fundamental question of investigation in interpersonal perception is the ability of people to know (perceive), as accurately as possible, the meaning of another person's traits, motives, intentions, behaviour, etc. This is fundamental since the accuracy or inaccuracy of perceiving, interpreting and understanding another person is the source of all love and hate, warmth and coldness, cooperation or opposition, inclusion or exclusion, and all the opposing dualities that one can think of, among human beings. The accuracy of perception is cause for harmonious and healthy interpersonal relationships which, in turn, is the *sine qua non* for one's success or failure in personal and professional life.

Given the above context, Paradise Splendour provides a golden opportunity for young graduates and postgraduates to thoroughly understand the perceptual process and its applications in one's personal and professional life. The case provides enough insight, and more, into the application of the perceptual process in one's day job and how perception impacts one's actions and reactions to interpersonal stimuli that all of us keep receiving right from the time we get out of our beds in the morning until the time we get into our bed at night.

Target audience

The case study is suitable for undergraduate and postgraduate students studying general management, hotel management and hospitality management. In particular it would be ideal for undergraduates in their final year of study just before they enter the job market.

Teaching methods and equipment

The case study can be taught as a class workshop using thought-provoking discussions. Equipment needed includes a printout of the case study, flipcharts, coloured highlighter pens, PowerPoint and access to Chapter 5 of Robbins and Judge (2019) (referred to in discussion question 2).

Teaching instructions

The time needed to teach the case study in class is 2½ hours. Tutors/instructors are free to use the case study as they wish in order to achieve the learning objectives. However, one possible suggestion is outlined here.

Prior to the class session students may be asked to read the chapter titled 'Perception' in Robbins and Judge (2019) *Organizational Behaviour*. Also students may be required to read a printout of the case study

Table 12.1. Clarification of what the concepts mean.

Concept	Brief clarification
Perception	Personalized **point of view** of reality by individual. When individuals receive stimuli from their environment, both within (i.e. internally) and without (outside their body), the sensory organs select the data from the stimulus and help them to make sense of it or provide meaning from it. This sense making results in creation of the point of view of reality.
Attitudes	An individual's favourable or unfavourable **feelings and reactions** about people, things, situations, etc. around them. They are a mixture of feelings, thoughts and action or behaviour.
Feelings	Positive or negative **sensations** that individuals experience while receiving a stimulus from a place, person or situation.
Factors influencing perception	In every perceptual process there are three characters: (i) **perceiver**; (ii) **target** (the thing that is being perceived); and (iii) the **situation** (the context of the target and the perceiver). There are several factors in these three characters that influence perception of any object, person, situation or phenomenon.
Attribution theory	This explains why we judge people the way we judge them. While observing behaviours of people with whom we interact, we tend to attribute meaning to those behaviours. This process of meaning creation is primarily driven by our personalized point of view of reality (perception) which we attribute to the immediate target resulting in judgement making about people and situations.
Attitude and perception formation	Our attitudes express themselves first, internally through our feelings, and then externally through our behaviour (demonstrated responses or reactions). Our reactions, therefore, tend to be evaluative in nature and are directed at a given object or target. The nature of attitude (i.e. positive/favourable, negative/unfavourable or neutral) influences our personalized point of view of reality (perception).

Table 12.2. Difference between perception, feelings and attitude. Authors's own data.

Characters	Organizational roles	Organizational relationship with respect to Palash	Perception	Feelings	Attitude
Palash Patil	Front Office Supervisor		Most of his peers had pedigree academic backgrounds accompanied by work experience and PP could quickly make out that getting accepted among his peers would be a definite struggle for him. 'In the meantime, try to stay away from perceiving that their behaviour is because of your humble pedigree.' After Manoj came back from his leave, PP observed that there was a marked change in the demeanour and behaviour of Manoj, Vidyut, Shefali and Jagat. As PP's time with Paradise Splendour increased his work was getting gradually appreciated and talked about. It was at this point that PP suggested to Shefali: 'What if the celebrity would have heard Bharat differently or just forgotten about the candid conversation that Bharat had with him on the phone? Or is it just possible that he promised his family the luxury of a Presidential Suite and, now since it is not available, is feeling embarrassed before his family?'	A couple of times when he tried to join them, he felt ignored since no one gave him the context of the topic under discussion. It was hurting. PP was feeling low, and all his excitement of a new job was lost in the first few weeks of his tenure. PP was finally getting to feel a sense of acceptance by his colleagues.	PP went about his job chores and work-related conversation with the very same colleagues who were ignoring him, as if nothing had happened. He was friendly and helpful while engaging with them in work-related conversations while continuing to join them during lunch breaks. PP said, 'Can I help you for a few days, by doubling up my job and yours too? I have some experience of the reservation's role during my days at the casino.' He immediately informed Rosaline (the Front Office Manager) and rushed to the top floor of the facility, where all suites were located.

			Probably, our guest has perceived the service at Paradise Splendour very differently, because of something that has put him on a spot, rather than because of a communication error from our end. 'so that he perceives that we have gone out of our way to do a service recovery and his family perceives the power of his celebrity status.'	To this, PP replied: 'Shefali, I am just looking at this situation from a different lens. Probably, our guest has perceived the service at Paradise Splendour very differently, because of something that has put him on a spot, rather than because of a communication error from our end. Perceived from the lens of the guest, our effort should be to pull him out his predicament and support him in standing tall before his family.'
Vidyut Rao	Front Office Supervisor	Peer		For instance, during lunch breaks he was not invited by his peers to join in. A couple of times when he tried to join them, he felt ignored since no one gave him the context of the topic under discussion. Whenever he tried to change the topic, he found that no one was sharing their perspective on the same, yet politely hearing him out, before going back to continue their discussion on the earlier topic. Come let us intervene and suggest this to the guest. Saying this, she almost dragged PP by his hand to join the conversation that Rosaline was having with the celebrity.

Continued

Table 12.2. Continued.

Characters	Organizational roles	Organizational relationship with respect to Palash	Perception	Feelings	Attitude
Manoj Sharma	Reservation Supervisor	Peer	Manoj was grim and glum and was trying to pacify Bharat by sharing his opinion about celebrity tantrums.		
Shefali Mishra	Assistant Manager Lobby	Manager	Shefali, his manager, was very much impressed with the way PP dealt with the guests of Paradise Splendour.	Shefali, Bharat, Manoj and PP were hurdled up at a distance and Shefali was visibly worried about the dent to the brand, should the celebrity post about this situation he and his family were going through on social media.	
			Shefali, Bharat, Manoj and PP were hurdled up at a distance and Shefali was visibly worried about the dent to the brand, should the celebrity post about this situation he and his family were going through on social media. The thoughts running in their minds was – 'Is PP taking sides?', 'Is he thinking that Bharat had not clearly communicated the non-availability of the Presidential Suite?'	Shefali's face lit up immediately. All the anxiety lines that were showing on her face were gone.	
Jagat Pandey	Assistant Manager CRM	Senior colleague			
Celina Ryan	Deputy Manager Customer Relations	Shift Manager			
Bharat Dubey	Deputy Manager Reservations	Senior colleague	Bharat was angry since he thought that he was being falsely targeted by the guest for no fault of his.	Visibly cowed down, Bharat said,Bharat was angry, since he thought that he was being falsely targeted by the guest for no fault of his.	

Rosaline Janson	Manager Front Office	Department Head		The temperatures cooled down, the face of the celebrity looked more welcoming and accommodating, the perspiration on Rosaline's forehead started to dry up	That night Rosaline gathered all her team members for an impromptu party to celebrate and raise a toast to the health of PP for …
Celebrity	Guest at Paradise Splendour	Customer	'This is the problem with you guys. You promise something and offer something else.'	However, upon arriving at the suite, the celebrity was visibly annoyed.	The celebrity was furious. Raising his voice he said…. Before Bharat could offer further clarification, the celebrity started throwing tantrums and his face turned red with anger. 'I would not like to talk to this person.' However, by that time, the celebrity was livid and was threatening to walk away from the resort. 'Yes, Mam, please go ahead', said he, showing no remorse for his anger or behaviour.
			'I would not like to talk to this person.'	Before Bharat could offer further clarification, the celebrity started throwing tantrums and his face turned red with anger. However, by that time, the celebrity was livid and was threatening to walk away from the resort. The irate guest turned his head to face Shefali with his eyes, red, as if they were spitting out fire. The temperatures cooled down, the face of the celebrity looked more welcoming and accommodating, the perspiration on Rosaline's forehead started to dry up.	

[a]CRM, Customer relationship management.

Table 12.3. Factors that influenced the process of perception formation in Palash's mind.

Factors in the perceiver – Palash	Case facts	Factors in Palash's self-image	Case facts	Factors in the situation –new job and size and scale of new organization	Case facts
Attitudes	Several instances in the case suggest that Palash was a person with a positive attitude, including: **1.** In spite of his peers ignoring him, 'PP [Palash] went about his job chores and work-related conversation with the very same colleagues who were ignoring him, as if nothing had happened.' **2.** Ignoring the behaviour of his peers Palash, 'was friendly and helpful when engaging with them in work-related conversations while continuing to join them during lunch breaks.' **3.** During the commotion that happened with the celebrity, Palash, on his own volition, 'rushed to the top floor of the facility, where all suites were located.' It may be noted that: (i) he was not in the team that was assigned to take care of the celebrity; (ii) the commotion was not happening in his work area, the lobby; and (iii) his boss, who was already with the celebrity, had not called for him.	Novelty	There were enough and more novelty factors that could have influenced the perception of Palash. Some of them are: **1.** Five-star property to which Palash had no prior exposure **2.** Working with colleagues who had pedigree academic backgrounds as well as work experience **3.** Customer milieu of a five-star facility by comparison with that of a casino in Goa **4.** Organizational structure of a hotel by comparison with that of a casino **5.** Transition from an individual contributor role of customer support staff to a supervisory role in the front office.	Time	No case facts available
Motives	There is no data in the case about Palash's motives. It is best not to make assumptions or hypotheses.	Motion	There is no data in the case on this factor.	Work setting	A fair assumption to make here would be the informal settings of a small organization like a casino versus the formal work settings of a large organization.

Interests	Though there is not much in the case that talks about Palash's interests, one thing that did interest Palash was building rapport with people. For example: **1.** For instance, once, Manoj Sharma, the Reservation Supervisor was completely off-colour while on duty. PP could notice that not all was well with Manoj. He walked up to him and asked, "Hope you are okay Manoj?"' **2.** 'Before, Manoj could continue, PP said, "Can I help you for a few days …"'	Sounds	The laughter of his peers while discussing topics over lunch, on which Palash had no clue at all.	Social setting	No case facts available
Experience	Palash's experiences, both academic and workplace, appear to have been quite humble, devoid of any glare and glitz of reputed educational institutions. For example: **1.** 'Academically, he had a postgraduate diploma in hotel management from a not-so-famous hotel management institute "down south".' **2.** 'His professional career began when he was picked from the campus by a casino in Goa.' **3.** 'Having been catapulted from a casino set-up to a five-star environment was no mean achievement for him.'	Size	The size and scale of operations at Paradise Splendour compared with the casino in Goa.		
Expectations	It is clear from the case that Palash's expectation was to make a mark for himself at Paradise Splendour and be accepted among his superiors and peers. For example: **1.** 'Palash had not quite settled into his new milieu' **2.** 'PP, as he was lovingly addressed by his friends and colleagues, had higher aspirations though.' **3.** 'As PP's time with Paradise Splendour increased his work was getting gradually appreciated and talked about. Shefali, his, manager, was very much impressed with the way PP dealt with the guests ...'	Background	No case facts on the family or social background of Palash.		
		Proximity	Closeness of Palash to his uncle who was his mentor and guide		
		Similarity	No case facts available		

(as described under the heading 'Case') before the class session, or time can be allocated for this during the class. The instructor may initiate the class discussion session with a brief lecture on the concepts that relate to the case study and what they mean (see Table 12.1). Following this the instructor may ask students questions about the details of the case as they understand it, writing the main points on the board. Once the instructor is satisfied that students understand the concepts and the case, students may be asked to answer the discussion questions. For this students may be asked to work in small groups to brainstorm their answers, which may be followed by a feedback session when each group presents their answers, followed by a class discussion of the issues. Finally, the instructor may conclude the session by briefly summarizing the case facts and outlining his/her analysis and answers to the discussion questions.

Analysis

1. Discuss in your group the difference between perception, feelings and attitude. Thereafter, identify the case facts where these three important concepts have been demonstrated by the pertinent characters of the case.

Details of the difference between perception, feelings and attitude in the behaviour shown by the main characters in the case are provided in Table 12.2.

2. What factors influenced the process of perception formation in the mind of Palash? Discuss in your group with reference to the factors cited by Robbins and Judge (2019) in their book, *Organizational Behaviour*, 18th edition, Chapter 5, 'Perception and Individual Decision Making'. Cite case data to justify your answers.

The factors that influenced the process of perception formation in Palash's mind are shown in Table 12.3.

References

Proust, M. (1999) *In Search of Lost Time, Volume V: The Captive and the Fugitive*. English translation of original 1923 (*The Captive*) and 1925 (*The Fugitive*) editions. Translated by C.K. Scott Moncrieff and T. Kilmartin, Revised by D.J. Enright. Random House, New York.

Robbins, S.P. and Judge, T.A. (2019) *Organizational Behaviour*, 18th edn. Pearson Education, Chennai, India.

Further Reading

Altman, S., Valenzi, E. and Hodgetts, R.M. (1985) *Organizational Behaviour – Theory and Practice*. Academic Press, Cambridge, Massachusetts.

Lotto, B (2017) *Deviate: the Science of Seeing Differently*. Hachette Books, Paris.

Robbins, S.P. and Judge, T.A. (2019) *Organizational Behaviour*, 18th edn. Pearson Education, Chennai, India.

13 Café Cherrapunjee: a Guest House with a Sustainable Approach

Kimberley Camelia S. Langstieh* **and Evarisa Mary Nengnong**
North-Eastern Hill University, Shillong, India

Abstract

Café Cherrapunjee is a guest house following a sustainable approach with a 125-year-old 'dak' bungalow converted into a restaurant located at the entrance of Sohra (Meghalaya), known for being the wettest place on earth. It is owned by Mr Alan West Kharkongor, a native of Sohra and a tourism enthusiast, who owns another guest house in the state. Besides being environmentally sustainable, they also employ locals without experience in hospitality but who are trained by the owner. The hotel industry accounts for around 2% of the global 5% carbon dioxide (CO_2) emitted by the tourism sector (UNWTO, 2022), stressing the need to promote a sustainable tourism model. Meghalaya, a state in India, is an ecotourism hot spot, which requires tourism businesses to follow a sustainable approach to preserve their natural resources. This case study will analyse Café Cherrapunjee as an example of sustainable tourism accommodation and how following a sustainable approach will help preserve the state's social and cultural environment.

Subjects: general management, operations management

Learning Objectives

At the end of the case study, the learner will be able to:

1. Explain the concept of the triple bottom line (TBL).
2. Identify the sustainability indicators.
3. Identify the World Tourism Organization (UNWTO) sustainable development goals.

Case

With its scenic terrain of rolling hills, meandering rivers, cascading waterfalls, lush forests, rich flora and fauna, and unique culture and heritage, Meghalaya, the 'Abode of the Clouds,' has a lot of promise for tourism development. In recent years, Meghalaya has rapidly grown in popularity among domestic and foreign tourists. Meghalaya has a lot to offer in terms of its natural scenic beauty, opportunities for adventure and ecotourism, and the availability of many trained human resources to support delivery systems. The government has chosen to foster tourism as a significant source of revenue for the state. Meghalaya actively promotes local participation, including community-based tourism; it emphasizes local training of youth and

*klangstieh@gmail.com

DOI: 10.1079/9781800621022.0013

tourism stakeholders in the state, and establishes relevant tourism practices. Further, Meghalaya is frequently described as one of India's most beautiful states. The state has considerable tourism substance and the potential to draw visitors with a wide range of interests. While growing tourist numbers, an effort is being made to develop the facilities sustainably.

Café Cherrapunjee, a 125-year-old 'dak' bungalow, is a guest house started by Mr Alan West Kharkongor, a resident of Sohra/Cherrapunjee, to promote local entrepreneurs and provide employment for the youth residing in the area. He owns two properties at Sohra /Cherrapunjee, which are both run following sustainable practices. Mr Alan West is a member of the Meghalaya Rural Tourism Forum, Khasi Heritage Conservation Forum and the 'Clean Sohra campaign'. As Chairman of the 'Clean Sohra campaign', he has taken many initiatives for cleaning drives in various tourist spots in Sohra (Cherrapunjee). He also conducts a cleanliness awareness programme in the village/ community. This case study reflects how sustainable practice by one property can be used by tourism organizations to preserve and promote sustainable practices.

Tourism in Cherrapunjee (famous for being the world's wettest place) was dilapidated until the early 2000s, with tourism businesses in the area earning little or no income. Seeing the tourism potential, Mr Alan West, a professional journalist eager to discover the state's hidden gems, conducted tours of undiscovered rural Meghalaya. In 2012 Sohra had only four local tourism practitioners; it was here that Mr Alan West saw the tourism potential and started Café Cherrapunjee, which, as its name indicates, started as a café to cater for tourists visiting the area. Established in a century-old building popularly known as *'iing dak'* or a dak bungalow, the main attraction of the guest house is the 125-year-old bungalow which now holds the restaurant. Dating back to the British Raj in India, it was one of the many bungalows set up by the British for changing official horses between Dhaka and Assam. An interesting fact about the bungalow is the reports of an unidentified flying object seen by locals over the property in the 1960s. Its credibility is unclear, but this adds to the property's history, mystery and attractiveness.

In 2014, Mr Alan West started setting up two glamping tents to accommodate tourists – these were the first in the state. Also four additional rooms were added to the bungalow, which already had two rooms. The luxury tents at the guest house are well furnished with all the essentials of a cosy home resting on a wooden deck. The tents include a bathroom, heating facility, hot water, a dressing table, and other necessities. The alpine camper tents are ideal for an overnight stay, as they are fitted with solar tent lights and mobile-charging units. Facilities and services provided by the guest house are: (i) housekeeping and room services; (ii) special tours organized to tourist destinations in the surrounding areas by local tour guides; and (iii) a cab-on-call service. Four expert guides are available in the guest house to fulfil the needs and requirements of the tourists visiting Meghalaya. Owing to its perfect setting one can also enjoy a range of recreational activities, including zip lining, scuba-diving, snorkelling, river rafting, riverside camping, abseiling, angling and trekking.

Wanting to promote local talent from the area and its surroundings, the guest house employs nine locals who are trained by the owner himself (he conducts training sessions to improve their skills so they can deliver a quality service) or by trainers invited by the owner from AVENUES and the Institute of Hotel Management (IHM) Shillong. Despite not being professionally trained workers, the feedback from guests tells us otherwise, where the service quality is appreciated. A commendable part of the operation is the sustainable practices, where the three pillars of sustainability, environmental, economic and social, are implemented, which is in line with the triple bottom line (TBL) theory and the UNWTO sustainable development goals. The first pillar of sustainability (i.e. environment) is practised here by installing a greywater treatment plant and a recyclable garbage disposal unit, run by the locals. The second and third pillars are reflected in the staff working at the café, where only locals are recruited and trained to provide service of the highest quality. The recruitment of local staff directly improves the livelihood of these individuals. Mr Alan West also ensures that only local guides are hired to assist the tourists, which acts as a revenue earner, and their real-life stories and experiences that are shared by the guides adds to the authenticity of place.

Mr Alan West owns another property, 'Kutmadan Resort' located in Sohra/Cherrapunjee that follows the same sustainable methods as Café Cherrapujee.

The owner of Café Cherrapunjee demonstrates TBL sustainability in three aspects:

1. Social sustainability: The owner of the business pays attention to the community, the social surroundings and internal resources as part of his corporate social responsibility. He only employs local people and the wages he pays them helps to support the local community.
2. Environmental sustainability: The owner's business practices foster the duty to conserve natural resources and maintain global ecosystems by promoting health and well-being today and in the future.
3. Economic sustainability: The business aims to be sustainable in terms of its financial performance and profitability. Also the owner is aware of his firm's impact on the general economy and its business growth and aims to contribute to these.

Café Cherrapunjee has been rated number one on TripAdvisor in ratings and reviews by guests who have stayed and used their services while visiting the area. Most travellers have given positive ratings, and of these 43 are excellent and 40 very good. One guest who visited the property commented:

> It's a beautiful hotel in Sohra. It has nice rooms and two tent rooms for people who want a tent experience but want luxury. The food here is excellent, and many local people also come here to eat. The rooms are well kept, have a heater, and are very neat and clean. Trekking areas can also be found around the vicinity for adventurous explorers. The setting is elegantly created.
> (guest at Café Cherrapunjee)

The guest house has made a mark in Sohra, which is reflected in the occupancy rate (see Table 13.1). Café Cherrapunjee has a total of six rooms that can accommodate 12 people. The total number of guests in 2016 was 1009, with the month of July having the highest occupancy of 140. The years that followed saw an increase in the occupancy rate in most of the months, in particular March and May of 2018 saw the highest occupancy of 253 guests.

Table 13.1. Occupancy at Café Cherrapunjee.[a] Authors' own table.

	Arrivals			
Month	2016	2017	2018	2019
January	44	72	85	50
February	13	193	34	40
March	64	96	*253*	132
April	63	120	119	121
May	94	*197*	*253*	*206*
June	101	166	153	178
July	*140*	167	118	126
August	92	96	112	74
September	73	122	86	84
October	138	141	160	163
November	96	146	141	105
December	91	129	159	149
Total arrivals	**1009**	**1645**	**1673**	**1428**

[a] The highest values for occupancy at Café Cherrapunjee for each year are set in italics.

Understanding the greywater treatment plant

The guest house has installed a greywater treatment plant to ensure the water consumed by the guest house is safe for disposal into the rivers. A greywater treatment plant purifies contaminated water through a series of steps involving filtering and purification resulting in pure, uncontaminated water.

The greywater treatment plant consists of four processes: (i) pre-treatment; (ii) primary treatment; (iii) secondary treatment; and (iv) tertiary treatment. Each process goes through three steps: anaerobic digestion, sand gravel filtration and a plant filter in which water flows through a reed bed. After completion of all four processes the result is pure, uncontaminated water (see Fig. 13.1).

At present, only four households/businesses in the state have set up greywater treatment plants, which is negligible when compared with the population of the state. The initial cost is 200,000 rupees.

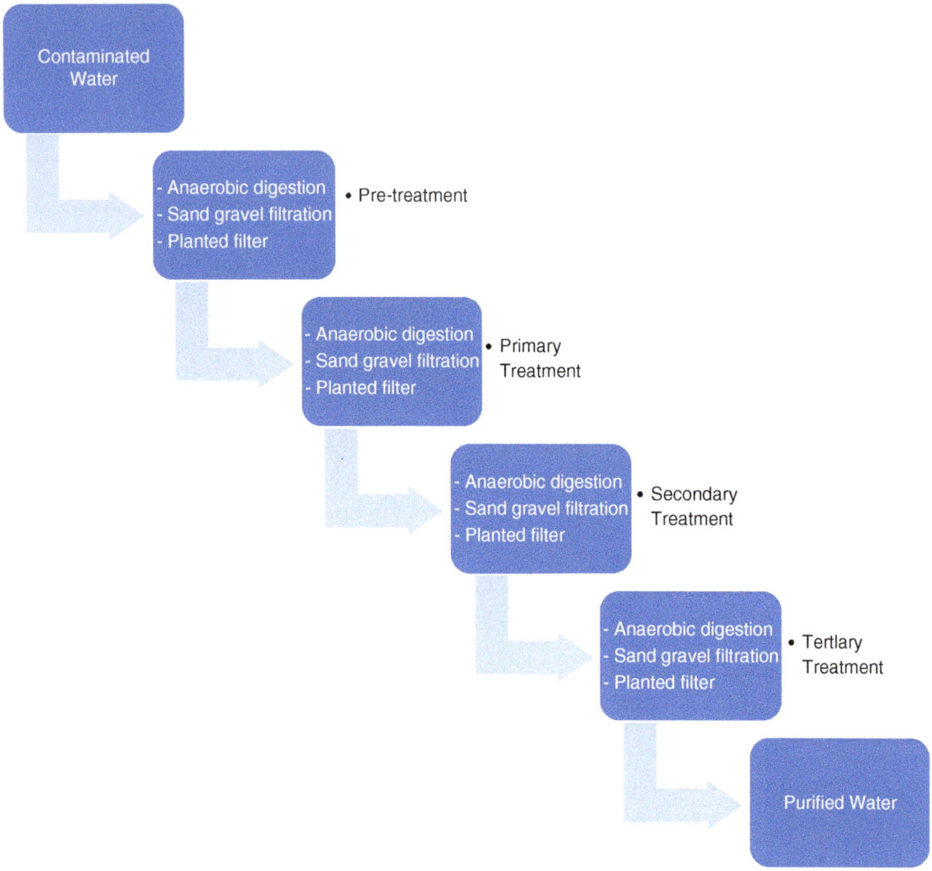

Fig. 13.1. Greywater treatment process. Authors' own diagram.

Understanding the triple bottom line (TBL)

Sustainable tourism is a practice of sustainable consumption of resources, which requires one to keep in mind future consumers in three main spheres: (i) environmental sustainability; (ii) social sustainability; and (iii) economic sustainability. Sustainable practices in the tourism industry are critical, whereby the UNWTO has also implemented the 17 sustainable goals laid out by the United Nations (UN). The case subject is analysed in line with achieving Sustainable Development Goal (SDG) 11 sustainable cities and communities and SDG 6 clean water and sanitation.

The concept of sustainability will be studied using the TBL framework incorporating the three pillars of environment, society and economy (UNWTO), otherwise known as the three P's: (i) people; (ii) profit; (iii) planet.

The concept of TBL was incorporated after three main historical events, which Elkington (2002) termed as three waves, limits, greens and globalization, falling under the timeline of the early 1960s, 1988 and 1999, respectively.

The theory of the ethical economy, the increased attention on society, and economic coordination require organizations to become more responsive to their environmental and social impact by employing more sustainable operational practices. Five sustainability indicators are used: (i) the organization's economic performance; (ii) its environmental performance; (iii) its social performance; (iv) its labour practices; and (v) its public policies and implementation measures (Stoddard *et al.*, 2012).

In the tourism industry, economic indicators such as heads in beds (for hotels), revenue per available room (RevPAR), or visitation (for attractions) are popular measures. Social performance is indicated by dimensions such as employee retention rates, job satisfaction levels, and investment per employee in illness and injury prevention. Environmental performance is indicated by measurements such as total energy use, use of recycled materials, and water sources significantly affected by an organization's use of water.

The term sustainable tourism (encouraged by both the UNWTO and the UN) is used as an umbrella concept under which terms such as ecotourism, heritage, cultural tourism, and geo-tourism may fall. Sustainable tourism and development find a balance between economic prosperity, environmental protection and social equity. The UNWTO developed 11 destination-specific core indicators to measure sustainable tourism impact, including: (i) site protection; (ii) stress; (iii) use intensity; (iv) waste management; (v) critical ecosystems (ecological); (vi) consumer satisfaction; (vii) local satisfaction; (viii) social influence (social indicators); (ix) tourism contribution to the local economy (economic indicators); (x) development control; and (xi) planning process (planning indicators) (Dymond, 1997).

Keeping in mind the TBL, Fig. 13.2 is a visual representation of sustainable practices by the guest house, including:

- segregation of waste ensuring environmental sustainability (A);
- the dak bungalow which now houses the main building and the café (B);
- accommodation units in the form of glamping tents (C);
- locals employed by the guest house ensuring social and economic sustainability (D); and
- a greywater treatment plant ensuring environmental sustainability (E and F).

Fig. 13.2. Café Cherrapunjee, Meghalaya. (A) Sign at the entrance to the grounds informing visitors about litter disposal; (B) reception and restaurant; (C) the two glamping tents; (D) the staff; (E and F) the water purification treatment plant. Photos courtesy of the authors.

Discussion Questions

1. Identify the various sustainable practices at Café Cherrapunjee by referring to the triple bottom line (TBL) model.
2. Identify sustainable approaches which tourism organizations can implement while expanding. Use data from the case on Café Cherrapunjee to support your answer.
3. Is an increase in salary proof of sustainable economic growth?
4. How will sustainable development processes be measured? Is an increase in occupancy proof of tourism development and in this case sustainable development?
5. How does sustainable development make sense for society? Is installing a greywater treatment plant enough to reduce a hotel's carbon footprint?

Teaching Notes

The purpose of the case is to familiarize learners with the concept of the TBL. The main objective is to equip bachelor's and master's degree students' as well as research scholars' programmes. Also it can be beneficial to tourism business owners and tourism policy makers who can participate in sustainable practices. Caring and giving importance to social and environmental matters by organizations and companies in the same manner as financial gains is what businesses term the triple bottom line (TBL). The primary use of the bottom line was for the financial matters of a company, calculating the company's profit and loss; when people and the planet are added to the equation along with profit is where the definition of TBL is made. The TBL is associated with the three leading primary aspects of a society, or the three P's: people, planet and profit. The concept of people refers to not just people belonging to a business but people of the society and community who are affected by the business both directly and indirectly, by considering the benefits of the company to the society. TBL ensures that workers are paid their dues, there are no human rights violations in the workplace, and helps the community through philanthropy. Keeping in mind the need for financial gains, the company under TBL also ensures that people and the planet are protected and given importance.

Sustainable tourism is a much needed concept given the current situation of mass tourism and depleting natural resources; sustainable tourism will ensure that tourism is available for future generations. It covers all tourism experiences, including environmental, economic and social aspects, with the need to improve tourist experiences and keep in mind the needs of the host communities. The UN sustainable development goals were introduced and adopted by the UNWTO; keeping Sustainable Development Goal (SDG) 11 and SDG 6 in mind, the study of Café Cherrapunjee was initiated.

SDG 11 which states:

> [to] accommodate everyone, there is need to build modern, sustainable cities. In order for everyone to survive and prosper, there is a need for intelligent urban planning that creates safe, affordable, and resilient cities with green and culturally inspiring living conditions.
>
> (Vaidya and Chatterji, 2020, p. 182)

SDG 6, water security and environmental sustainability are now considered important by international organizations since its adoption as a UN sustainable development goal. The goal of SDG 6 is for the management and availablility of water and sanitation in a sustainable manner.

Companies following a sustainable approach understand the need to care about society and the environment, which in some cases can bring profit to the organization because of its sustainable approach, thereby attracting customers who care about not just profit but people and the planet. TBL can be measured through renewable energy, a sustainable approach, improving the three R's (reduce, reuse, recycle), and the efficient use of natural resources. The current case study demonstrates how sustainability is aligned with the business of a hospitality organization. It will help learners understand how hospitality units can promote livelihoods, give gainful employment opportunities, and strengthen the rural economy in Meghalaya.

Target audience

Apart from students in their bachelor's, master's degree and research programmes, the study is also beneficial for tourism business owners and tourism policy makers who can use this case as an example of participation in sustainable practices.

Teaching methods and equipment

The concept of sustainability and its impacts can be understood by first evaluating each of the three pillars of sustainability (social, environmental and economic) through examples of local establishments, analysing the results and making a connection with the sustainable development goals. Group activities such as group discussions can be used to start up a conversation and bring about ideas for sustainable practices. Analysing research papers can justify and support claims made by the students.

Teaching instructions

The case can be taught in a 2-hour class using the following teaching instruction plan. Learners should be familiar with the concept of sustainability and the TBL model. However, at the beginning the instructor or lecturer should evaluate the learners' understanding and familiarity with the concept by interacting with the students, asking them questions about the TBL and their experience of it (e.g. examples they are aware of).

Follow-up work *after* the case study may include the lecturer instructing the students to collect information in the state or region of their interest in order to identify a business venture that they can investigate with regards to the TBL model, and in particular the basic strategies adopted by the business with regards to its sustainability practices.

Teaching instruction plan

Discussion point	Time required (min)	Teaching strategies
Introduction of the case	10	The instructor briefly introduces the case study, outlining what it is about, and finding out if the students are familiar with the destination. The instructor asks the students to raise their hands if they want to ask or answer questions.
Briefing about the concept of sustainability	10	The instructor introduces the concept of sustainability. The instructor asks the students to provide examples regarding sustainability in the state or region, in particular focusing on the three main spheres, environment, social and economic. To answer or ask questions the learners raise their hands.
Concept of TBL	15	The instructor briefly discusses the background information about the TBL, including the three pillars and three P's, and how the TBL model can be used as a framework of indicators for studying sustainability in a tourism business.
Summary of the case	10	The instructor briefly discusses the background information about the case, introducing students to specific details about Café Cherrapunjee, in particular in relation to the development of sustainable practices in the business. Students are asked to get into small groups or teams to answer the discussion questions.
Question 1	15	**Identify the various sustainable practices at Café Cherrapunjee by referring to the TBL model.** Now, the instructor allows the groups/teams to identify the pillars of sustainability at Café Cherrapunjee.

Continued

Teaching instruction plan Continued.

Discussion point	Time required (min)	Teaching strategies
		After receiving the teams' responses, the instructor needs to write the answer given by each team on the board and display the chart created. Instructors ask each group to justify the rationale behind their answers referring to the TBL model. If there is any variation in the response between teams, then specific teams need to clarify their decision further.
Question 2	15	**Identify sustainable approaches which tourism organizations can implement while expanding. Use data from the case on Café Cherrapunjee to support your answer.** The instructor asks the team to identify what are the ways a tourism business can expand keeping in mind sustainable practices, citing examples from the existing business.
Question 3	15	**Is an increase in salary proof of sustainable economic growth?** Instructor supplies students with data on salary increase at Café Cherrapunjee (see Table 13.3 in 'Analysis' section) and asks them to use this, as well as referring to the TBL model, as the basis to answer this question.
Question 4	10	**How will sustainable development processes be measured? Is an increase in occupancy proof of tourism development and in this case sustainable development?** Students are asked to give examples by referring to the TBL and table on occupancy (Table 13.1).
Question 5	10	**How does sustainable development make sense for society? Is installing a greywater treatment plant enough to reduce a hotel's carbon footprint?** The instructor asks students how to measure the development of sustainability in the tourism sector. Also to identify other ways to reduce waste and carbon emissions, which can be used by accommodation units (e.g. in the city where they live).
Closing remarks and discussion	10	The instructor concludes the case analysis session with his/her analysis and summary of the answers to the TBL model discussion questions.

Duration

120 min

Analysis

1. Identify the various sustainable practices at Café Cherrapunjee by referring to the TBL model.

As per the case study, the students examine the practices of sustainability followed by Mr Alan West Kharkongor at Café Cherrapunjee in terms of the three P's (people, planet and profit as shown in the diagram in Fig. 13.3) also known as the three pillars, that is social, environmental and economic sustainability.

People/social

People play an essential role in ensuring the sustainable growth of any business. Sustainability can be reflected in the improved standard of living through employment generation, affecting the social environment through the involvement of the locals.

Planet/environment

In any business, ensuring the environment is protected is a positive indicator. Setting up a greywater treatment plant, encouraging

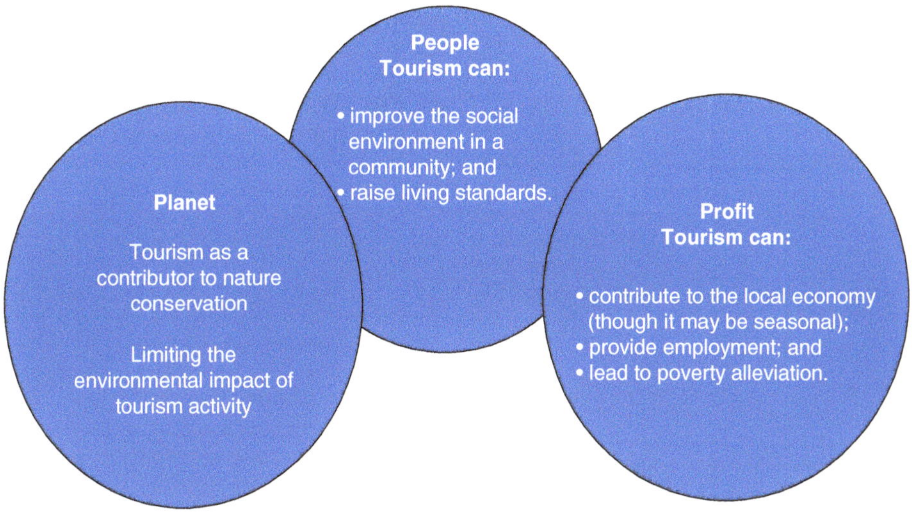

Fig. 13.3. Tourism and the three P's of sustainability (planet, people and profit). Authors' own diagram.

proper waste disposal through the 'Clean Sohra campaign', and ensuring adequate segregation and recycling of the waste within the property are ways that the Café are involved in saving and ensuring sustainable environmental growth.

Profit/economic

Any business establishment would expect an economic outcome. Ensuring this outcome is sustainable would mean having a positive impact on the livelihood of the locals. Café Cherrapunjee ensures this by recruiting only locals with minimal experience or no experience. In turn, they are trained by the guest house owner, therefore providing a livelihood for themselves and their families. The setting up of Café Cherrapunjee also brought a positive impact on tourism businesses in Sohra, a tourist place where locals were not involved in tourism. After Cafe Cherrapunjee was set up there was an increase in tourist inflow, and the years 2015–2017 saw a drastic growth in the number of accommodation units, rising from five to 87 in Sohra alone.

2. Identify sustainable approaches which tourism organizations can implement while expanding. Use data from the case on Café Cherrapunjee to support your answer.

Table 13.2. Increase in the number of employees at Café Cherrapunjee. Authors' own table.

Year	No. of employees
2013	5
2015	7
2017	9

The increase in the number of employees at Café Cherrapunjee between 2013 and 2017 indicates the positive impact of the guest house on society (see Table 13.2). Beginning with only five employees in 2013, the guest house currently has nine employees.

Growth in the number of employees would mean expansion and an increase in tourist inflow. An increase in number of employees could mean more money for locals to support themselves and their families perhaps raising living standards in the local community.

3. Is an increase in salary proof of sustainable economic growth?

A salary increase among all the employees indicates an improvement in the standard of living, with salary among some employees increasing by two- to threefold (see Table 13.3).

An increase in salary is a sign of growth which would also mean improvement in livelihood which is sustainable economic growth.

4. How will sustainable development processes be measured? Is an increase in occupancy proof of tourism development and in this case sustainable development?

Table 13.3. Salary increase over the years at Café Cherrapunjee. Authors' own table.

	Salary (rupees)	
Position	Past	Current
Manager	8,000	50,000
Chef	10,000	25,000
Housekeeping	4,000	10,000
Food and beverage service	6,000	15,000

Sustainable development can be measured by keeping in mind the economic, social and environmental aspects, and the impacts that tourism activities can bring which can be both positive and negative. One aspect that can be analysed here is the tourist inflow.

The inflow of visitors since the Café Cherrapunjee guest house started operations as a café in 2012 saw a gradual increase, with March and May 2018 seeing the highest occupancy of 250+ guests and a 63.03% increase in occupancy in 2017 compared with 2016 (see Table 13.1). Though some years saw a decline or a slow growth, this is due to the increase in the number of accommodation units that

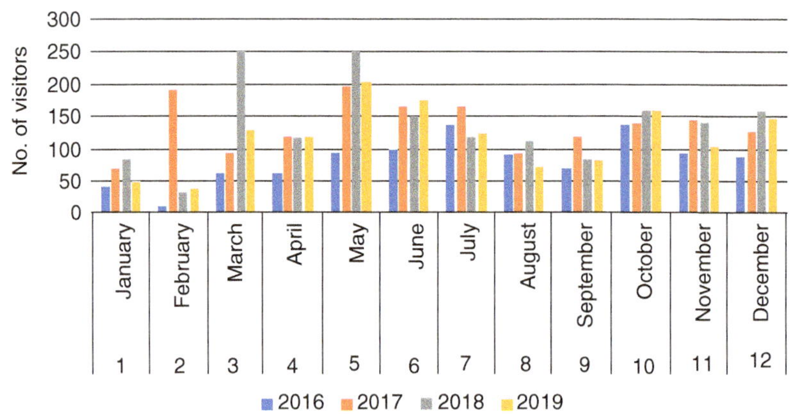

Fig. 13.4. Inflow of visitors to the Café Cherrapunjee guest house. Authors' own diagram.

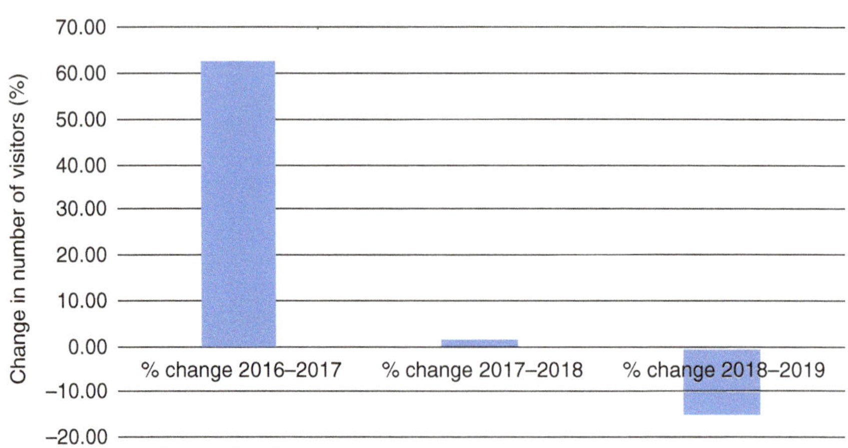

Fig. 13.5. Percentage change in number of visitors to Café Cherrapunjee. Authors' own diagram.

became available in the region, reflecting the positive impacts of tourism in Sohra, where it has grown to be a viable source of income for the locals. Refer to Figs 13.4 and 13.5.

Increasing occupancy should be done keeping in mind local policies and making sure that the positive impact of increasing tourist numbers is also seen by the local community while protecting natural attractions and should not lead to mass tourism.

5. How does sustainable development make sense for society? Is installing a greywater treatment plant enough to reduce a hotel's carbon footprint?

A greywater treatment plant is one way to ensure rivers and oceans are not contaminated by tourism business. But being the sole business with this practice is not enough; there is a requirement for both businesses and households to install these treatment plants to see its effect.

References

Elkington, J. (2002) Corporate strategy in the chrysalis economy. *Corporate Environmental Strategy* 9(1), 5–12.

Dymond, S.J. (1997) Indicators of sustainable tourism in New Zealand: a local government perspective. *Journal of Sustainable Tourism* 5(4), 279–293.

Stoddard, J.E., Pollard, C.E. and Evans, M.R. (2012) The triple bottom line: a framework for sustainable tourism development. *International Journal of Hospitality & Tourism Administration* 13(3), 233–258. DOI: 10.1080/15256480.2012.698173.

UNWTO (World Tourism Organization) (2022) Hotel Energy Solutions (HES). Available at: https://www.unwto.org/hotel-energy-solution (accessed 15 December 2021).

Vaidya, H. and Chatterji, T. (2020) SDG 11 Sustainable cities and communities. In: Franco, IB., Chatterji, T., Derbyshire, E. and Tracey, J. (eds) *Actioning the Global Goals for Local Impact: Towards Sustainability Science, Policy, Education and Practice*. Springer Nature, Singapore, pp. 173–185. https://doi.org/10.1007/978-981-32-9927-6

Further Reading

Goodwin, H. (2017) The challenge of overtourism. *Responsible Tourism Partnership Working Paper 4*. Available at: https://haroldgoodwin.info/wp-content/uploads/2020/08/rtpwp4overtourism012017.pdf (accessed 15 December 2021).

McGrath, G.M., Lockstone-Binney, L., Ong, F., Wilson-Evered, E., Blaer, M. and Whitelaw, P. (2020) Teaching sustainability in tourism education: a teaching simulation. *Journal of Sustainable Tourism* 29(5), 795–812.

Mihalič, T. (2000) Environmental management of a tourist destination. A factor of tourism competitiveness. *Tourism Management* 21, 65–78. DOI: 10.1016/S0261-5177(99)00096-5.

Pereira, V., Silva, G.M. and Dias, Á. (2021) Sustainability practices in hospitality: case study of a luxury hotel in Arrábida Natural Park. *Sustainability* 13, 3164. https://doi.org/10.3390/su13063164

UNWTO (World Tourism Organization) (2004) *Indicators of Sustainable Development for Tourism Destinations: a Guidebook*. UNWTO, Madrid.

14 The Twinhill Hotel: Revenue Management at a Family-owned Hotel

Rohan Jugran*
Manipal Academy of Higher Education, Manipal, India

Abstract
This case intends to familiarize learners with the concept of forecasting, pricing and displacement. The main objective is to equip undergraduate and postgraduate students and research scholars of hotel management to identify the difference between assumption and calculative thinking and subsequently apply a higher level of knowledge. Since revenue management is a concept-based learning, it becomes very important for students to understand about the market, competitors and analyse decisions before arriving at the conclusion.

Subjects: revenue management, family business, hospitality management

Learning Objectives

At the end of the case study, a learner will be able to:

1. Understand the business lead (i.e. an individual or organization who is interested in the product or service you sell, and has the potential to become a future customer), how to qualify a lead, how to study the market and how the lead will benefit the hotel in achieving its budget.
2. To compare hotels in terms of the services, facilities and location from the perspective of revenue management.

Case

The Twinhill Hotel is a fully serviced hotel in New Delhi. The hotel is a ten-storey, 300-room property, originally built in 1990. The current management took it over in 2005 with an annual fee of US$28 million and invested another US$14 million in property improvements in 2015. The hotel is located in the heart of Delhi, the capital city of the Republic of India, generally surrounded by a mix of commercial, retail and residential uses. Due to the variety of the surroundings, the hotel attracts demand from all segments, namely corporates, groups (both leisure and

*rohan.90r@gmail.com

DOI: 10.1079/9781800621022.0014

corporates), meetings, incentives, conferences and exhibitions (MICE) and leisure travellers. The predominant market segment for the property is corporate travellers.

The hotel's guest rooms include 180 'Deluxe' rooms, 100 'Executive' rooms and 20 suites, distributed over nine floors. The public space features two restaurants, one pool, side bar and one indoor bar with discotheque, five meeting rooms (of which the largest can accommodate up to 500 guests), an outdoor pool, business centre, executive lounge and spa with salon and state-of-the-art gymnasium. The city's economic strength comes from its corporate hubs, factories located in the outskirts of the city, regional healthcare cluster and a broad retail sector. Additionally, being the national capital, the city draws millions of visitors including students, families and friends to the famous institutes and colleges. The leisure guests include tourists from both domestic and international destinations visiting the famous attractions and international sports facilities. The hotel is also a ideal location to cater for guests visiting any other event happening in Delhi throughout the year.

Hotel revenue management

The Twinhill Hotel is a family-owned hotel but handed over to third-party professional management company that is responsible for day-to-day operations. Ram is a newly recruited Revenue Manager, who is also heading the Reservation Department with a team of three. Before Ram joined, the hotel did not have a specific pricing strategy and they used to follow a traditional method for pricing (i.e. one rate per season and that rate was proposed on the intuition of old employees and passed to the General Manager for approval). But now as a new revenue manager, Ram is also in-charge of pricing, inventory allocation and other issues pertaining to bookings. He works closely with sales and marketing and other departments to understand the operational cost, and also with online travel agencies as well as other distribution channels to understand the market. In order to get engaged with all the departments, Ram holds a morning meeting with the Sales Department and a weekly meeting with others including the hotel's General Manager, unit financial managers and all other stakeholders. For every MICE query, Sales have to seek approval from Ram for a price decision. The owners and the management of the hotel have just spent a great amount of money on its renovation and now expect to see an increase in revenue.

Geographical setting

New Delhi is the capital of India with a mix of domestic and international travellers visiting for various purposes such as leisure, business and education. The city has dedicated areas for embassies, corporate offices, factories, banks and others, creating a never-ending demand for hotels. It also has one of the busiest airports in the world which also assist neighbouring cities (i.e. many passengers transit through New Delhi). The hotels in the city cater for the demand, efficiently balancing it out across different districts as per customer need.

Competitive hotel

Ram has defined five main competitors for the hotel out of the nearby hotels, given their similarity in services, quality and market orientation, backed by some research like SWOT (strengths, weaknesses, opportunities and threats) and other market-relevant data. After conducting some research about the past 2 years' performance of the competitive set or comp set (a group of hotels that compete with your property for business, selected to benchmark your performance against the competition), Ram found that the Twinhill Hotel has been performing a little below the rest of the market. The comp-set supply features 1700 guest rooms, with an estimated average occupancy of 85.6% and an estimated average daily rate (ADR) of US$85.43

for the year 2019/20. At the moment the hotel falls in the least expensive end with its competitors, while the hotel's occupancy is above the market average. The comp set comprises five hotels, details of which are as follows.

Hotel 1

This hotel is run by a European brand which has more than 75 hotels across the globe. The brand manages two hotels in India and planned to open a few more but they had to shift the focus to Hotel 1 as it has been under a lot of pressure to perform given the strong competition, steep budget, high management expectation and moderate customer reviews. The hotel has facilities like an indoor heated swimming pool, spa and gym, two meeting rooms (60 persons capacity in each), two board rooms with 20 capacity, two banquet halls with 600 persons capacity (theatre style) and one small banquet hall with 200 persons capacity (theatre style) to offer. Like the Twinhill Hotel, they also enjoy high demand from Monday to Friday and an average demand at weekends (Saturday and Sunday).

Hotel 2

The hotel is run and managed by an American brand (with one of the highest number of keys across the globe). With good presence in the market and a strong membership programme, the hotel has a better market mix than its competitors and good direct booking. The hotel only deals in the upscale market in India and good support from the global team keeps it busy throughout the year. The hotel offers both an indoor and outdoor pool with free access to the spa and yoga facilities for all in-house guests. The hotel has one of the best gyms, which attracts popular personalities. The hotel has limited meeting rooms and one banquet hall with capacity for 600 persons (theatre style) and one lawn for 200 guests (but rarely uses it for any gatherings). With back-to-back events and high demand, the hotel attracts non-residential guests to utilize banquet spaces.

Hotel 3

This select-service hotel offers well-equipped rooms for avid travellers with less focus on MICE. The hotel keeps rates lower than other hotels in the comp set and focuses on short-stay guests. This gives the hotel an edge over all the four-star hotels in the nearby vicinity and it acquires business from a lot of five-star hotels. The ratings on online distribution partners are not that great for the hotel but it gets bookings due to its low-rate strategy approach.

Hotel 4

The hotel is run and managed by an Indian brand (one of the best luxury hotels in India). With good presence in the market and a strong membership programme, the hotel has a great share of international travellers. The hotel only deals in the upscale market in India and gets good support from the well-known travel agencies across the globe. This helps them focus on luxury instead of keeping lower rates for domestic travellers. The hotel is usually busy throughout the year and offers an indoor heated swimming pool with access to steam, sauna and spa for all in-house guests. The hotel, similar to Hotel 2, has one of the best gyms and attracts popular personalities in the city. The hotel is the first choice for all travellers seeking luxury. Due to this, the hotel remains busy during the year with back-to-back VIP movements/arrivals. Since the main focus of the hotel is not meeting rooms, they have fewer employees managing banquet sales in comparison to any other hotel in the city. The hotel has limited meeting rooms and only one banquet hall with a capacity of 600 persons (theatre style) which can be divided into three halls as and when required.

Hotel 5

This boutique hotel attracts all segments of guests but mostly gets filled by spillover from neighbouring hotels. Due to the complex market approach, the hotel keeps competitive dynamic prices on online platforms and competitive offline rates for corporates and travel agencies. The sales team finds it hard at times to fill the hotel with this approach as they look for spillover business, however, the banquet attracts a large volume of MICE in the city and this compensates as it fills up the base occupancy.

Historical data

The historical data from 2011 to 2019 for hotel market growth in the comp set (Table 14.1) reflects the overall healthy growth of the comp-set hotels in the area, which convinces Ram that an increase in revenue for the Twinhill Hotel is achievable.

Table 14.2 shows the ADR for all hotels in the comp set.

Determining the optimal price for the hotel rooms

Ram holds a daily revenue meeting at 09:15 a.m. Ankit, one of the event sales representatives, arrived with good news as he got a call from the secretary of the Boxing Association requesting a proposal for a group requirement of 60 rooms as they are hosting the National Junior Boxing Tournament. The event is scheduled from 20 to 31 October and they are looking for accommodation from 19 October to 1 November for officials and chief guests. Due to the nature of the event, the guests will be occupied with official work and require only breakfast and dinner in the hotel. This is the first time they have approached the Twinhill Hotel and Ram knows for sure that they are shopping at hotels in their comp set as well. Comp-set data is the most important secondary data used by the hotel. The comp set will consist of a set of five to ten hotels which are considered as direct competitors to the target hotel, or it is a list of hotels a guest will compare the target hotel against while booking.

Ram is also aware about the amount of pressure that he will get from the hotel owners and management, as the hotel has shown negligible growth since renovation. During the meeting he decides that the hotel's pricing strategy needs a change to respond to the market demand and to drive more bookings. Before the meeting could end, Ram announces that from today onwards, the hotel is employing new room rates for all segments.

On getting back to the office, Ram looks at the hotel's performance and starts preparing the new pricing strategy to obtain the best possible revenue per available room. Segment-wise occupancy (e.g. guests booking for leisure, group bookings) and variance for the period 24 September–2 November is presented in Table 14.3.

Table 14.1. Historical hotel market growth 2011–2019 for the competitive set (comp set).[a,b] Author's own table.

Year	Occupied rooms	Available rooms	Occupancy (%)	ADR (US$)	RevPAR (US$)
2011	456,068	620,500	73.5	71.89	52.84
2012	462,295	622,200	74.3	74.69	55.50
2013	469,719	620,500	75.7	76.11	57.62
2014	482,749	620,500	77.8	77.84	60.56
2015	495,159	620,500	79.8	79.18	63.19
2016	505,226	622,200	81.2	80.20	65.13
2017	513,774	620,500	82.8	82.10	67.98
2018	519,979	620,500	83.8	83.78	70.20
2019	531,148	620,500	85.6	85.43	73.13
Average			79.39	79.03	62.90
CAG	1.71%			1.94%	3.68%

[a]ADR, average daily rate; CAG, compound annual growth; RevPAR, revenue per available room.
[b]US$ to Indian rupee (INR) = 71.52.

Table 14.2. Average daily rate (ADR) of hotels in the comp set. Author's own table.

Hotel	No. of rooms	Hotel type	ADR 2020 (US$)	ADR 2019 (US$)
Hotel 1	320	Upscale	88.66	88.38
Hotel 2	350	Upscale	88.76	88.34
Hotel 3	350	Select service	78.87	78.17
Hotel 4	280	Upscale	90.45	89.92
Hotel 5	400	Boutique	83.76	82.37

Discussion Questions

1. Considering the current market factors such as supply and demand and historical data provided in Tables 14.1–14.3, propose the room rate for the transient market segment for the Twinhill Hotel for the period 19 October–1 November (when the Boxing Association wants to make a group booking for 60 rooms) in order to capitalize during the demand period.
2. What strategy should be followed for transient business for the requested dates? Include the length-of-stay restrictions to be applied during these dates, and provide an explanation for your decisions based on the data provided.

Teaching Notes

This case study is intended to familiarize learners with the concept of forecasting, pricing and displacement. The main objective is to equip undergraduate and postgraduate students and research scholars of hotel management to identify the difference between assumption and calculative thinking and subsequently apply a higher level of knowledge. Revenue management is a concept that is complex as various factors have to be considered, so it is important for students to understand about the market, competitors and analyse decisions before arriving at the conclusion. The case is designed for a minimum of two students (a pair of students) to understand and discuss the assumptions and calculations. This will help students to take corrective steps in acquiring business leads (i.e. individuals or organizations who are interested in the product or service being sold by the business, and who have the potential to become future customers) for their respective business/hotel. The maximum number of students could extend to six, where each student analyses the queries for their respective hotels. The instructor in the case is expected to take the role of the owner/management. If there are more groups of students in the class, the tutor may instruct the student team to identify the problem within different hotels using different factors, for example a lead-time study, a booking trends study, and/or a no-show and cancellation study.

Target audience

This case can be used in undergraduate and postgraduate level courses of hotel revenue management and introduction to pricing and distribution.

Teaching methods and equipment

The case can be studied through scenario analysis and group discussions. Analysis of year-to-year growth and demand needs to be done first, so students understand the background to the case before answering the discussion questions. The decision and answer to each discussion question should be based on simple concept and growth analysis calculated through Microsoft Excel. The strategies should be supported by understanding of the various factors that contribute to revenue management (e.g. market conditions, competitors' offers). The instructor may ask the students to read relevant research articles and books on group displacement (group displacement is when a hotel accepts a group booking, it may have to turn down some individual or non-group bookings as a result).

Table 14.3. Segment-wise occupancy and variance at the Twinhill Hotel from 24 September to 2 November. Author's own data.

	17 Sept	24 Sept		17 Sept	24 Sept				Last year				
Date[a]	Leisure bookings (no. of rooms)	Leisure bookings (no. of rooms)	Week-to-week variance[b]	Group booking (no. of rooms)	Group booking (no. of rooms)	Week-to-week variance[b]	Rooms sold	Occupancy rate	Rooms sold	Occupancy rate	Year-to-year variance	Rooms available	Last year no-show
24 Sept	137	167	30	23	53	30	220	73%	187	62%	33	80	0
25 Sept	127	120	–7	10	55	45	175	58%	200	67%	–25	125	1
26 Sept	114	112	–2	10	55	45	167	56%	246	82%	–79	133	1
27 Sept	83	83	0	57	57	0	140	47%	155	52%	–15	160	2
28 Sept	77	87	10	109	109	0	196	65%	289	96%	–93	104	5
29 Sept	63	82	19	44	56	12	138	46%	298	99%	–160	162	0
30 Sept	92	110	18	58	88	30	198	66%	165	55%	33	102	2
01 Oct	93	120	27	44	89	45	209	70%	175	58%	34	91	3
02 Oct	74	89	15	34	34	0	123	41%	267	89%	–144	177	0
03 Oct	66	76	10	53	89	36	165	55%	148	49%	17	135	1
04 Oct	59	63	4	200	200	0	263	88%	183	61%	80	37	3
05 Oct	52	87	35	113	123	10	210	70%	199	66%	11	90	4
06 Oct	54	68	14	170	170	0	238	79%	199	66%	39	62	2
07 Oct	54	59	5	10	10	0	69	23%	244	81%	–175	231	3
08 Oct	72	81	9	39	39	0	120	40%	281	94%	–161	180	0
09 Oct	81	97	16	41	41	0	138	46%	182	61%	–44	162	5
10 Oct	89	104	15	29	29	0	133	44%	169	56%	–36	167	1
11 Oct	71	73	2	131	131	0	204	68%	158	53%	46	96	2
12 Oct	67	70	3	175	175	0	245	82%	135	45%	110	55	6
13 Oct	61	64	3	165	165	0	229	76%	174	58%	55	71	2
14 Oct	61	74	13	7	7	0	81	27%	161	54%	–80	219	4
15 Oct	56	71	15	2	2	0	73	24%	175	58%	–102	227	3
16 Oct	55	68	13	61	61	0	129	43%	111	37%	18	171	0
17 Oct	63	56	–7	96	116	20	172	57%	138	46%	34	128	3
18 Oct	55	49	–6	85	105	20	154	51%	139	46%	15	146	3
19 Oct	57	41	–16	121	141	20	182	61%	280	93%	–98	118	2
20 Oct	68	44	–24	112	132	20	176	59%	270	90%	–94	124	1
21 Oct	50	68	18	110	110	0	178	59%	177	59%	1	122	3

Continued

Table 14.3. Continued.

	17 Sept	24 Sept		17 Sept	24 Sept				Last year				
Date[a]	Leisure bookings (no. of rooms)	Leisure bookings (no. of rooms)	Week-to-week variance[b]	Group booking (no. of rooms)	Group booking (no. of rooms)	Week-to-week variance[b]	Rooms sold	Occupancy rate	Rooms sold	Occupancy rate	Year-to-year variance	Rooms available	Last year no-show
22 Oct	38	52	14	133	133	0	185	62%	176	59%	9	115	4
23 Oct	52	51	–1	156	116	–40	167	56%	172	57%	–5	133	2
24 Oct	44	46	2	127	117	–10	163	54%	156	52%	7	137	1
25 Oct	52	73	21	167	110	–57	183	61%	219	73%	–36	117	4
26 Oct	34	53	19	140	93	–47	146	49%	299	100%	–153	154	5
27 Oct	48	73	25	131	91	–40	164	55%	282	94%	–118	136	4
28 Oct	40	40	0	66	66	0	106	35%	123	41%	–17	194	2
29 Oct	44	49	5	0	15	15	64	21%	183	61%	–119	236	1
30 Oct	50	39	–11	0	15	15	54	18%	269	90%	–215	246	7
31 Oct	54	26	–28	0	15	15	41	14%	187	62%	–146	259	6
01 Nov	59	55	–4	65	65	0	120	40%	235	78%	–115	180	3
02 Nov	56	37	–19	70	70	0	107	36%	133	44%	–26	193	2

[a]Sept, September; Oct, October; Nov, November.

[b]Week-to-week variance refers to the number of rooms that have increased or decreased in comparison to 17 Sept.

Teaching instructions

The case can be taught in a 2-hour class using the following teaching instruction plan. The suggested plan assumes that learners are familiar with the concept of group displacement. The instructor should evaluate the learners' understanding of and familiarity with the concept by interacting with them. If the students are not very familiar with the concept, then the tutor could begin by asking learners about their experience of evaluating a business. The discussion will help the tutor to ensure a common understanding of what the case is about. The case may be discussed in two phases, so that after an initial phase of the case study, it may be extended to another session to discuss the strategies and the approach to the situation. The tutor may instruct students to look at the booking pattern in comparison to same time in the previous year. Students need to analyse the group displacement and decide whether to take the group booking for the hotel.

Teaching instruction plan

Discussion point	Time required (min)	Teaching strategies
Introduction	20	The instructor begins the case study by describing it to the learners and also by asking them questions about their understanding of the case. The instructor randomly asks students to volunteer themselves to narrate their understanding and decide on the impact of the group booking on the hotel. For a voluntary response from students, the instructor asks the students to raise their hands. The instructor writes the points in the board. Also the instructor asks a few questions to verify the students' understanding of reading materials supplied before the case analysis.
Briefing on concepts related to group displacement	30	The instructor introduces the concepts of group displacement, lead-time study, booking trends study, no-show and cancellation study. The instructor asks randomly selected students to provide an understanding of each study. If anyone volunteers to answer, the instructor asks them to raise their hand.
Summary of the case	20	The instructor briefly discusses the background information, the scenario, the protagonist, and the dilemma or decision to be taken, and discussion questions. The instructor also discusses the adaption of the spreadsheet to calculate the group displacement. After summarizing the case, the instructor creates multiple teams to answer the discussion questions.
Assignment question 1	25	**Propose the room rate for the transient market segment for the Twinhill Hotel for the period 19 October–1 November (when the Boxing Association wants to make a group booking for 60 rooms) in order to capitalize during the demand period.** The instructor asks the students to look at the hotel's capacity and break it into segments to understand the number of transient rooms displaced by the group booking for the period. Then the students need to: (i) check the number of transient bookings displaced on the shoulder dates; and (ii) run a comparison with the same period last year in order to analyse the type business expected during the period. The instructor must tell the teams not to disclose their results to other teams. After receiving each team's response, the instructor needs to write the answer given by each team on the board and display the Excel table or spreadsheet and report created by the students based on their understanding. The instructor then briefs the students about the different pricing method used in the period.

Continued

Teaching instruction plan Continued.

Discussion point	Time required (min)	Teaching strategies
Assignment question 2	25	**What strategy should be followed for transient business for the requested dates? Include the length-of-stay restrictions to be applied during these dates, and provide an explanation for your decisions based on the data provided.** The instructor now facilitates the teams to debate on their choice with their rationale. Each team must defend their choice using the data and give reasons for their choice.

Duration

120 min

Analysis

The students will be asked to read the case study and then break the case study into different parts as follows:

- introduction of the hotel;
- roles and responsibilities of the Revenue Manager;
- introduction of the group lead;
- discussion of the business's bookings data (as shown in Table 14.3);
- prepare a table for group displacement;
- compare room bookings in the current year versus last year;
- calculate:
  - business on the books in terms of room nights (as per given figures in Table 14.3);
  - number of expected leisure room nights during the requested period (i.e. 19 October–1 November);
  - number of room nights expected to be lost in case the group booking is confirmed;
- read about the competitive set and understand the positioning of all the hotels; and
- conduct a discussion on all hotels including a clear concept of positioning.

Process

1. We can clearly observe that the hotel has rooms available for the said demand period (i.e. from 19 October to 1 November). Also, the percentage change in ADR is 1.94% from 2011 to 2018 (Table 14.1).

2. With growth in ADR from previous years, we can look at increasing the room rates to push the ADR. The hotel will follow a pricing strategy based on a mix of: (i) cost-based pricing; (ii) budget-based pricing; and (iii) dynamic pricing (i.e. practice of changing prices in relation to changes in market conditions, especially increasing prices when there is high demand).

3. The hotel has picked-up fewer transient room nights in comparison to 17 September, thus keeping a high price may result in losing higher priced transient room nights.

4. Category selling (i.e. breaking the hotel rooms on offer down into discrete groups or types which will be suitable for different types of customer, and priced accordingly) will be key during the period.

5. The hotel needs to prepare a rate grid as shown below (i.e. one for when group booking is not picked up and another for when the group booking is picked up).

A. When the booking for the group is not picked up

- The hotel is at an average occupancy of 51% for the month of October and 46% for the said period (as calculated from data shown in Table 14.3).
- The average comp-set ADR for 2019 is US$85.43 and for 2020 is US$86.10 (as calculated from data shown in Table 14.2).
- The Twinhill Hotel's ADR has increased over time; hence, a similar percentage

Table 14.4. Scenario-based pricing if group booking is not accepted.[a,b] Author's own data.

Occupancy/tier	Scenario I	Scenario II
50–55%	$79.03	$79.03
55–60%	$80.56	$82.09
60–65%	$82.12	$85.28
65–70%	$83.72	$88.59
70–75%	$85.34	$92.02
75–80%	$87.00	$95.59
80–85%	$88.68	$99.30
85–90%	$90.40	$103.16
90–95%	$92.16	$107.16

[a]Last rooms sold as per availability and based on the comp set.
[b]Scenario I is based on a 1.94% change in ADR and Scenario II on a 3.88% change in ADR (i.e. double 1.94%). All prices are in US dollars (US$).

change will be used to increase pricing as per occupancy as displayed in Table 14.4.

- The hotel will work on two scenarios: (i) at 1.94% change in ADR (based on CAG from Table 14.1); and (ii) at 3.88% (i.e. double 1.94%). Similarly the room price will increase with increasing occupancy levels.
- The average of 9-year ADR is US$79.03 (see Table 14.1) which will be a good base rate to build-up on pricing (5% = 15 rooms).

Table 14.5. Scenario-based pricing if group booking is accepted.[a, b] Author's own data.

Occupancy/tier	Scenario I	Scenario II
70–75%	$85.43	$92.03
75–80%	$88.74	$95.60
80–85%	$92.19	$99.31
85–90%	$95.76	$103.16
90–95%	$99.48	$107.17

[a]Last rooms sold as per availability and based on the comp set.
[b]Scenario I is based on a 1.94% change in ADR and Scenario II on a 3.88% change in ADR (i.e. double 1.94%). All prices are in US dollars (US$).

B. When the group booking is picked up

- The hotel is at an average occupancy of 51% for the month of October and 46% for the said period.
- The group requirement is for 60 rooms which is equivalent to 20% of Twinhill Hotel's guest rooms, which will increase the occupancy to 71%. As the group booking is picked up to boost occupancy, the pricing can be followed as shown in Table 14.5.
- The average comp-set ADR for 2019 v US$85.43 and for 2020 is US$86.10 (as calculated from data shown in Table 14.2).

Further Reading

Szende, P. (2020) *Hospitality Revenue Management: Concepts and Practices.* Apple Academic Press, Point Pleasant, New Jersey.

15 To Stay or Not to Stay: How to Convince the Very First Guest of a Resort Property

H. Kader Şanlıöz-Özgen*
Özyeğin University, Istanbul, Türkiye

Abstract
Hotel openings in resort destinations have several challenges to overcome and manage, such as: (i) the completion of the hotel project; (ii) preparation of hotel operations for the customer experience; (iii) pre-opening tasks and procedures to test the technical and functional performance of all the service units; and (iv) organization of site visits to marketing and sales partners. The conflicts between the hotel operation and project teams and the delays in openings due to planning defects or unexpected problems may prolong this period with multiple issues for managing executives to deal with. This case addresses the possible challenges of such a period and helps learners comprehend and analyse the situational aspects to seek the best possible solutions to overcome these challenges.

Subjects: tourism and hospitality industry, operations management, hospitality marketing, tourist decision making, customer experience

Learning Objectives

At the end of the case study, a learner will be able to:

1. Discuss the challenges of the hotel opening period.
2. Analyse the decision-making process of potential hotel customers.
3. Develop actions to help customers make their decisions based on knowledge about decision making and customer experience.

Case

Antalya (Türkiye) is one of the most attractive holiday destinations of the Mediterranean region in Southern Europe. The destination welcomes millions of tourists from various countries (8.7 million tourists in 2021; 14.5 million tourists in 2019, the highest number before the pandemic) (Republic of Turkey Ministry of Culture and Tourism, 2021). Leading nationalities are German, Russian and other European populations in addition to the markets of the Middle East. Sillium Sun Beach and Golf Resort is a large property located in the eastern part of the destination. The property featured 1000 acres of land with 500 accommodation units and plenty of facilities. The opening was in 2008 with many difficulties after a 2-year project period. The expected opening was originally

*kader.sanlioz@ozyegin.edu.tr

 Teaching Cases in Tourism, Hospitality and Events (eds. S.K. Dixit and S. Piramanayagam)
DOI: 10.1079/9781800621022.0015

the year 2007; however, for several reasons there was a 1-year postponement which led to deep financial problems for the company which owned the property. Therefore, it was extremely critical to open the resort at the beginning of the 2008 summer season despite the fact that the full construction was not completed.

As the key players of the tourism industry, particularly in resort destinations, tour operators' marketing was so critical for the property that large numbers of tourists would stay in the property in the coming months to ensure a certain level of business and revenues for the company. The resort management, therefore, faced a very tough challenge to convince the very first guest of the resort to stay, and the marketing team of the resort and the tour operator would announce the opening of this splendid resort despite very limited facilities being in service and many others still being under construction. The Marketing Director, who was also responsible for other properties owned by the company, needed to revise the opening plans and promotional content of the resort several times due to the extension of the construction period. Therefore, the Director relied extensively on the competence of the resort operation team, including the General Manager and the Guest Relations Manager to handle this critical situation.

The Marketing Director, tour operator marketing team and General Manager had several meetings before the arrival of the very first guest to the property. Finally, a day before the arrival of the guests, the Regional Operation Director of the tour operator came to inspect the resort to decide if they would bring the guest to the property. The Operation Director observed the availability of some different room types with sea views which could be offered according to the guests' preference, the main restaurant with a considerable variety of food, a main bar with some night events, very limited sports facilities (a beach volley pitch and a tennis court) and a small part of the beach with shades and sun loungers. The Operation Director also explained that the guests were a senior-aged couple and the female guest was an individual with access requirements as she was sitting in a wheelchair. In addition, the Director declared that they would bring the guests to the resort first, in order to give the guests the opportunity to decide where to stay, as they would make a second tentative reservation in a similar resort in case the guests would not accept staying in an uncompleted holiday resort that was still partially under construction.

Depending on the feedback from the Operation Director of the tour operator, the General Manager and Guest Relations Manager discussed the multiple aspects of this crucial subject in order to decide how to treat the guests when they arrived at the resort. The General Manager was an experienced professional who was aware of the common problems of hospitality as an international industry. Physical infrastructure and facilities are indispensable for foreign visitors in a highly competitive environment in terms of destinations and properties. The resort was suffering from all the aspects of mass tourism as the resort had a large number of rooms to fill, therefore its business was highly dependent on tour operators. The delay of hotel openings is another common feature for properties for several reasons such as: (i) lack of competence in project management of the owners and managers; (ii) delays in the procurement process; (iii) delays in employment of the hotel managers and other team members; and (iv) poor execution of opening activities and licence/permit application. Persuading the guests to stay at the resort the next day was a crucial issue for the property and the company, and both managers as well as other members of the management team were under great stress to find ways to make this happen.

From another point of view with a customer focus, both managers also recall some essentials of the guest experience and decision-making process. The ultimate aim of a resort team as a collaborative effort was to offer unforgettable experiences to guests and this experience incorporates both physical factors (e.g. the design, layout, upkeep and ambience of the resort) and human environment factors (e.g. staff on hand to serve the guests and ensure they have an enjoyable holiday) to offer the value that the guests can perceive. In addition, three major elements were required to design the guest experience: (i) the environment, including various dimensions of experiences (cognitive, emotional, sensory, relational, behavioural); (ii) making guests participate in these experiences; and (iii) the need to surprise, amaze, challenge and encourage self-improvement. The guest, on the other hand, would decide to stay in a resort based on: (i) the perceived image of the property based on emotional and cultural aspects; (ii) their attitude based on prior information and beliefs; (iii) the perceived risk to make a judgement about the product; and (iv) cognitive dissonance (questioning the decision). Keeping all this background knowledge in mind in relation to the opening of a resort and the decision-making process of the guests, the General Manager and Guest Relations Manager decided on an action plan for the following day, and shared this with other members of the management team, to welcome the very first guests to the resort and convince them to stay.

Discussion Questions

1. Describe the challenges that must be overcome by the managerial executives of holiday resorts at the pre-opening and opening stage.
2. Identify and discuss the essential roles played by managing executives of a property (hotel or resort) at the pre-opening and opening stage.
3. Explain how the elements of the tourist decision-making process may affect the decision of the very first guests to stay in the resort described in the case study.
4. Discuss why it is critical to open the resort on the predetermined date from various aspects including marketing, sales, finance, operational functions such as front of office, food and beverage, housekeeping and other administrative functions.
5. Propose how resort management may act to convince the very first guests with regard to the distinct features and elements of the customer experience in a resort property.

Teaching Notes

This case study aims to make students identify and discuss the challenges of opening a new resort property which is a business highly dependent on global tour operator operations. Based on their knowledge from prior subjects in the academic curriculum, undergraduate students of tourism/hospitality/hotel management will be able to recognize and discuss various aspects and dimensions of the tourism system and industry, including: (i) the distribution structure; (ii) the role of marketing function for new properties in the resort destinations; (iii) economic and financial issues to deal with; and (iv) the operational actions of the opening stage to ensure customer satisfaction. In this respect, students will be required to make use of their theoretical knowledge in the relevant courses to analyse the situation in the case, regarding the features of the organization and the sector, roles and functions of individuals and customer expectations.

The instruction method around the case study can be designed as a process of two stages including group discussion and role play. First, students in the class (maximum 20 students) can discuss the situation from various perspectives and answer the discussion questions. Secondly, a group of volunteer (or assigned) students will role play the continuation of the case. A group of five to seven students may act as the management team of the resort (General Manager, Guest Relations Manager, Front Office Manager, Food and Beverage Manager, Executive Housekeeper, Entertainment and Sports Manager, etc.) to find ways of convincing the very first guests to stay in the resort through an action plan. The conversation between the resort management and the guests can be simulated too, by the participation of additional students to role play the guests. Students should remember the details given in the case in order to cater for the specific needs of the guests with the available resort facilities. After the role play, other students in the audience can discuss the effectiveness of the solutions proposed and simulated, elaborate the proposed solutions and actions and offer alternative ways.

Target audience

The case will be beneficial for undergraduate students in the tourism/hospitality/hotel management programmes as a part of their hotel/resort/club management, hospitality marketing or operations management courses.

Teaching methods and equipment

The case is designed to facilitate group discussion and role playing of students so that they will feel for themselves how it is to be prospective managing executives handling similar situations. Course instructors may ask students to review certain articles or academic/sectoral content (some are given in the references) in order to provide rationale and justify their proposed actions. In addition, if the higher education institution includes an internship programme to provide on-the-job work experience, contribution from the students with prior work experience will make the discussions sparkle with possible real-life experiences or observations. Instructors may also motivate students to think and imagine the post-stay stage of the guests' experiences upon departure in terms of unforgettable experiences, loyalty, and word-of-mouth recommendation to others if they are convinced. Seating arranged in a semi-circle will be convenient to facilitate group discussion among the students. The role-playing group may be placed in a smaller semi-circle of seats in the front, facing the larger semi-circle, so that all the students in the audience will be able to follow the conversations and interactions easily. Flipcharts or boards in the classroom to note important remarks of discussions will be effective. The final action plan of how to treat the very first guests on their arrival at the resort may also be displayed in a tabular form by writing on the board, including the details and answers to who, when, where, what and how questions.

Teaching instructions

The case is designed for a class lasting a maximum of 2 hours (with an additional 30 min preparatory reading before the class). The teaching instruction plan that follows provides details of the process, including the preparatory review of the literature before the case discussion in class.

Teaching instruction plan

Discussion point	Time required (min)	Teaching strategies
Before the case discussion	30	Students will be required to review the following papers in order to acquire essential knowledge on specific issues related to the hotel opening period: Chan (1993), Ingram (1995), Claver-Cortes *et al.* (2007) and Noordzy and Whitfield (2014). Therefore, students will be aware of several issues and challenges to deal with by managing executives at the hotel opening period.
Introduction	10	Instructor asks students if they have had prior work experience in any hotel/resort property either as an intern or any other position. Depending on the responses, students and instructor will try to identify various challenges of different departments to overcome in a property. If any observation or knowledge about a hotel opening experience is available from students (e.g. may be a student has met a manager who experienced the opening of a property), then this example will be discussed to recognize specific issues of the hotel opening period.

Continued

Teaching instruction plan Continued.

Discussion point	Time required (min)	Teaching strategies
Summary of the case	10	Instructor asks students to summarize the case with essential points; a different student will address an issue or detail in the case each time. Each contribution will be written on the board by the instructor and relevant issues will be interconnected or contradicting ones will be highlighted so that all the students will remember the fundamental issues in the case for further discussions. In addition, they will also observe how to analyse situational factors and elements of a case.
Theoretical background	10	Based on the given literature or additional papers selected by the instructor, the general content and dimensions of the specific theoretical concepts will be explicated to the students. Students need to remember the tourist decision-making process, and dimensions and elements of the customer experience in a hotel/resort.
Discussion question 1	10	**Challenges of the opening and pre-opening period for the managerial executives of holiday resorts** The instructor will open a discussion on challenges based on prior readings and possible past experiences of students. Those challenges can be classified in various categories such as internal and external, physical or human, operational or administrative, supply- or demand-related ones. Students also need to identify the primary and secondary departments in a hotel to deal with such challenges. All the challenges will be noted on the board by the instructor and any relations or classifications will be marked, once the students have reached a consensus from the matters discussed. Note: *Pre-opening* refers to the period where there are no customers but invited stakeholders (owners, suppliers, project team members, travel agents, company representatives, etc.) or associates visit to test the hotel facilities and rooms. *Opening* refers to the period where customers of the property start their experience and stay in the hotel/resort.
Discussion question 2	10	**Roles of managing executives at the opening and pre-opening stage** In relation to the answers of the previous question, departments or managing executives to deal with challenges will be determined by the students taking part in a discussion that is directed by the instructor. One challenge may be under the responsibility of multiple departments. The coordination of the General Manager is inevitable. Leading roles for specific challenges can also be identified. The instructor can develop a drawing method to identify potential linkages among challenges, departments and roles (using different colours, types of lines, shapes, etc.). For example, primary roles could be shown in a circle, secondary departments to deal with a challenge could be shown with dotted lines, etc.
Discussion question 3	10	**Decision-making process of the very first guest to stay in the resort** The instructor will ask students to consider themselves in the situation of those customers. Therefore, each student will imagine how they would make their decisions through the image of the property they perceive in the conditions given in the case, emotional, cultural and cognitive aspects, perceived risk and cognitive dissonance. Students should remember the details given in the case such as: (i) the senior age of the customers; (ii) the access requirements of the female customer; (iii) the limited number of facilities available; (iv) that some parts of the resort are still under construction; and (v) the tentative reservation in another resort.

Continued

Teaching instruction plan Continued.

Discussion point	Time required (min)	Teaching strategies
Discussion question 4	10	**Timely opening of the resort** This question is a complementary one to question 1 in terms of understanding how delays may make challenges even tougher or how timely openings may have an affect (positive/negative/neutral) on operations, marketing and team members of the property. Students will be free to share their opinions in relation to challenges or potential opportunities of a timely opening. However, in order to comprehend the issues better, the instructor will classify them in categories such as those in relation to marketing, operational functions such as front of office, food and beverage, housekeeping and other administrative functions.
Role play question 5	30	Discussions in the first four questions will make students analyse the situation and conditions of the case in a comprehensive manner to understand the critical points to develop solutions or actions. A group of volunteer (or assigned) students will be selected to role play the management team of the Sillium Sun Beach and Golf Resort. Each student will take the role of the one of the following managers: General Manager, Guest Relations Manager, Front Office Manager, Executive Housekeeper, Food and Beverage Manager, Entertainment and Sports Manager. If more students are available, then additional roles such as the Marketing Director, Finance Manager and Human Resources Manager may also be involved in the role play. A simple meeting-room setting will be prepared with a rounded seating in the middle of the classroom or in front of the board so that all the students in the audience will easily observe the management team. The General Manager will be the leader of the meeting to discuss the situation in the case and develop solutions as a team. The Guest Relations Manager will remind the meeting of the customer-related details given in the case and of the feedback from the Operations Director of the tour operator. The managers at the meeting are responsible for developing an action plan as a process for the customer experience from the very first moment they arrive at the property to the time where the guest will decide whether they wish to stay or not. Students need to remember that an action plan requires answers to the questions of Who?, When?, Where?, What? and How? for each action to take throughout the entire process. The action plan will be written on the board and amended with any additional remarks or comments. Students in the audience will not interfere in the discussions of the management team.
Discussion question 5	10	When the action plan is completed with the final consensus of the management team, it will be opened to a discussion among the students in the audience. Students are expected to criticize the expected outcome of each action on the plan and question if all the required aspects have been included or any missing or unnecessary matter exists. Students from the management team can make some explanations to the points addressed if they feel necessary. The final version of the action plan will be determined upon a consensus of all the students in the classroom. The instructor's role at this stage is essential to facilitate the open communication and discussion among the students.
Concluding remarks and student feedback	10	The instructor will ask the students about their feedback of the case discussion and role play in terms of: (i) its contribution to their knowledge; (ii) its effect on their perspective of the industry; and (iii) their understanding of a critical issue of hospitality marketing and management.

Duration

2 hours (and an additional 30 min to review papers before the class)

Analysis

Answers to the discussion questions can be elaborated by the insight and opinions from students. However, some key considerations are stated below to guide the instructor in order to showcase the situation in the most effective way. As the case is based on a real-life experience of the author, some hints from what really happened are indicated.

1. Describe the challenges that must be overcome by the managerial executives of holiday resorts at the pre-opening and opening stage.

Market-related challenges

There is high dependence on sales from tour operators. Therefore, opening is in reality the decision of the tour operator rather than the resort management. It is a competitive environment with plenty of resorts to fill the gap created by the resort that fails to open on time. Also there is difficulty for marketing, sales and promotional efforts if the opening is not possible on time (at the final stage, openings can be postponed by a few weeks which require substantial marketing efforts to manage), and it is difficult to produce promotional works and materials (visuals, photographs, films in a world where experiences should be reflected in the marketing content). Customer satisfaction is also affected in the case of possible technical or operational problems leading to poor relations with the tour operators (tour operators may not take the risk of bringing more customers to the property if the problems are repeated and remain unsolved).

Resort-related challenges

Incomplete structure and layout of the resort property create difficulties for offering services at a property where there is ongoing construction. Also it is difficult to manage resort team members and other project-related workers, as there are potential security problems in having customers, resort team members and construction workers at the same place together. Further, employment procedures need to be handled and revised based on postponed dates of opening. In addition, it is difficult to find qualified team members in an uncertain period caused by delayed openings to manage the handover process of each unit of the property from the project (construction) team and to deal with unexpected technical and operational problems which require immediate remedies. There may be cash-flow problems to manage procurement and limited financial sources to run the hotel to satisfy customers. Finally there is increased stress for the management team in managing issues among the owners, customers and tour operators/marketing team, and difficulties in motivating team members in order to overcome the unexpected difficulties of the opening period.

Owner-related challenges

At the opening stage there may be a lack of financial resources to complete the project and open the hotel, the majority of financial resources having been spent on the construction leaving limited funds for the operation of the resort. The stress of managing loan repayments can lead to tensions between the owners and the management team, resulting in a lack of confidence in the management team, and owner's concerns about the return on the investment if opening is delayed.

2. Identify and discuss the essential roles played by managing executives of a property (hotel or resort) at the pre-opening and opening stage.

General Manager

The General Manager is the bridge between the owner and the management team, and the representative of the property to the local community and organizations. They are responsible for: (i) licences, permits and security; (ii) coordinating the opening activities; (iii) managing and motivating members of the management team; and (iv) customer-related issues that require immediate action.

Marketing/Sales Manager

The Marketing/Sales Manager is responsible for: (i) marketing and promotion activities; (ii) marketing relations; (iii) sales contracts, pricing and revenue management; (iv) customer satisfaction issues; and (v) reputation management in the social media.

Chief Engineer

The Chief Engineer is responsible for: (i) takeover of the units with all the required technical specifications; (ii) floor and emergency plans; (iii) the work-order management system; and (iv) follow-up of construction-related issues with the project team.

Human Resources and Training Manager

The Human Resources and Training Manager recruits the opening team members and plans the staffing budget in line with the expected occupancy level. Also they handle local situations for employment, and organize orientation and training programmes for team members.

Food and Beverage Manager

The Food and Beverage Manager is responsible for: (i) takeover of the units and outlets (e.g. bars, restaurants, kitchens and stockrooms) in coordination with the engineering department; (ii) preparation of outlets (e.g. bars/restaurants) for customer service; (iii) purchase and delivery of items and products; (iv) preparation of work plans and checklists; (v) training the serving team members; (vi) stock management; and (vii) customer-satisfaction issues.

Executive Housekeeper

The Executive Housekeeper is responsible for: (i) takeover of the guest rooms and other units (e.g. laundry facilities, stores for cleaning materials and equipment) in coordination with the engineering department; (ii) preparation of rooms and other places for customer service; (iii) follow-up of unexpected technical problems in coordination with the engineering department; (iv) preparation of work plans and checklists; (v) training of housekeeping team members; and (vi) customer-satisfaction issues.

Finance Director

The Finance Directory is responsible for: (i) managing cash flows; (ii) loan payments; (iii) procurement payments; (iv) insurance procedures; and (v) stock-control procedures for effective purchase orders.

3. Explain how the elements of the tourist decision-making process may affect the decision of the very first guests to stay in the resort described in the case study.

Communication, coordination and cooperation with the tour operator are critical for this process. The resort management team and the tour operator need to decide how they will inform the customers about the current situation of the property. This knowledge may affect the perception of the customer in terms of the resort image. However, the decision of the tour operator to bring the customers to the resort first is very helpful in order to give the customers the opportunity to compare their expectations with their observations and make their final decision based on their assessment of the perceived risk. Following this valuable action by the tour operator, it is critical that the resort management team act in order to reflect the best experience of staying in the resort and to construct a positive first impression based on emotions and other human-related factors in addition to the physical environment elements.

4. Discuss why it is critical to open the resort on the predetermined date from various aspects including marketing, sales, finance, operational functions such as front of office, food and beverage, housekeeping and other administrative functions.

Persuading the very first guest to stay in the resort will start the opening stage of the resort, announcing it to the relevant stakeholders including the tour operators, owners, suppliers and team members. If the customers are satisfied with their experience at the resort, positive word of mouth will also have an effect on future sales from tour operators. An increase in sales will also raise the occupancy levels to generate revenue and business for the resort which will facilitate all the relevant procedures of the financial considerations, payments and procurement. Also team members in the operation departments will become motivated to welcome customers and serve them, meeting their needs and requests, to create an experience together.

5. Propose how resort management may act to convince the very first guests with regard to the distinct features and elements of customer experience in a resort property.

This question is the most critical part of the case discussion as all the points addressed in the previous questions will be possible if the very first guests are convinced to stay in the hotel at the resort. Therefore, the resort management team should use all their expertise to make the best possible action plan to conduct the process without problems so that the potential customers will decide to stay. The emotional touch is very critical at this stage and this orientation should be enhanced with the ultimate level of personalization offered to those first customers. First of all, the resort management team should be honest and explain to the customers what they may face during their experience and what services will be available for them. Despite several challenges, this experience of being the very first guest of a property is also a once-in-life experience for customers. Therefore, the resort management can emphasize this fact and ensure that all the team members will be at their service to serve them personally and meet their needs and requests. Resort management can also offer this holiday to those customers and refund the payment made for this stay. Customers may also be given a VIP status so that they will be treated in a special manner whenever they come back to the resort. The resort management team and tour operator should also ensure that the guests have the right to change to another resort whenever they like in case of any inconvenience.

It may also be helpful for the instructor to know what happened in the real case, which was experienced by the author of this case study. The customers, a senior-aged couple and female customer in a wheelchair, arrived at the resort as planned. The General Manager and Guest Relations Manager welcomed the customers together to discuss the issue at the bar while having some refreshments. They were frank and explained the situation at the hotel with both negative and positive aspects. They also reminded the customers about the options for a refund or changing to another resort. They underlined the opportunity to offer a hyper-personalized experience to the customers as the resort looked forward to welcoming them and had a few reservations coming up. They offered a site inspection of the resort so the customers could see the facilities. The Guest Relations Manager showed them all the rooms available, including food and beverage units and facilities. The Executive Housekeeper prepared a selection of guest rooms including suites and showed all the possible options to the customers. However, the female customer opted for a standard room with full sunshine all day long on the terrace with direct access to the pool which would be convenient for her husband while taking care of her. The resort management team offered the customers a lunch in the open buffet restaurant where all the team members were eating together. The Guest Relations Manager was with them all the time and introduced them to the team. All the team members greeted them with a warm welcome as they were all aware of the importance of those very first customers. After their lunch, the customers announced that they would stay in the resort and all the people in the restaurant were so pleased that they applauded them for their decision. The customers had a memorable experience of 2 weeks at the resort. The male customer was a sporty individual. When he wanted to play tennis or beach volleyball, immediately team members got ready to play with him. Although they were happy with the meals, their special requests were also fulfilled. They were offered a few spa treatments with special discounts. Finally, the customers were so delighted with their experience that they paid for a special cocktail event where they invited all the team members to say how much they had appreciated their efforts. They kept coming back to the resort (and to other properties of the company). One final remark is that the male customer was an engineer and he was very helpful in identifying some problems or defects of the resort during his first stay. He prepared a detailed report about these issues and was very happy when the management took his report into consideration, acted immediately to resolve the issues and give him feedback (some of the issues were resolved quickly and the customer could observe that his feedback was taken into consideration for correction and improvement).

References

Chan, W.W. (1993) Managerial roles of hotel pre-opening teams in developing countries such as China. *International Journal of Hospitality Management* 12(2), 155–161.

Claver-Cortes, E., Molina-Azorin, J.F. and Pereira-Moliner, J. (2007) Competitiveness in mass tourism. *Annals of Tourism Research* 34(3), 727–745.

Ingram, H. (1995) Hospitality and tourism: international industries experiencing common problems. *International Journal of Contemporary Hospitality Management* 7(7), 44–54.

Noordzy, G. and Whitfield, R. (2014) Root causes of hotel opening delays in China. *Cornell Hospitality Report* 14(10), 1–24.

Republic of Turkey Ministry of Culture and Tourism (2021) Border Statistics 2021. Available at: https://yigm.ktb.gov.tr/TR-249702/sinir-istatistikleri.html (accessed 2 September 2022).

Further Reading

MacLean, M.A. (2018) *Consumer Psychology of Tourism, Hospitality and Leisure*. Society Publishing, Burlington, UK.

Mason, P. (2010) *Tourism Impacts, Planning and Management*. Routledge, Abingdon, UK.

Sotiriadis, M. and Sarmaniotis, C. (2016) Collaborating to provide attractive hotel guests' experiences. In: Sotiriadis, M. and Gursoy, D. (eds) *The Handbook of Managing and Marketing Tourism Experiences*. Emerald Publishing, Bingley, UK, pp. 175–194.

16 Mayday, Mayday: Falling Off a Legacy

Senthilkumaran Piramanayagam[1]* and Saurabh Kumar Dixit[2]
[1]*Manipal Academy of Higher Education, Manipal, India;*
[2]*North-Eastern Hill University, Shillong, India*

Abstract
Family-owned businesses play a significant role in nations' economies worldwide. They have become an integral part of economic development by providing employment generation and wealth creation. However, the family-business image has been tarnished by issues such as the informal culture and structure, lack of professionalism, lack of planning and nepotism. It is evident that the majority of family-run businesses fail to survive beyond three generations. This hypothetical case study is about the issues faced by a family-owned hospitality business organization called the Raymond Hotel Company (RHC). The problems faced by the RHC are common among family-owned business firms that affect their survival and profitability. The case study will help students become familiar with issues related to professionalism and nepotism in managing a family-owned hospitality business.

Subject: family business, nepotism, leadership, role of managers

Learning Objectives

At the end of the case study, the learner will be able to:

1. Understand the challenges in managing family-owned businesses.
2. Identify and diagnose the issues related to lack of professionalism, issues of nepotism in the family business and its subsequent impact on the organization.

Case

The Raymond Hotel Company (RHC) is a new hospitality business arm run by Mr Edmund. Mr Edmund and his family members have been running various businesses for the last nine decades. Edmund's father, the Late Mr Raymond was a first-generation entrepreneur. He established his business empire through his hard work and sincerity and founded the most-respected business house, Raymond & Sons. He was also known for his business ethics and philanthropy. Raymond & Sons was a success story among the new generation of entrepreneurs in the country who would like to emulate the family business model of Raymond & Sons. The company consistently attained and exceeded the turnover and profit targets. Raymond & Sons received many awards, accolades and recognition from both federal government and industrial associations for

*senthil.kumaranp@manipal.edu

 Teaching Cases in Tourism, Hospitality and Events (eds. S.K. Dixit and S. Piramanayagam)
DOI: 10.1079/9781800621022.0016

their innovative business model and business performance.

After his father's retirement, Edmund took on the role of Managing Director of the company, as he is the eldest among his siblings in the family. Mr Edmund created the RHC in 2015 to spread the wings of Raymond & Sons into the booming hospitality sector in the country. Mr Edmund is head of the whole business that belongs to the family, including his four brothers. In terms of capital structure, the major portion of the capital is obtained through bank loans. Edmund always invests about one-third of capital from their own resources and retained earnings from the other businesses operated by the family in any new business the family would venture into. His father's goodwill and impeccable repayment record has always helped Edmund to raise funds from both public and private-sector banks.

Edmund entered into the hospitality business considering the current economic growth, increased standard of living and economic affluence, increased tourist flow, rising number of international events, changing lifestyle and increased numbers of luxury-loving millennials in the country. The RHC is at the forefront in delivering a unique hospitality experience to high-end corporate customers in a luxury landscape. Currently, it operates Raymond's Palace, a five-star classified property in the capital city. Raymond's Palace was built between 1818 and 1824, a magnificent palace that reflects the unique heritage, culture and handicraft of the city and the art collection of the local royal family. Mostly white statuario marble from Carrara in Italy was used for its construction. The launch of Raymond's Palace was a high-profile event, with invitees from both the corporate and the political world. Now the palace also houses a museum and a well-appreciated art gallery, worth millions.

Recently, Mr Edmund has made a cash transaction, for an undisclosed amount, to purchase, outright, a 180-year-old castle of a nobleman in the city. The castle, now Raymond's Castle, was an attractive proposition as it is located on a hill which offers an exotic view of the city. It is accessible as it is only a 20-min drive from the airport, is very close to the main shopping area and in the middle of the most elite neighbourhood in the city. A rumour among the business houses in the city is that RHC committed about two-thirds of their assets to buy this castle, as against the family's policy of investing one-third of capital.

Raymond's Castle is a kind of tribute to the Late Mr Raymond, who started his reign in a humble way but with an audacious goal to become a role model for young entrepreneurs. Mr Raymond is the one man in their entire family who made history and dreamed of a more enduring and timeless business legacy. He was a visionary, unmatched strategist and a great appreciator of arts and luxury. When Raymond embarked on his first business venture, he had a seed capital of just US$150. But in his lifetime, he built a billion-dollar business empire known for its business ethics and the great respect and devotion of its employees.

After the outright cash purchase, 46 rooms in the castle have been converted to guest rooms and the courtyard in the castle is now restored as the lobby for the hotel. An architect known for transforming palaces and old castles into beautiful hotels was appointed as a consultant to convert the castle into a luxury hotel. Edmund allocated more than 50% of the total funds committed for the hotel business to renovate and refurbish the castle into a luxury hotel.

Raymond's Palace has become the city's most sought-after luxury hotel with its aristocratic background. It is no surprise that it has attracted many corporate houses that intend to offer 'royal' hospitality to its top-notch professionals, chief executives, foreign and most-valued business partners. Corporate houses have now become the most important source of business for RHC. The current customer segments of RHC and their contribution to annual business turnover is presented in Table 16.1.

In terms of total contribution to the annual business turnover, the high-end corporate customers contribute the most to the business. After recognizing the typical requirement of corporate customers, Raymond's Palace, operated under RHC, would like to continue to focus more on corporate customers as it was able to customize many

Table 16.1. Customer segments of Raymond Hotel Company (RHC) and their contribution to annual business turnover. Authors' own table.

Customer segment[a]	Contribution to annual business turnover (%)
Corporate	72.8
OTA	18.6
Walk-in guests	8.6

[a] OTA, online travel agencies.

of its services. Unlike online travel agencies (OTA) and walk-in guests, most of the corporate customers are looking for unique hospitality offerings. The corporate houses never hesitate to spend money on any special and unique requirement. Now with a significant contribution, corporate customers are the main source of RHC's profitability and success. The current occupancy rate at Raymond's Palace stands about 74% which is 10% higher than the current city's average of 64.5%, thanks to the corporate customers of RHC. In terms of customer service, Raymond's Palace has always been rated greater than 4.5 out of 5. Raymond's Palace has been lauded for its service, supportive employee behaviour, food, amenities and overall experience on online platforms. Mr Edmund has always insisted on the sales force building a long-term relationship with its corporate clients to sustain the success and these high ratings. It is expected that RHC will achieve break even (i.e. a point where the business profits equal the costs) within 4 years, considering the current occupancy rate. Witnessing the prestige and success of the newly started hotel business, Mr Edmund has planned to expand his business to three more metropolitan cities in the country.

In terms of human resources, the number of employees recruited at Raymond's Palace is about 72. Mr Lewin (age 41), Mr Edmund's youngest brother, is currently managing the hotel taking the position of Managing Director. Lewin is a graduate in arts, having studied in the same locality. Immediately after graduation, as per the family tradition, he was inducted into the family business. He is very interested in cooking, loves to travel, and is a foodie. Considering his interest in food, Mr Edmund instructed Mr Lewin to manage the affairs at Raymond's Palace. Other top executive positions in the hotel are given to either Edmund's cousin or to the nieces of Edmund's wife, Cecilia. Mr Rotham, one of Edmund's nephews, has been appointed as an executive manager to assist Mr Lewin. Mr Rotham is a graduate in political science from a reputable university waiting for admission to a PhD fellowship. Mr Edmund used to evaluate and monitor the performance of the hotel periodically. He has instructed the Chief Accountant to submit the financial reports for Raymond's Palace every quarter.

A bad morning

On a busy day, Mr Edmund has received a call from Mr Melinton, head of a popular business house in the city involved in information technology-related services. The company managed by Melinton is a major client for Mr Edmund's hospitality business as he used to send his international clients to Raymond's Palace. After the usual interaction, Melinton informed him that he is no longer willing to send his clients to Raymond's Palace and also indicated that he may be forced to cancel the agreement made between the business house and RHC on the lease of ten rooms during the current financial year if the issues are not addressed effectively. Melinton hinted that he had received many complaints from his most valuable international clients who have stayed at Raymond's Palace on: (i) the misbehaviour of employees with guests; (ii) a quarrel among the employees in front of a guest; (iii) the poor quality of the food; and (iv) the poor maintenance of guest rooms and public areas. Mr Edmund was shocked as he listened to the complaints, but he promised to take the necessary action and conveyed his deep regrets to Mr Melinton. When he urgently called Lewin to discuss the issue, Edmund received the message that the person was away from network connectivity. Edmund later understood that Lewin was on his 1-month culinary experience road trip to tribal areas, which is located far away in the north-eastern part of the country.

Feeling desperate, Mr Edmund immediately called the Chief Accountant to ask him to submit the financial reports for Raymond's Palace. On evaluating the third-quarter result, Mr Edmund was shocked to see that the business is moving towards loss. He has called the external social-media management firm that handles the Raymond's Palace social media and OTA portals to verify the customer rating. Another upset for Mr Edmund is that ratings were abysmal and Raymond's Palace now has been rated even below a shabby business-class hotel. The review is filled with complaints and destructive narratives on employee behaviour, food, maintenance, and overall guest experience. He has written a strongly worded message full of discontent to the chief of the media management firm complaining about their poor professional practice and the behaviour of the accounts manager. However, in her reply, the chief of the media handling firm denied any lack of professionalism and reporting of the facts. She informed that the firm's accounts manager had submitted daily, weekly and monthly summary social media reports and analytics to Lewin, with suggested managerial recommendations to address the customer complaints. She has also attached the acknowledgement letter that Lewin signed to prove her claim. When Edmund asked about Mr Rotham, he learned that he was on leave for a month and had gone to meet a professor at a university for preparing a synopsis for his doctoral research.

Now, along with Chief Accountant, Mr Edmund has visited the hotel. He was greatly disappointed with the current state and maintenance at Raymond's Palace. He found that many corporate clients had stopped sending their international and most-valued domestic customers. He also found that costly kitchen and equipment in the public area is under repair. The menu card was torn, and the menu list was similar to the roadside eateries. The executive chef humbly informed him that as all the professional chefs had left their jobs, he had to modify the menu as he was managing the kitchen with only a few kitchen stewarding staff.

It is disheartening to see that a 150-year-old antique clock in the lobby is not functional and stood as the victim of poor maintenance and housekeeping practice. Half of the fused bulbs in the big chandelier in the lobby had not been replaced because an imported mechanized ladder is undergoing repair. When Mr Edmund enquired about the absence of Ms Maria, Lewin's cousin, who is the Executive Housekeeper of the hotel, employees informed him that Maria was accompanying Lewin on the long road trip. The Chief Accountant also narrated that when the loss in the business was communicated to Mr Lewin, he replied that the loss may have arisen due to the competition and wanted to reduce the room tariff to compete with other hotels. Mr Lewin also assured the accountant that everything would be all right after his road trip to the tribal area. Now looking at the damaged antique gold crystal chandelier imported from Belgium, Mr Edmund has understood that RHC is in a big crisis.

Discussion Questions

1. What is your general perspective on appointing a family member in the leading role of the business?
2. Do you believe that appointing Mr Lewin, a foodie with a passion for travel, as a Managing Director of Raymond's Palace is right?
3. If you have been appointed as a consultant to provide solutions for the problems at RHC, what could be your response?

Teaching Notes

The case study aims to help students become familiar with the issues related to professionalism and nepotism in managing family-owned hospitality business firms. This is important as many of the students of hospitality opt for a career in a family-owned hospitality business, and while some students may be familiar with and able to manage the issues highlighted in the case study, others will benefit from the exercise. This case study helps the students to apply a complex theoretical concept and model to solve a day-to-day challenge in a family business. The case has been taught to postgraduate students of hospitality, tourism and event management in classroom settings and has proved to be an effective training tool to teach entrepreneurship and challenges in a family-owned business.

The case is suitable for both group and individual learning, but is best done as group work. Learners may be divided into groups of four–five students to address the discussion questions and present their various solutions with appropriate analysis. Based on the presentation, a tutor may extend the case by using other theories in the domain (e.g. Max Weber's bureaucratic management theory or Ludwig von Bertalanffy's systems theory). Alternatively, individuals can be instructed to answer each question with relevant analysis and the support of theoretical concepts.

Target audience

The case study is suitable for undergraduate and postgraduate hospitality students. Prior knowledge of various concepts and theories related to nepotism, the role of manager, functions of management and professionalism in business will be an added advantage but is not mandatory. A tutor may provide a short lecture and reading materials on nepotism, professionalism and Henry Mintzberg's (1989) role of managers to familiarize students with these concepts before discussion of the case.

Teaching methods and materials

As already stated, it is best if students work in groups of four–five, and the learners are instructed to assume the role of a management team or a team of management consultants responsible for solving the issues faced by the RHC. Each team (group) should address the issues phase by phase (see teaching instructions below). The team will identify the symptoms, underlying causes, and impact of the challenges arising from nepotism, lack of professionalism and poor management of the hotel's day-to-day operation, customer relationship management, brand image and organizational performance. The students must emphasize the theory and make concrete and measurable recommendations. Each team's response will be evaluated based on the relevance, effectiveness and quality of the explanations. The tutor has to develop a structured rubric to evaluate the teams as per the expected learning outcomes, and the students should be made aware how they will be assessed at the outset.

Teaching instructions

The tutor who aims to incorporate this case study in their course must introduce the basic elements such as: (i) the definition of professionalism; (ii) the definition of nepotism; (iii) their relevance in business; and (iv) typical challenges in managing a family business with real-time examples.

The case can be discussed in a phased manner as shown in the teaching instruction plan that follows. In the first phase, prior to the classroom session, each learner should read the content of the case and prepare some basic ideas on professionalism and its relevance in business. Then, in the classroom, the tutor will engage with the students by asking questions about their experience/understanding of family-owned businesses, professionalism and their overall understanding of the case. In the second phase, students will work in groups of four–five and each team (group) of learners will address the issues raised in discussion questions 1 and 2. In the third phase, the groups answer discussion question 3 by developing structured, concrete and measurable managerial recommendations. As already mentioned, the case may be extended to three or four sessions with additional inputs.

Teaching instruction plan

Discussion point	Time required (min)	Teaching strategies
Phase 1		
Introduction	15	The tutor begins by asking the question about the difference between managing a family-owned business and a professionally operated business. The tutor may randomly select a few students to ask about the advantages and disadvantages of a family-owned business. The question may be extended further to ask whether the students are familiar with any family-owned businesses in the region. The tutor writes the information gained from the students' answers on the board concerning: (i) the difference between family-owned and professionally operated companies; (ii) the advantages and disadvantages of family-owned business; and (iii) a few examples of family-owned business in the region.
Briefing on professionalism in family business	15	The tutor needs to give a brief lecture on the importance of professionalism and present empirical evidence on the impact of nepotism, lack of professionalism in family business, and its impact on business performance. At the end of the lecture, the tutor may ask a few students selected randomly to provide their understanding of these concepts.
Summary of the case	15	The tutor may ask for volunteers to raise their hand to summarize the case and list the case facts (students will have read the case prior to the classroom session). As the students answer, the tutor should write a list of the main points on the board with the aim of providing an overview of the case. Students should be asked to get into groups (teams) of four–five in preparation for Phase 2.
Phase 2		
Assignment question 1	15	**What is your general perspective on appointing a family member in the leading role of the business?** The tutor asks the students to answer this question by participating in a poll using an online app or a Google Form link. After finishing the poll, the tutor shows the output as a percentage. The tutor must tell the teams not to disclose and discuss their arguments for or against appointing a family member with other teams. For postgraduate students, the tutor may instruct the teams to support their arguments with empirical evidence.
Assignment question 2	15	**Do you believe that appointing Mr Lewin, a foodie with a passion for travel, as a Managing Director of Raymond's Palace is right?** Before the students answer this question, it would be helpful if the tutor provides a short lecture on the roles of managers as suggested by Henry Mintzberg (1989) and discusses with the students what is meant by nepotism. Then, the tutor should ask students to discuss their answers to question 2, giving their reasons for both why Mr Lewin should be appointed, and why not.

Continued

Teaching instruction plan Continued.

Discussion point	Time required (min)	Teaching strategies
Phase 3		
Assignment question 3	15	**If you have been appointed as a consultant to provide solutions for the problems at RHC, what could be your response?** Students working in their teams (groups) should discuss their suggestions for solutions between themselves. Then, to avoid overlapping answers, the tutor must ask each team to submit their suggestions before starting the class discussion. The tutor now allows the teams to present their suggestions to the class providing a strong rationale for each recommendation. The tutor must write the main points on the board to assist the discussion.
Closing remarks and discussion	15	The tutor needs to instruct the teams and students to raise any questions they have about their discussion. The tutor must emphasize that 'messiness' is expected for this type of case study. Finally, the tutor concludes the session with analysis and answers for each discussion question using theories related to professionalism, role of managers and nepotism.

Duration

1 hour 45 min

Analysis

1. What is your general perspective on appointing a family member in the leading role of the business?

Empirical evidence has established that family members in the leading role of a business are desirable as they tend to show higher commitment and desire to sustain the business (Basly and Saunier, 2020). There is an interplay between ownership, family, and management of the business. Family members who own a large part of a business have a strong positive impact on a firm's performance. A family's long-term outlook on a business is vital for business survival as they consider the business an asset to be passed on to the succeeding generation. Family members in an organization assume three simultaneous roles: (i) owners; (ii) relatives; and (iii) managers. As family members are primarily interested in return on investment, they take the role of owners; when taking more concern for welfare and unity of the family, they assume the role of relatives; and on taking the role of managers they are working towards operational effectiveness of the business. However, the existing literature also reveals that many family-run businesses lack professionalism, foster nepotism and have poor management practices. Many founders make the mistake of occupying the dual role of chief executive officer (CEO) and business leader or chairman. While the family business board chairman needs to be viewed as an adviser, the CEO should focus on operating the business. Professionalism is the process that intends to transform the business that is dependent on a single key person or a promotor, to a business with a formalized organizational structure where the owner-manager delegates the actual decision-making authority to a group of highly competent managers. The professionally run family business is more sustainable.

2. Do you believe that appointing Mr Lewin, a foodie with a passion for travel, as a Managing Director of Raymond's Palace is right?

An argument against the appointment of Lewin

Nepotism is interpreted in different ways by different authors. Nepotism denotes the psychological and social processes related to family membership and its influence in any public affairs or business. Nepotism is a form of unfair workplace procedure and process that favours family members or friends over others and supports their behaviour. An example is hiring family members for an executive role that is not based purely on their skill, experience and knowledge. In the current context, the recruitment of Mr Lewin is an act of nepotism as he has been appointed for a top management position for his passion and because he belongs to the family that own the business, not for his skill and experience in management. It is also a surprise that there is no professional approach towards managing the business as the business's day-to-day affairs have been handled very carelessly. It indicates the need for experienced managers to run the business.

Supporting the appointment of Lewin

There is nothing wrong with appointing family members in top management roles in a business. It is common for founders of businesses to appoint family members as they believe that it is essential to retain the control and identity of the business. The professionalization of business is not intended to remove the family and its members from an active role in a business. Boston Consulting Group (BCG), a well-known management consulting firm, narrates that professionalism will take different shapes at different stages of business growth (Bhalla and Orglmesiter, 2017). On its establishment, an organization brings in few outsiders. During the second phase, the organization hires external talents for important positions and once the organization has become a high-performing enterprise, it attracts top talents with a variety of backgrounds and there is a corporate culture. The main challenge for successful professionalization is integrating the inside long-serving members and the recruits. Considering the phases of professionalization of a family business, it is the right decision to involve family members in the business.

3. If you have been appointed as a consultant to provide solutions for the problems at RHC, what could be your response?

Suggestions to overcome the current crisis are as follows:

- *Recruitment of external members in managing the business*: Professionals who have education and experience in managing luxury hotels need to be appointed.
- *Management training*: Passion is not sufficient to manage the business, so all the family members need to be trained for the role of manager. This includes training in: (i) the functions of management; (ii) the significance of managing customers; and (iii) service quality and its implications in business outcomes.
- *Standard operating procedures (SOPs)*: Mishandling of equipment is a major reason for repair of machines and equipment. Developing SOPs on how each machine, process and work has to be handled is necessary. Undertaking different levels of maintenance (i.e. preventive, breakdown and periodic maintenance) is essential to keep the equipment readily available for when it is needed.
- *Board of Management*: A board consisting of external and internal family members to oversee and provide guidance will help in overcoming many business-related challenges.
- *Training*: Employee training and motivation is another important aspect of management to facilitate the smooth day-to-day running of the business.

References

Basly, S. and Saunier, L.S. (2020) Family members' commitment to the firm and family business continuity: investigating the mediating role of family-to-firm identity fit and emotional attachment. *Journal of Small Business & Entrepreneurship* 32(1), 9–32. DOI: 10.1080/08276331.2018.1551458.

Bhalla, V. and Orglmesiter, C. (2017) *A Founder's Guide to Professionalizing a Family Business*. Boston Consulting Group, Boston, Massachusetts. Available at: https://www.bcg.com/publications/2017/family-business-people-organization-founders-guide-professionalizing-family-business (accessed 21 February 2022).
Mintzberg, H. (1989) *Mintzberg on Management: Inside our Strange World of Organizations*. Simon and Schuster, New York.

17 Learning from Hilton Worldwide – Successful Corporate Social Responsibility Strategies

Rachel Hyunkyung Lee* and Tiffany S. Legendre
University of Houston, Texas, USA

Abstract
Corporate social responsibility (CSR) has been high on the agenda of the hospitality and tourism industry. Despite the emphasis on firms' CSR activities, it is not clear how companies implement CSR initiatives throughout their core business models and how they show the practical impact and positive changes to their stakeholders. In addition, teaching and learning materials for industry practitioners and the next generation are very scarce. From the case study on Hilton Worldwide, this case study explores ongoing CSR efforts and provides a direction for future changes in hospitality industries' CSR endeavours.

Subjects: corporate social responsibility (CSR), hotel CSR initiatives, Hilton Worldwide, CSR pyramid model, ethical CSR, altruistic CSR, strategic CSR

Learning Objectives

At the end of the case study, a learner will be able to:

1. Define what corporate social responsibility (CSR) is in the hospitality and tourism industry.
2. Explain and compare Carroll's CSR pyramid model and Lantos's archetypes CSR model.
3. Apply the three archetypes of the CSR model to analyse how Hilton Worldwide executes CSR initiatives.
4. Investigate how Hilton Worldwide minimizes its environmental footprint and maximizes its social impact on society.

Case

Introduction

Corporate social responsibility (CSR) has become an important agenda for the hospitality and tourism industry. Hospitality firms have been criticized by the public due to their wasteful characteristics. For example, the overall industry has a heavy dependence on single-use disposable plastics, energy, water resources and food waste (Bohdanowicz and Zientara, 2008). With global warming and changes in political and legal environments, there has been a major push for increasing the social roles of corporations in the form

*hlee59@cougarnet.uh.edu

(eds. S.K. Dixit and S. Piramanayagam)
DOI: 10.1079/9781800621022.0017

of their civic duties. The shifts in values of younger generations (i.e. Millennials and Gen Z) and their desires for social justice, protection for the environment, and business ethics has made the hospitality industry concerned about better engagement with CSR activities (Guzzo *et al.*, 2020). Many hospitality organizations recognize the importance of balancing profits with maintaining a positive corporate image through environmental, ethical and social responsibility; however, it is not an easy task to complete due to the conflict of interests among many different stakeholders and governance issues (Lee and Heo, 2009).

Despite the practical and theoretical importance of proper implementation of CSR initiatives, the resources surrounding CSR teaching and learning for the industry and the next generation are scant (Kim and Austin, 2020). With Millennials and Gen Z gradually taking up the consumer market share and increasing their disposable income, hospitality businesses have made big changes in terms of how consumers socialize, relax and do business (Rank and Contreras, 2021). Moreover, given that Millennials and Gen Z are tech- and web-savvy, and concerned about environmental and social causes (Tsai *et al.*, 2012), CSR will be highly valued by these generations, not to mention that information about CSR activities will be reaching more consumers via social media platforms. Hence, the primary goal of hospitality corporations should be to actively pursue CSR (Submittable, 2020). The effort should be genuine and transparent, as inadequately implemented CSR campaigns can taint a corporate reputation with negative evaluations of those campaigns (e.g. greenwashing, diversity washing).

In a rapidly changing external environment, success in CSR implementation requires catering for all the internal and external stakeholders involved. However, as the nature of CSR implementation is sophisticated and complex, it is necessary to learn from successful cases. Thus, this case study analyses Hilton Worldwide's CSR activities, the challenges involved, and how they overcame those challenges. Reflecting on Hilton Worldwide's CSR engagement, this case study also explores past and present CSR efforts and provides a direction for future changes in hospitality. Based on the insights from Hilton Worldwide's CSR endeavours, this case study helps instructors address the CSR activities and challenges in the hospitality industry.

CSR efforts in the hospitality industry and the need for sustainability

There is considerable research interest in CSR activities in relation to: (i) consumer communication and stakeholder engagement (Janković and Krivačić, 2014); and (ii) employee commitment and job satisfaction in the hospitality industry (Lee *et al.*, 2012, 2013). Although the definitions of CSR differ, the generally accepted definition is: 'CSR involves the conduct of a business so that it is economically profitable, law-abiding, ethical and socially supportive. ... The CSR firm should strive to make a profit, obey the law, be ethical, and be a good corporate citizen' (Carroll, 1999, p. 286). Glavas and Kelley (2014) define CSR as 'caring for the well-being of others and the environment with the purpose of also creating value for the business' (p. 171), which is connected to a company's economic performance.

In terms of what constitutes CSR, Carroll (1979) proposes a pyramid model that comprises four levels of responsibility: (i) economic; (ii) legal; (iii) ethical; and (iv) philanthropic. It is proposed that CSR should be carried out from the bottom to the top – from economic responsibility to philanthropic responsibility – for the best CSR results. At the first level, economic, organizations have a responsibility to produce products and services by satisfying customers' needs and maximizing profits (Carroll, 1979). Second, the business has a legal responsibility to comply with the law, which is regulated by the national or regional administration (Carroll, 1979). Third, ethical responsibility indicates moral deeds expect of business entities, such as engaging in ethical behaviours, respecting people, and/or preventing social harm. Although consumers, employees, shareholders and community members expect companies to follow ethical social norms, it is difficult to reflect on the ethical component of CSR activities, because of its blurry definition (Carroll, 1979). Finally, the philanthropic responsibility to society is also called voluntary or discretionary responsibility and includes philanthropy, social welfare and financial aid (Carroll, 1979).

Although Carroll's CSR pyramid model is designed to conceptualize and demonstrate critical progress in CSR research, it could not clearly distinguish between mandatory and voluntary responsibilities for philanthropic duties (Jamali, 2007). Therefore, building upon Carroll's CSR pyramid model, Lantos (2002) identified three archetypes of CSR: (i) ethical CSR; (ii) altruistic CSR; and (iii) strategic CSR. Lantos's (2002) model divides philanthropic responsibilities into altruistic and strategic CSR and combines economic, legal and ethical responsibilities into 'ethical CSR' (see Table 17.1). In this case study, Lantos's archetypes CSR model rather than Carroll's CSR pyramid model is applied for further analysis, because Hilton Worldwide's CSR is comprehensive and they currently pursue exemplary philanthropic responsibilities beyond economic and legal responsibilities.

Ethical CSR

Lantos's (2002) research, which focuses on the ethics of CSR (Holcomb, 2010), separates Carroll's philanthropic responsibilities into altruistic and strategic responsibilities. Ethical CSR are moral responsibilities that go beyond fulfilling a company's economic and legal obligations, by avoiding the company's damage or public blame (Lantos, 2002). In other words, ethical CSR is not only required at a legal and a moral level but is also actively keeping businesses away from potential social harm. If these requirements are not met, the company's reputation could be damaged or, in a worse case, be subject to legal prosecution (Lantos, 2002). An example of ethical CSR is provided in Box 17.1.

Table 17.1. Corresponding classification of corporate social responsibility (CSR). Authors' own table.

Carroll's four CSR pyramid model[a]	Lantos's three archetypes CSR model[b]
Philanthropic responsibility	Strategic CSR Altruistic CSR
Ethical responsibility Legal responsibility Economic responsibility	Ethical CSR

[a] (Carroll, 1979).
[b] (Lantos, 2002).

Hilton Worldwide's TMRGs took critical action to protect the key business entities and team members in response to the agile impact of the COVID-19 pandemic. By supporting human rights and diversity, Hilton Worldwide is morally responsible for group members that may inflict actual or potential harm in a particular course of action. Recently, CSR initiatives are not only needed to fulfil legal/economic responsibilities but also require companies to include ethical responsibilities. Moreover, considering that the majority of Hilton Worldwide's team members are Millennials and Gen Z, ethical CSR can inspire their pride as team members. Thus, ethical CSR, which encompasses economic responsibilities relating to the fair treatment of employees and legal responsibilities that expect to fulfil the mission within legal boundaries, is most evident in Hilton Worldwide's TMRGs.

Altruistic CSR

According to Lantos (2002), altruistic CSR is philanthropic CSR that embodies charitable and beneficial activities for social stakeholders who contribute to the company's sacrifice by giving up their own profits and covering the costs of such activities. Altruistic responsibilities explain corporations' efforts in going above and beyond their ethical responsibilities and resolving social issues that the company is not directly responsible for. Altruistic CSR is the fulfilment of an organization's charitable responsibility to help mitigate public welfare deficits beyond preventing possible harm, whether this is advantageous to the corporation or not. Although altruistic CSR may not directly benefit the company – at least the benefit to

Box 17.1. Team Member Resource Groups (TMRGs) and human rights.

As the coronavirus disease 2019 (COVID) pandemic intensified in April 2020, Hilton Worldwide decided to temporarily give frontline employees' time off and decrease the salaries of corporate employees, executive committee members and chief executive officers (CEOs) until the end of the year. The company put its team members first during the pandemic by supporting healthcare coverage and a portion of employees' pay while providing online training materials. Hilton Worldwide also provided financial and mental care to team members and their families directly affected by the pandemic. Through its Team Member Assistance Fund (TMAF), Hilton Worldwide sought to rebuild its workforce by providing financial support to manage and franchise hotel and corporate team members directly impacted by disasters or pandemics. Hilton Worldwide partnered with Global Impact to implement this fund, providing nearly $500,000 to more than 130 team members and their families to cover medical expenses, food and family care for those diagnosed with COVID-19. Hilton Worldwide also allowed team members to provide monetary support via Hilton Honors Points donations so that they could help their colleagues directly affected by COVID-19.

Also Hilton Worldwide is committed to attaining global gender equality and 25% more racial diversity at the US corporate leadership level, along with regional diversity targets, by 2027. Hilton Worldwide announced plans to expand gender and ethnic diversity as employment accelerated within the hospitality industry (Hospitality Net, 2021). Specifically, Hilton Worldwide launched Team Member Resource Groups (TMRGs), which build relationships with a variety of non-profit strategic partners, such as the National Association for the Advancement of Colored People (NAACP), to provide more opportunities for employees and increase their engagement. TMRGs, sponsored by Hilton senior leaders, share business insights into internal business practice or external market factors with these senior leaders. TMRGs also develop career management, personal branding and emerging industry trends and engage local diverse communities through service projects while supporting corporate partners. More than 8000 team members participate in TMRGs at corporate and hotel properties, and 27% of corporate team members are members of one or more TMRG. This participation rate is growing every year. To provide many events and activities for team members while promoting TMRG membership, employees focus on career development, community support and innovation. Hilton Worldwide launched its Women's TMRGs globally in 2017, as well as its military TMRGs, African American TMRGs, LGBTQ (lesbian, gay, bisexual, transgender and questioning) and Friends TMRGs, and Millennials TMRGs in 2016 to increase its commitment and provide leaders with insights on community outreach strategies.

the business will not be the primary focus – most CSR practices have a positive long-term effect on the company. Altruistic CSR is an effort to decrease public issues such as poverty, environmental problems and lack of medical treatment, marking the corporation's attempt to develop society's welfare and enhance the quality of life. Distinct from ethical CSR, which is based on morally mandated obligations, altruistic CSR is considered to be a high-minded and virtuous action. An example of altruistic CSR is provided in Box 17.2.

Hilton Worldwide's charitable and beneficial activities administered through the Hilton Effect Foundation described in Box 17.2 are a good example of altruistic CSR, as their philanthropy is aimed at helping communities resolve social needs and cope with global environmental problems. Further, given that CSR is essential for the younger generation, Hilton Worldwide's altruistic CSE, which in many ways emphasizes its philanthropic side, appeals to the younger generation.

Strategic CSR

Strategic CSR includes the philanthropic activities of a caring company to achieve its strategic business goals. It also involves a company's participation in CSR practices that increase not only social well-being but also benefit the company while achieving business goals such as energy-cost reduction and brand management. Strategic CSR is the core of a company's value-creation activities and enables companies to choose activities that will enhance their competitive advantages. Unlike altruistic CSR, strategic CSR is

Box 17.2. Hilton Effect Foundation.

To celebrate Hilton Worldwide's 100th anniversary in 2019, the Hilton Effect Foundation announced its first Hilton Effect Grant as one of its philanthropic missions to make the world a better place to travel. In particular, the foundation awards grants within three sectors: (i) the annual Hilton Effect Grant; (ii) disaster response and resiliency; and (iii) signature partnerships. It awarded $580,000 in grants to community partners in 2019 and 2020, as well as more than $1 million to grant recipients in food security, sanitation, economic security and clean air and water. Hilton Worldwide has focused its grants on promoting inclusive recovery for a society severely impacted by COVID-19. The foundation supports double investments to make social impacts by investing in short-term relief and long-term resiliency to support communities in need while building resilience and supporting organizations fighting the spread of infection. More than 20 communities have received awards from the Hilton Effect Foundation, positively influencing global environmental and social needs. The foundation contributes monetary and in-kind charitable giving to 23 organizations addressing the most urgent humanitarian needs, recognizing the severe impact the COVID-19 pandemic had on marginalized communities. Grants to organizations are also used to help vulnerable populations severely affected by natural disasters and hardship.

The foundation also supports key partnerships with the International Youth Foundation and the World Wildlife Fund, which drive positive environmental impact. For example, Habitat for Humanity, one of the foundation's grantees, installs toilets and handwashing facilities and establishes roadside market shelters for people in need in Fiji. Meanwhile, Love Beyond Walls provides portable handwashing stations for homeless people who had no access to handwashing facilities during the pandemic (Love Beyond Walls, nd). The International Rescue Committee in Jordan supports financial recovery and development for refugees and vulnerable host communities, and the National Environmental Education Foundation (NEEF) in the USA supports communities that need to maintain green spaces and waterways. The Hilton Effect Foundation invests in organizations responding to the urgent needs of people in hardship.

used to achieve a company's strategic goals in response to social needs. Strategic CSR strikes a balance between the economic and social value of the company and enhances stakeholder relationships with the company. Also, strategic CSR recognizes corporate threats and evolves sustainable business initiatives. Thus, strategic CSR is performed only if it is expected to advance both the company and the stakeholder groups. While the motive of altruistic CSR is that a company's financial profit is not the main goal, in the case of strategic CSR, the company believes that contributing to its stakeholders is the best way to ensure financial benefits since CSR practices are already philanthropic (Lantos, 2002). An example of strategic CSR is provided in Box 17.3.

As a strategic CSR, Hilton Worldwide has set energy conservation goals through environmentally sustainable implementation and cost reduction. In particular, Hilton Worldwide established goals and the LightStay system to achieve strategic business goals along with ethical CSR and altruistic CSR. Specifically, in line with the Travel with Purpose 2030 Goals, the LightStay system is dedicated to halving environmental impacts across the value chain by describing and tracking all Hilton hotels' energy, carbon, water and waste reduction goals. This strategic CSR maximizes CSR activities that align with the interests of Hilton Worldwide and its stakeholder groups, promoting overall win-win outcomes.

Conclusion

Through this case study, Hilton Worldwide's CSR initiatives are analysed according to Lantos's three archetypes of CSR. As an ethical CSR, Hilton Worldwide has fulfilled its corporate moral role by upholding human rights and diversity and both legal and ethical values through TMRGs. Economic responsibility may not be directly derived from TMRGs, but in the long run, the economic benefits that come from acknowledging regional and ethnic diversity for team members cannot

Box 17.3. Travel with Purpose 2030 Goals and 'LightStay'.

Hilton Worldwide is investing in projects that preserve the resources of hotels in the most beautiful areas, reduce the environmental footprint of their supply chains and work with communities to cut their environmental impact in half. At a United Nations (UN) summit in 2015, world leaders adopted the UN Sustainable Development Goals of the 2030 Agenda for Sustainable Development; in the same year, 2015, Hilton Worldwide introduced its CSR initiatives which it refers to as the Travel with Purpose 2030 Goals. With a promise to double its social investment and halve its environmental impact by 2030, Hilton Worldwide utilizes its Travel with Purpose 2030 Goals to pursue efficient operations and supply chains as well as engage customers, owners and employees by adopting 'LightStay', an award-winning environmental, social and governance (ESG) management system.

Through Hilton's LightStay, hotels assess and operate waste reduction, recognize priority areas, discover and share the best initiatives, and report on current sustainability practices. Using LightStay to monitor and predict their consumption, Hilton team members consider internal benchmarking to reduce waste consumption. As a result, Hilton Worldwide has decreased environmental consumption over the last few years. Since 2008, Hilton Worldwide has reduced cumulative utility costs by more than $1 billion by using LightStay systems to reduce carbon emissions by 36%, waste intensity by 44%, energy intensity by 26%, and water use by 20% (CDP, 2020). By understanding Hilton Worldwide's waste consumption and efforts to manage waste over time, Hilton Worldwide has gathered more than 10 years of data, which it has used to encourage greater efficiencies and establish the best initiatives and training for both customers and team members.

be ignored. As an altruistic CSR, the Hilton Effect Foundation delivers to people in need from all walks of life and forms partnerships where physical help is required to help the community develop. In terms of strategic CSR, Hilton's Travel with Purpose 2030 Goals and LightStay system provides justification for the business to fulfil its philanthropic responsibilities. In particular, by subdividing the philanthropic CSR responsibilities of Carroll's model into altruistic CSR and strategic CSR, and by clearly specifying the ambiguous boundaries of Carroll's model, Lantos's archetypes CSR model shows that the company's philanthropic responsibilities also have economic benefits. This is not simply an action that a company shows while being conscious of society's gaze, but instead the company is providing genuinely sustainable initiatives that take care of all moral and economic benefits.

Discussion Questions

1. Identify corporate social responsibility (CSR) in the hospitality and tourism industry.
2. Identify the three archetypes of the CSR model in the case of Hilton Worldwide.
3. What is the impact of Hilton Worldwide's CSR initiatives on their employees and customers?

Teaching Notes

Summary

Hilton Worldwide has recently started Travel with Purpose 2030 Goals, which taps into Hilton's environmental, social and governance (ESG) strategies that generate positive environmental and social influences throughout its global operations (Hilton, 2020). The company is also focusing on halving its environmental impact and doubling its investments in social impact through its Travel with Purpose 2030 Goals (Hilton, 2020). Since 2008, Hilton has saved more than $1 billion by reducing carbon emissions by 36%, waste by 44%, energy by 26% and water consumption by 20% to increase operational efficiency (CDP, 2020). Hilton has established an environmental protection system called 'LightStay' to measure hotels' environmental and social impact on sustainable development. Through the system, the environmental protection performance of each hotel is measured and continuously managed. It seems clear that Hilton Worldwide's CSR initiatives are going some way to meet the current challenges that the company faces, including newly emerging policies (e.g. single-use plastic ban introduced in New York and California) and the changing customers' values of Millennials and Gen Z.

This case study is intended to familiarize learners with CSR in the hospitality and tourism industry, and apply Lantos's archetypes CSR model to analyse Hilton Worldwide's CSR initiatives. It provides a teaching resource about CSR that is useful for both the industry and the academic community.

Target audience

This case study is suitable for practitioners in the hospitality and tourism industry looking to educate their managers about CSR activities and how these may be implemented. Also it is suitable for undergraduate students studying hospitality and tourism marketing.

Teaching methods and equipment

Students should work in groups of four to five individuals in order to answer the discussion questions. This may be followed by a feedback session in which it may be helpful if the instructor wrote the main points on a board or flipchart to facilitate the class discussion. Students will also require access to computers in order to conduct online research to find: (i) statistical data to support their answers, for example Hilton Worldwide's annual ESG report (Hilton, 2020); and (ii) CSR initiatives of other tourism and hospitality companies so these can be evaluated.

Teaching instructions

This case may be used in a 90-min class, using the teaching strategies described in the following teaching instruction plan. It is assumed that the learners will read the case in advance and understand Carroll's CSR pyramid model and Lantos's archetypes CSR model. Instructors should interact with learners to confirm that they have understood the concepts and summarize them. In particular, instructors can start the class by asking students about their satisfaction, experiences or impressions of Hilton Worldwide hotels. After a brief explanation of the concepts of Carroll's and Lantos's CSR models and summary of the case of Hilton Worldwide's CSR strategies, instructors can divide the students into small groups of four or five individuals, in order to answer the discussion questions. Students should be given about 10 min to work in their groups to answer each discussion question, before the instructor asks groups to feedback their answers to the class (perhaps writing the main points on a board/flipchart) in order to lead a class discussion about their responses. Depending on the students' level of understanding, instructors could ask students to access the Hilton website to see if there are additional CSR initiatives and how these can be applied to the CSR model (e.g. regarding answers to discussion question 3).

Teaching instruction plan

Discussion point	Time required (min)	Teaching strategies
Introduction	5	The instructor demonstrates concepts of CSR by asking questions about the learners' recent experiences of CSR initiatives in the hospitality and tourism industry. The instructor then discusses: (i) the positives and negatives of those CSR experiences from the consumer's perspective; and (ii) the benefits of CSR activities from the business's perspective. The instructor introduces the Hilton Worldwide case and explains the company's CSR philosophy.

Continued

Teaching instruction plan Continued.

Discussion point	Time required (min)	Teaching strategies
Briefing on concept of Carroll's CSR pyramid model	10	The instructor introduces Carroll's CSR pyramid model. The instructor asks learners to provide definitions and examples for each element of the CSR pyramid. Examples can be shared during this phase to encourage discussions (e.g. What is going to happen if businesses only care for economic benefits?; What will be the likely consequences when businesses only perform what is required by the law but do not go beyond that?)
Briefing on concept of Lantos's archetypes CSR model	5	The instructor introduces the concept of Lantos's archetypes CSR model and explains why Lantos's CSR model is more suitable for the Hilton Worldwide case than Carroll's pyramid model. The instructor also encourages students to compare and contrast the two CSR models.
Summary of the Hilton Worldwide case	10	The instructor briefly summarizes Hilton Worldwide's CSR strategies. Then the instructor could provide an example of a CSR initiative from the Hilton case, and ask the learners about what level it falls under in Carroll's model and Lantos's model. Learners can be encouraged to think of further initiatives that Hilton Worldwide could adopt to fulfil greater CSR outcomes. In preparation for answering the discussion questions, the instructor should ask learners to get into groups of four to five students.
Discussion question 1	15	**Identify corporate social responsibility (CSR) in the hospitality and tourism industry.** The instructor asks learners to find other examples of CSR initiatives in the hospitality and tourism industry by visiting the website of any hospitality corporation. These practices can then be compared with those from Hilton Worldwide's case. If other companies' CSR initiatives appear to be implemented poorly, learners can be asked to suggest how they could be improved.
Discussion question 2	20	**Identify three archetypes of the CSR model in the case of Hilton Worldwide.** As already stated, learners should be familiar with the case of Hilton Worldwide having read the case prior to the classroom session. The instructor should ask the students questions about the case in order to establish that they understand and can identify each of the three archetypes of the CSR model (ethical, altruistic and strategic), and can provide examples of Hilton Worldwide's CSR initiatives for each archetype.
Discussion question 3	20	**What is the impact of Hilton Worldwide's CSR initiatives on their employees and customers?** Having shown that students have understood the general idea of Hilton Worldwide's CSR initiatives (by answering discussion question 2), the instructor can encourage learners to think about the impact Hilton Worldwide's CSR initiatives have on their employees and customers. To support their arguments students should provide evidence from secondary data gained from a web search (e.g. quantitative data from Hilton Worldwide's annual ESG report (Hilton, 2020)).
Closing remarks	5	The instructor concludes the session by summarizing the answers to the discussion questions, including the instructor's own observations about CSR in the hospitality and tourism industry.

Duration

90 min

Analysis

1. Identify CSR in the hospitality and tourism industry.

Since the concept of CSR in the hospitality and tourism industry is broad, it is easier for students to understand it with examples. Therefore asking the learners to find other examples of CSR initiatives by tourism and hospitality companies, and perhaps comparing these with Hilton Worldwide's CSR activities, will be beneficial.

2. Identify three archetypes of the CSR model in the case of Hilton Worldwide.

Answering this question should show that students have understood the general idea of Hilton Worldwide's CSR initiatives, in particular in relation to the three archetypes of Lantos's CSR model consisting of ethical, altruistic and strategic CSR. Further, students should be able to demonstrate, using specific examples, how these initiatives aim to minimize Hilton Worldwide's environmental impact and maximize its social impact by helping marginalized communities and vulnerable populations affected by natural disasters and hardship.

3. What is the impact of Hilton Worldwide's CSR initiatives on their employees and customers?

Students should think about Hilton Worldwide's CSR initiatives that have been described in the case, and for each initiative they should note down: (i) how it impacts the employees; and (ii) how the customers may view it. More specific Hilton Worldwide CSR initiatives can be found in the Hilton Worldwide ESG report (Hilton, 2020) and these should also be considered. Quantitative data, for example from this ESG report, should be used as evidence to support the students' arguments which each group should present to the class in the feedback session (instructor could briefly list the main points for each initiative under the headings 'Employees' and 'Customers') followed by a class discussion.

References

Bohdanowicz, P. and Zientara, P. (2008) Corporate social responsibility in hospitality: issues and implications. A case study of Scandic. *Scandinavian Journal of Hospitality and Tourism* 8(4), 271–293.

Carroll, A.B. (1979) A three-dimensional conceptual model of corporate performance. *Corporate Social Responsibility* 4(4), 37–45. DOI: 10.5465/amr.1979.4498296.

Carroll, A.B. (1999) Corporate social responsibility: evolution of a definitional construct. *Business and Society* 38(3), 268–295. DOI: 10.1177/000765039903800303.

CDP (2020) *Hilton Worldwide, Inc – Climate Change 2020*. CDP Disclosure Insight Action. Available at: https://cr.hilton.com/wp-content/uploads/2020/02/Hilton-2019-CDP-Climate-Change-Questionnaire.pdf (accessed 24 May 2022).

Glavas, A. and Kelley, K. (2014) The effects of perceived corporate social responsibility on employee attitudes. *Business Ethics Quarterly* 24(2), 165–202. DOI: 10.5840/beq20143206.

Guzzo, R.F., Abbott, J. and Madera, J.M. (2020) A micro-level view of CSR: a hospitality management systematic literature review. *Cornell Hospitality Quarterly* 61(3), 332–352. DOI: 10.1177/1938965519892907.

Hilton (2020) Hilton 2020 Environmental, Social and Governance (ESG) Report. Available at: https://cr.hilton.com/wp-content/uploads/2021/04/Hilton-2020-ESG-Report.pdf (accessed 28 November 2021).

Holcomb, J.L. (2010) General managers' perceptions of corporate social responsibility in Florida hotels. Doctoral dissertation, University of Central Florida, Florida. Available at: http://purl.fcla.edu/fcla/etd/CFE0003466 (accessed 28 November 2021).

Hospitality Net (2021) Hilton Ranked #1 on DiversityInc's Top 50 Companies for Diversity List. Hospitality Net. Available at: https://www.hospitalitynet.org/news/4104298.html (accessed 28 November 2021).

Jamali, D. (2007) The case for strategic corporate social responsibility in developing countries. *Business and Society Review* 112(1), 1–27. DOI: 10.1111/j.1467-8594.2007.00284.x.

Janković, S. and Krivačić, D. (2014) Environmental accounting as perspective for hotel sustainability: literature review. *Tourism and Hospitality Management* 20(1), 103–120. DOI: 10.20867/thm.20.1.8.

Kim, S. and Austin, L. (2020) Effects of CSR initiatives on company perceptions among Millennial and Gen Z consumers. *Corporate Communications* 25(2), 299–317. DOI: 10.1108/CCIJ-07-2018-0077.

Lantos, G.P. (2002) The ethicality of altruistic corporate social responsibility. *Journal of Consumer Marketing* 19(3), 205–232. DOI: 10.1108/07363760210426049.

Lee, C.K., Song, H.J., Lee, H.M., Lee, S. and Bernhard, B.S. (2013) The impact of CSR on casino employees' organizational trust, job satisfaction, and customer orientation: an empirical examination of responsible gambling strategies. *International Journal of Hospitality Management* 33(1), 406–415. DOI: 10.1016/j.ijhm.2012.10.011.

Lee, S. and Heo, C.Y. (2009) Corporate social responsibility and customer satisfaction among US publicly traded hotels and restaurants. *International Journal of Hospitality Management* 28(4), 635–637. DOI: 10.1016/j.ijhm.2009.02.007.

Lee, Y.K., Kim, Y.S., Lee, K.H. and Li, D. (2012) The impact of CSR on relationship quality and relationship outcomes: a perspective of service employees. *International Journal of Hospitality Management* 31(3), 745–756. DOI: 10.1016/j.ijhm.2011.09.011.

Love Beyond Walls (nd) Available at: https://www.lovebeyondwalls.org/lovesinksin/ (accessed 6 September 2022).

Rank, S. and Contreras, F. (2021) Do Millennials pay attention to corporate social responsibility in comparison to previous generations? Are they motivated to lead in times of transformation? A qualitative review of generations, CSR and work motivation. *International Journal of Corporate Social Responsibility* 6(1). DOI: 10.1186/s40991-020-00058-y.

Submittable (2020) Millennials, Gen Z, and the Rising Demand for Corporate Social Responsibility. American Marketing Association, 6 July. Available at: https://www.ama.org/2020/07/06/millennials-gen-z-and-the-rising-demand-for-corporate-social-responsibility/ (accessed 3 September 2022).

Tsai, H., Tsang, N.K.F. and Cheng, S.K.Y. (2012) Hotel employees' perceptions on corporate social responsibility: the case of Hong Kong. *International Journal of Hospitality Management* 31(4), 1143–1154. DOI: 10.1016/j.ijhm.2012.02.002.

18 Researching to Solve the Problem of Developing a Skilled Workforce in Tourism and Hospitality: a Case of the Sultanate of Oman

Mohit Kukreti,[1] Amitabh Mishra[1]* and Aarti Dangwal[2]
[1]*University of Technology and Applied Sciences, Ibri, Oman;* [2]*Chandigarh University, Gharuan, Mohali, India*

Abstract
The Sultanate of Oman has targeted tourism as one of the main economic diversification initiatives to boost the national economy and employment generation. The government has fully realized the significance of tourism and hospitality education for providing trained and skilled employees in the tourism industry to improve the quality of services and boost Oman's tourism industry's overall competitiveness. However, the tourism and hospitality jobs are not acceptable to Omani students, and they are hesitant to pursue tourism and hospitality careers. This impediment has severely impacted the prospects for tourism development in the country and hindered efforts to give the tourism workforce a national character as currently more expatriates than Omanis work in the sector. The purpose of this case study is to elicit information on how qualitative research can contribute towards solving the national-level problem of developing a skilled workforce in tourism and hospitality. This case explores: (i) the Omani citizens' motives for studying tourism and hospitality; (ii) their expectations and attitudes about working in the business and their perceptions of tourism and hospitality as a career option; (iii) what parents, relatives and acquaintances of students think about their children/relatives/acquaintances working in tourism; and (iv) the impediments that the Omani government confront in the tourism industry.

Subjects: career impediments, tourism and hospitality employment, tourism and hospitality education, students' career behaviour

Learning Objectives

At the end of the case study, the learner will be able to:

1. Explain how simple qualitative research can be utilized to solve a national problem of developing a skilled workforce in tourism and hospitality.
2. Identify the impediments to career acceptance faced by tourism students.
3. Explain the cultural and societal issues affecting the tourism workforce.

*amitabh.ibr@cas.edu.om

(eds. S.K. Dixit and S. Piramanayagam)
DOI: 10.1079/9781800621022.0018

Case

The Sultanate of Oman has placed emphasis on tourism and hospitality as a means of economic diversification and job creation for Omani nationals. To support this economic diversification initiative, the Ministry of Higher Education, Research and Innovation (MoHRI) had started tourism and hospitality majors/specializations in various higher education institutions (HEIs) in Oman. Tourism and hospitality management education is interdisciplinary in nature and aims to prepare students for operational and management positions. The trained and skilled employees in the tourism industry can improve the quality of its offerings, thereby boosting Oman's tourism industry's overall competitiveness. Fewer Omani graduates are employed in the tourism and hospitality sector than expatriates. The government bodies in Oman understand the significance of tourism and hospitality education as a backbone that supports the industry and requires continuous government support. As many universities offering tourism and hospitality programmes require trained and experienced teachers, one of the famous universities recruited Dr Alex Macdonald to teach its tourism and hospitality students. Dr Alex was very interested in tourism development in the Middle East and willingly accepted the job at the university. He commenced teaching the tourism and hospitality students in high spirits but soon he realized that many Omani students are hesitant about continuing to pursue tourism and hospitality careers in the future. This career impediment factor is negatively impacting the students' motivation to continue tourism and hospitality education. Dr Alex discussed this issue with his Faculty Dean. The Faculty Dean invited Dr Alex to prepare a report on the issue. Dr Alex roped in two more researchers, namely Dr Sweetline and Dr Asma, from the university as his team members and they started work on preparing the report.

Dr Alex, Dr Sweetline and Dr Asma met with several stakeholders such as present students, alumni, tourism and hospitality teachers, industry leaders, industry associations, parents and government officials and collected their views in order to develop the report. Dr Alex and his team conducted a random survey of the current tourism and hospitality students studying on the certificate, diploma and bachelors' degree programmes to analyse their future job perceptions, motivations, beliefs and attitudes. Dr Alex and his team believe that the findings will serve as a first step towards establishing a quality tourism and hospitality education programme at the university to attract talented Omani youth to build their career prospects in the industry.

Also Dr Alex and his team conducted the survey questionnaire for all tourism organizations. This included a wide range of tourism corporations such as airlines, hotels, restaurants, travel agencies, tour operators, tourist attractions, sports and recreational facilities (Table 18.1). The survey included focus group meetings with the local males, females and student groups pursuing tourism majors at the university. All

Table 18.1. Types of tourism businesses and number of business organizations. Compiled by the authors.

Types of tourism businesses	No. of business organizations
Hotels and resorts	120
Tour operators	50
Travel agencies	45
Food and beverage corporations	40
Transportation companies	20
Sports and recreation companies	15
Tourism guides	32
Cultural industry organizations	5
Tourist equipment rental companies	10
Miscellaneous tourism services businesses	15
Total	352

survey respondents took part on a voluntary basis and their anonymity was guaranteed. The data were analysed quantitatively using statistical software.

The main objective of this research was to acquire first-hand information regarding the local population's views of the tourism industry as an employer. The student groups were surveyed about their knowledge of the industry and careers, attitudes, career objectives and plans. The researchers conducted focus group meetings in the various regions of Oman based on their present stake in tourism development.

The team started the research with the aim of using the findings to help solve the national-level problem of developing a skilled workforce in the tourism and hospitality industry. Specifically they tried to answer the following questions through the research:

- What are Omani citizens' motives for studying tourism and hospitality?
- What are their expectations and attitudes about working in the business and their perceptions of tourism and hospitality as a career option?
- What do parents, relatives and acquaintances of students think about their children/relatives/acquaintances working in tourism and hospitality?
- What obstacles does the Omani government confront in the tourism industry?

Findings in the report prepared by Dr Alex, Dr Sweetline and Dr Asma

HEIs offering tourism education

Dr Alex and his team surveyed other competing HEIs offering tourism and hospitality courses/qualifications and came up with the following information:

- *Sultan Qaboos University* – BA in tourism, hospitality management, tourism guidance (tour guiding);
- *Oman Tourism College* – International Diploma in tourism and hospitality management, Professional Certificate in tour guiding, front office, housekeeping, culinary, service and restaurant management, travel management and International Air Transport Association (IATA) programmes in food safety and hygiene, business languages (English, French and German), customer services/ guest relationship;
- *Colleges of Applied Sciences (CAS)* – BBA with majors in tourism;
- *GU Tech* – BSc in sustainable tourism and regional development;
- *Gulf College* – BA in travel and tourism management;
- *Sur University College* – BBA in hotel management and tourism;
- *Majan University College* – BA in business administration with pathway in tourism;
- *Al Buraimi University* – bachelor's degree in tourism and leisure management; and
- *National Hospitality Institute (NHI)* – Diploma in international hotel management for existing tourism employees, certificate in vocational courses, and a diploma for cabin crew.

The survey revealed that most of the programmes listed above are offered by the Business Department in HEIs. Since Oman ranks low in global competitiveness reports, the perception about the quality of tourism programmes is poor. A broad review of the curriculum is necessary to improve the quality of the tourism curriculum.

Analysing the size of the tourist market and its human resource needs

To analyse the size of the tourism market in the country, Dr Alex, Dr Sweetline and Dr Asama collated the following data from the various publications of the National Statistical Centre for Information (NSCI) Oman.

DATA ON TOURIST ARRIVALS. The future of the tourism industry is quite promising if tourism numbers increase year on year as they did in the period 2011–2020 (Table 18.2), with more than 3.5 million tourist arrivals in 2019 and US$3.08 billion as receipts. This was approximately 3.5% of gross national product (GNP). The income per tourist arrival was almost US$878.

EMPLOYMENT IN TOURISM. Dr Sweetline conducted the survey and collected data on

Table 18.2. Tourism indicators (inbound and outbound tourism) for Oman from 2011 to 2020. Compiled by the authors.

Tourism indicator	Year									
	2020	2019	2018	2017	2016	2015	2014	2013	2012	2011
Inbound tourism										
Inbound visitors - (000)	869	3,506	3,242	3,178	3,207	2,634	2,225	1,923	1,714	1,393
Tourists (overnight visitors)	622	2,500	2,301	2,316	2,335	1,909	1,611	1,392	1,241	1,018
Inbound tourism expenditure (000 RO)[a]	244,592	620,079	679,175	532,264	427,341	364,753	306,489	238,473	192,993	158,614
Outbound tourism										
Outbound visitors - (000)	1,657	6,633	5,975	6,368	5,902	5,424	4,727	4,301	3,972	3,341
Tourists (overnight visitors)	1,098	4,224	3,350	4,473	4,167	3,838	3,358	3,103	2,888	2,446
Outbound tourism expenditure (000 RO)[a]	319,925	941,106	955,278	642,969	542,530	481,242	435,192	397,311	364,697	285,811

[a] RO, Omani rial.

Table 18.3. Numbers of workers employed in tourism in 2019. Compiled by the authors.

Employment sectors	From survey	Estimates for the whole country[a]
Accommodation sector 1–5 star	7,000	9,000
Tour operator/travel agency	1,400	3,500
Restaurants	1,000	3,000
Dive centres	60	170
Car rentals	150	600
Airlines	4,700	6,800
Tourist attractions	45	400
Water sports	120	200
Clubs	50	50
Others	1,500	8,000
Total	16,025	31,720

[a]Estimated numbers for the whole country were based on the data collected from the survey.

employment in tourism. Based on this, Dr Alex and the team estimated the number of workers employed in tourism in the whole country. Table 18.3 shows that tourism was estimated to employ a total of approximately 31,720 workers, both locals and expatriates.

The survey data reveals that since management jobs are well paid (Table 18.4), Omani employees are willing to work in managerial positions.

Dr Asma and Dr Sweetline identified the following skill gaps where future training should be focused: (i) management skills; (ii) hospitality skills; (iii) information technology; (iv) foreign languages, especially English; (v) customer service skills; (vi) sales and marketing skills; (vii) finance; (viii) ticketing and reservations; (ix) tour guiding; (x) sport and recreation; (xi) golf course management; (xii) cultural and heritage interpretation; (xiii) handicrafts and souvenirs; and (xiv) meetings, incentives, conferences and exhibitions (MICE) organizing skills.

Table 18.4. Average salary in various tourism jobs in 2019. Compiled by the authors.

Tourism job	Average salary (RO)[a]
Management	1000
Sales and marketing	650
Culinary	700
Accounting and finance	530
Public relations	450
Ticketing	680
Front office	650
Engineering and maintenance	500
Food and beverage	480
Housekeeping	450
Tour guiding	500
Drivers	350
Security	450

[a]RO, Omani rial.

Students' motivations, beliefs and inclinations

Only a few students had expectations regarding landing a well-paid job with a tourism degree. However, many students felt that once they opted for a tourism and hospitality course it would be simple to study and pass the course. Certain students truly found tourism to be an enthralling subject to study. Dr Alex and his team investigated whether students' first choice of study was a tourism programme and reported that, surprisingly, the majority of students did not choose tourism as their first course of study. Around half of the students said they wanted to study something different at first. It also appeared that students who had not chosen tourism and hospitality education as their first option had enrolled in the programme because it was the only one available to them or for which they had been awarded a scholarship.

Dr Alex and his team found that the students were motivated to pursue tourism and hospitality education because the available job opportunities were attractive. Some students were motivated by the impression that this education had multiple options and would help them in building an international career. Some students had a genuine interest in the tourism and hospitality subjects. Many students, based on the quality of their internship experience, were motivated and inclined to join the tourism and hospitality industry. The general attitude towards the tourism and hospitality profession appeared to be neither positive nor negative.

Omani students' views towards working in a hotel compared with work as a tour guide

Dr Sweetline compared students' views of working in hotels and restaurants to working as a tour guide, in order to determine the viability of two distinct career opportunities for Omani students. In the Omani tourism industry, tour guiding is a vocation that has been covered under 'Omanization' or the 'nationalization' of the workforce (i.e. giving the workforce a national character). However, tourism students are required to pass an oral examination and obtain a 'Tour Guide Licence' from the government and this impedes the students' choice of a tour guide career. Some students opt to complete their tourism and hospitality management education first before applying for the tour guide licence. This maximizes the chances of obtaining the tour guide licence easily. A career as a tour guide is acceptable to the local community and families of the students. Work in hotels and restaurants, on the other hand, is frequently viewed as depraved, with associations such as exposure to alcohol and interaction with 'strangers', particularly among women. Hotel jobs are not culturally suitable for women. Many students agreed with these impressions, believing that the prevailing opinion in Omani society was that hotel and restaurant jobs are more suitable for expatriates rather than Omani citizens.

Career growth and development

Dr Alex and his team found that most of the students had limited knowledge about the nature of tourism and hospitality jobs and career prospects available in MICE, ecotourism, hotels and resorts, restaurants, catering, travel agencies and tour operations, or airlines, cruises and car rentals. However, students had a positive perception about the attractiveness of the salary in the industry. Regarding professional growth possibilities in the tourism and hospitality business, the majority of students anticipated that the industry would offer them good opportunities for career progress. However, regarding employment security, students were a little skeptical believing that private-sector jobs in the industry do not promise a secure future. Dr Alex highlighted strengthening the image of tourism and hospitality employment to attract a good quality workforce. In the report he emphasized the importance of providing proper career guidance in the field of tourism and hospitality.

Parents, relatives and aquaintances' views about their children working in tourism

Dr Asma, being an Omani national, conducted the research in the community and presented the perspectives of other nationals, including parents, relatives and aquaintances' views of tourism and hospitality as a career choice. She found that in Oman, relatives, particularly parents, substantially affect their children's professional and educational choices, and their input is valued in the decision-making process. Without the parents' guidance, students become confused about making career-related decisions. The students form their career opinion and perception based on the interaction with their parents and relatives whereas the teachers, academic counsellors and friends are less significant career influencers. The conservative parents are heavily influenced by the society where they live and continuously make the children aware about the consequences of their career choices and the subsequent impact on their families. The society sees a 'hotel' as a place where culturally and ethically unacceptable practices are carried out. The word 'hospitality' is perceived positively by the society and is compatible with the hospitable Arab culture. With this cultural sensitivity in mind, Dr Asma recommended considering that the title of the degree programme should be renamed 'Tourism and Hospitality Management' instead of 'Tourism and Hotel Management'.

Students' future career plans after graduation

Regarding the students' future career plans after graduation, including where they want to work, the data shown in Fig. 18.1 was collected and analysed by Dr Alex and his team. They found that the majority of students (55%) showed an interest in working in the tourism industry jobs in the public sector. Around 16% of students wanted to continue their studies, while 20% hoped to work in private-sector jobs in the tourism and hospitality industry. Only 4% of students wanted to pick up jobs in other sectors such as government, banking and the telecommunications sector and 5% of students opted to stay at home.

Challenges faced by government

Despite advancing better than other industries, tourism easily gets affected by deteriorating ecological, economic, social, political and health issues. The tourism industry confronts many global challenges. Numerous internal and external elements impact the tourism business, including climate change, global warming, political conflicts, health and financial crises. Internal problems include the lack of a tourism information system, the nature of the tourism offering, the seasonality effect, accommodation facilities, food crisis, transportation and accessibility issues.

Oman's tourism sector has just recently gained importance, and the current challenge is to strengthen infrastructural development in order to fulfil the country's anticipated tourism demand. Oman is establishing a tourism business that can sustain and promote Oman's tradition and culture to outside visitors through marketing, infrastructure development, human resource training, and improving visitors' attractions.

Dr Alex and his team found that according to the *Times of Oman* newspaper, the priority areas for 'Omanization' (nationalization) are tourism, logistics and the industrial sector under the economic diversification plan for the country. In 2020, the Omanization rate in the tourism sector was 44.1%, 20% in the logistics sector and 35% in the industrial sector (Table 18.5).

The 'Omanization' or nationalization of the tourism and hospitality sector workforce is one of the main challenges faced by the government. In Oman, education connected to the tourism and hospitality business is offered at the higher diploma and bachelor's degree levels in various institutions and universities. Only a few graduates from Omani universities join the tourism and hospitality employment market, which could lead to barriers for Omani employees in the industry (as expatriates may favour expatriate employees over Omani nationals). The expatriate workforce percentage will be high in the hospitality sector. The government believes that many countries send millions of tourists to Oman each year, and these visitors have the potential to contribute to the Sultanate of Oman's long-term growth. As the tourism industry expands, many organizations will have the chance to respond to increased demand

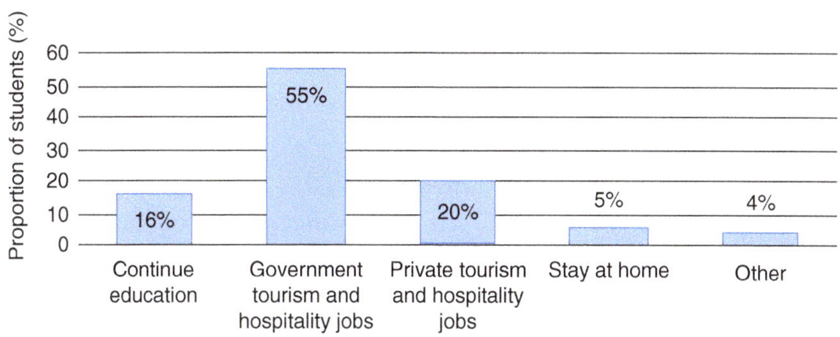

Fig. 18.1. Students' career plans after graduation. Author's own compilation

Table 18.5. 'Omanization' (nationalization)[a] in tourism, logistics and industrial sectors. Compiled by the authors.

Year	Travel and tourism sector (%)	Logistics sector (%)	Industrial sector (%)
2017	41.1	14	32.5
2018	42.1	16	33
2019	43.1	18	34
2020	44.1	20	35

[a]'Nationalization' refers to giving the workforce a national character.

for recruiting, employment, training and accommodation and tourist services. Although the number of visitors to Oman grows year after year, most tourist attractions are seasonal. The above challenges impact Oman's economy and have an impact on tourism and hospitality students' careers. As a result, many students are hesitant to pursue these careers in the future. The government is trying to tackle these issues by introducing many policies and programmes, but the implementations still need attention. Individuals, students, employees and others should be made more aware by the government of the significance of tourism to the economy. Dr Alex and his team realized the need for a specific curriculum for teachers and practitioners in local schools to teach students about the value of tourism. Through its funding, Dr Alex and his team also emphasized that government could excel in tourism and hospitality education by promoting innovative and creative tourism management skills, entrepreneurship development, and research work in tourism.

Briefing the Faculty Dean about the report

In their briefing to the Faculty Dean about the report, Dr Alex and his team explained the significance of developing human resources for the travel and tourism industry. They explained the following significance of the report to the university:

- more enhanced demand for tourism and hospitality education and training courses;
- increased student motivation;
- increased training opportunities; and
- stronger links with tourism and hospitality enterprises.

They emphasized the need to know what future workers' values and attitudes were regarding human resource development in the tourism and hospitality industry. Before joining the tourist and hospitality industry, students tend to have an unfavourable opinion of this profession, which affects their motivation to pursue careers in tourism and hospitality after graduation. Pessimistic attitudes of tourism jobs may affect their motivation to work in the tourism and hospitality industry after graduation. They further informed the Faculty Dean that only a few graduates from Omani universities join the tourism and hospitality job market, which will lead to difficulties in recruiting Omani employees in the industry. The efforts and investment by the government will be wasted if tourism graduates fail to follow tourism and hospitality careers.

It is crucial to understand the policy implications of developing the relationship between tourism and hospitality education and industry in order to satisfy the growing emphasis on labour nationalization and the future human resource requirements of the tourism and hospitality sector in Oman. Consequently, prospective students should have a better understanding of the tourist employment market and more realistic career expectations. The tourism industry's issues require more attention, both in tourist management and planning and in tourism development research in Oman, because a well-functioning tourism workforce is vital for a successful tourism business that is primarily based on human connection.

Discussion Questions

1. How do the tourism students feel about pursuing a career in tourism?
2. What are the motivations of the Omani students to pursue an education in tourism and hospitality?
3. What do the parents think about their children working in tourism?

4. What are Omani students' perceptions of working in hotels, restaurants and as tour guides?
5. What are students' aspirations for their future careers, including where they want to work?
6. What are the types of tourism career-related impediments and concerns students face? How could research solve this problem?

Teaching Notes

This case study intends to familiarize learners with the career impediments faced by tourism and hospitality students. The case examines Omani students' motives for studying tourism and hospitality, their expectations and attitudes about employment in the tourism and hospitality industry, and their impressions of tourism and hospitality as a career option. The case explores: (i) what Omani tourism students think about tourism; (ii) what factors inspire and encourage Omani students to pursue tourism as a career; (iii) what parents, relatives and acquaintances of students think about tourism and hospitality as a field of study; (iv) what positions Omani tourism students want to work in, and where they want to work; and (v) what impediments the Omani government confronts in the tourism industry. This information would result in students thinking beyond the classroom discussion and real-time application of the concepts that could facilitate in making informed career-related decisions.

Target audience

The case is suitable for undergraduate students studying tourism and hospitality management.

Teaching methods and equipment

The learners pursuing tourism and hospitality management education programmes can attempt this case study individually. It focuses on the career research of tourism management students and their views on the job assistance, career impediments and resources provided by the university to help them prepare for a successful career in tourism. The learners will identify motivations for studying tourism and hospitality, expectations and preferences for a future career, in particular the aspects they felt were important in a future job, and the general perception and reputation of the tourism sector as an employer. The learners must not accord significance to the theoretical aspects but make specific and significant recommendations; therefore, learners will be evaluated based on the explanation's applicability, usefulness and quality. Along with the case solutions, each learner could make a PowerPoint presentation to showcase their case findings (additional time would be required for this which would vary according to the number of students in the class). The case facilitator/instructor should develop a rubric to assess the learners based on the intended learning outcomes (e.g. see Table 18.6).

Teaching instructions

The case can be taught in a class lasting 2 hours using the following teaching instruction plan. Prior to the class session students are required to have read the case so they are familiar with it; the 'summary of the case' shown in the plan is an overview to highlight the main points to be considered. After a brief introduction, the facilitator/instructor starts the class session by introducing the significance of using simple qualitative research to better understand and find solutions to a problem such as development of a skilled workforce. The instructor then continues by encouraging students to think about career prospects in the tourism and hospitality profession and the impediments that they may face, moving on to summarize the case. Students will then work individually to answer the discussion questions, submitting their responses to the instructor at the end of the session for assessment. Learners will learn about the issues students face in their tourism and hospitality careers, the hurdles they face, and how they can overcome them.

Teaching instruction plan

Discussion point	Time required (min)	Teaching strategies
Introduction	10	The instructor begins the case analysis by asking questions about the learners' knowledge about the career options in tourism and hospitality sector. The instructor randomly asks a few students about their future career plans and likely impediments students feel they might face. The instructor writes the points on the board.
Brief lecture on the significance of research	15	The instructor should highlight the significance of research and should explain: (i) how a research study in the field of tourism workforce planning and analysis could be conducted; and (ii) how Oman effectively utilized qualitative research in solving the national-level problem of developing a skilled workforce in the tourism and hospitality industry.
Briefing on career prospects in the tourism and hospitality industry	20	The instructor introduces the concept of career prospects in the tourism and hospitality industry. The instructor can ask the students how and why should one ascertain the future growth of tourism industry? The students can calculate the percentage year-to-year change in the annual growth rate by using the tourist arrival data in Oman (Table 18.2) with the help of computer spreadsheets or MS Excel. A promising tourism growth in the country can ensure better employment opportunity for the skilled workforce in the tourism industry. The instructor may ask the students to refer to Table 18.1 and Table 18.3 in the case to get an idea about the employment categories available in tourism.
Summary of the case	10	The instructor may ask volunteers to raise their hand to summarize the case and list the case facts (students having read the case prior to the class session). The instructor may list the main points providing a brief overview of the case on the board. After summarizing the case, the instructor asks learners to write down their answers to the discussion questions.
Assignment question 1	10	**How do the tourism students feel about pursuing a career in tourism?** The instructor allows the individual learners to answer this question. The instructor must instruct the learners not to disclose their answers to others. After receiving each individual's response, the instructor summarizes the answer given by learners and compares it with the situation of the Omani students reported by Dr Alex and his team.
Assignment question 2	10	**What are the motivations of the Omani students to pursue an education in tourism and hospitality?** The instructor now facilitates the learners to find their own motivations. After receiving each individual's response, the instructor summarizes their answer and subsequently compares it with those of the Omani students.
Assignment question 3	10	**What do the parents think about their children working in tourism?** The instructor can enquire from the students how other individuals, such as their parents, relatives and friends, felt about tourism and hospitality as a career option. The students' response can then be compared with the Omani parents' views to identify the cultural differences and gaps.
Assignment question 4	5	**What are Omani students' perceptions of working in hotels, restaurants and as tour guides?** The instructor asks the students to discuss the question and identify their perception towards working in hotels, restaurants and working as tour guides. After receiving each individual's response, the instructor then compares the responses of the learners with the Omani students.

Continued

Teaching instruction plan Continued.

Discussion point	Time required (min)	Teaching strategies
Assignment question 5	10	**What are students' aspirations for their future careers, including where they want to work?** The instructor asks the students to list their aspirations for their future careers and subsequently compares them with the Omani students.
Assignment question 6	10	**What are the types of tourism career-related impediments and concerns students face? How could research solve this problem?** The instructor asks the students to write down the types of career impediments they think might arise. After receiving each individual's response, the instructor summarizes the learners' responses and compares them with the Omani students. Then, the instructor briefly discusses how the effective utilization of research can solve the national-level problem of skilled workforce development in the tourism and hospitality industry, and help to address dilemmas or impediments in a student's tourism career. The instructor outlines how the use of the qualitative survey of tourism corporations (airlines, hotels, restaurants, etc.) and focus group meetings with local males, females and tourism student groups could facilitate in analysing career-related impediments. This would help to understand the issues so that suitable solutions could be proposed.
Closing remarks and discussion	10	The instructor concludes the case analysis session with his/her analysis and answers for the discussion questions.

Table 18.6. Example of a rubric to assess learners based on the intended learning outcomes of the case. Compiled by the authors.

Grade	Criteria
A	An excellent response: (i) demonstrates relevant reading; (ii) shows and applies suitable examples; (iii) develops logical analysis and argument; and (iv) concludes appropriately.
B	As for A grade but falls short in comparison, either in terms of argument depth, example relevance, or logic and conclusion. Proof of relevant reading must be demonstrated.
C	Demonstrates a thorough understanding of some of the fundamental principles and theories; but limited analysis, assessment and research.
D	Eligible to pass but: (i) weak in terms of depth, logic and conclusion to the argument used; (ii) no/poor examples and illustrations; (iii) more descriptive rather than analytical; and (iv) limited evidence of relevant reading.
F	Not deserving of a pass, it does include some relevant reasoning.

Duration

The learner will require 2 hours to complete the case and the students' responses could be collected and assessed by the instructor.

Analysis

1. How do the tourism students feel about pursuing a career in tourism?

The 'answer' to this question will depend on the students' opinions in the class.

2. What are the motivations of the Omani students to pursue an education in tourism and hospitality?

The students have expectations regarding landing a well-paid job with a tourism degree. Many students feel that it would be simple to study tourism and hospitality courses and tourism would be a fascinating subject to study. Dr Alex and his team studied whether tourism was the students' initial choice of study and found that, rather surprisingly, half of the students had not picked tourism as their first choice. Around half of the students said they wanted to study something different at first. It also appeared that students who had not chosen tourism education as their first option had enrolled in the programme because it was the only one available to them or for which they had been awarded a scholarship.

3. What do the parents think about their children working in tourism?

In many Asian countries, relatives, particularly parents, substantially affect their children's professional and educational choices, and their input is valued in the decision-making process. Without the parents 'guidance' students become confused in taking career-related decisions. The students form their career opinion and perception based on the interaction with the influencers such as parents and relatives whereas teachers, academic counsellors and friends are less significant career influencers. The conservative parents are heavily influenced by the thinking of the society in which they live and continuously make the children aware about the consequences of their career choices and subsequent impact on their families. Omani society sees 'hotel' as a place where culturally and ethically unacceptable practices are carried out. The word 'hospitality' is positively perceived by the society as it is compatible with the hospitable Arab culture. Dr Alex and his team recommended considering cultural sensitivity with regard to the title of the degree programme which they suggested should be renamed 'Tourism and Hospitality Management' instead of 'Tourism and Hotel Management'.

4. What are Omani students' perceptions of working in hotels, restaurants and as tour guides?

In the Omani tourism industry, tour guiding is a vocation that has been covered under Omanization or nationalization of the workforce. The condition that tourism students are required to pass an oral examination and obtain a 'Tour Guide Licence' from the government is viewed by the students as an impediment to choosing the tour-guide career option. Some students opt to complete their tourism and hospitality management education first before applying for the tour guide licence. This maximizes the chances of obtaining the tour guide licence easily. The tour guide career is acceptable to the local community and families of the students. Work in hotels and restaurants, on the other hand, is frequently viewed as depraved, with associations such as exposure to alcohol and interaction with 'strangers', particularly among women. Hotel jobs are not culturally suitable for women. Many students agreed with these impressions, believing that the prevailing opinion in Omani society was that hotel and restaurant jobs are more suitable for expatriates rather than Omani citizens.

5. What are students' aspirations for their future careers, including where they want to work?

Concerning the students' career intentions after graduation, including where they wish to work, the majority of students (55%) reported interest in working in tourism industry positions offered by public-sector government employment. Around 16% of students wanted to continue their studies, while 20% hoped to work in private-sector jobs in the tourism and hospitality industry. Only 4% of students expressed the choice of picking up jobs in other sectors such as government jobs, banking and the telecommunications sector, and 5% of students opted to stay at home.

6. What are the types of tourism career-related impediments and concerns students face? How could research solve this problem?

Effective research could solve the tourism career-related impediments and the concerns faced by the students. The survey of tourism corporations such as airlines, hotels, restaurants, travel agencies, tour operators, tourist attractions and sports and recreational facilities and focus group meetings with the local males, females and student groups pursuing tourism majors at the university could facilitate in analysing career-related impediments. Such analysis could help in understanding: (i) concerns regarding student's future job perceptions, commitment factors and jobs for tourism students; (ii) attitudes, motivations, beliefs and inclinations about pursuing a tourism degree in preparation for a future career; (iii) the views of parents about their children working in tourism; (iv) the characteristics that students consider vital in a future career; and (v) the tourist industry's overall perception and reputation as an employer.

The focus of research on the career-related impediments highlights the pressing need for HEI policy makers to establish a quality joint education programme between academia and the industry. The Omani universities have distinct admittance requirements, which comprise three basic university programmes, a professional distribution policy, and students' national yearly examinations. The university admissions standards are far less demanding than university requirements for the course (meaning that university candidates may be admitted onto a course relatively easily but the demands of the course may be challenging particularly for less able or less committed students). The research suggests that by allowing university candidates more freedom in choosing a university topic that matches their interests could solve this problem. Applicants may be needed to approve any courses provided by the school depending on their education attainment level and eligibility to be admitted to an elected university.

In Oman's history, tourism management has long been a subject of education. Students from various social, economic and cultural backgrounds like pursuing higher degrees in tourism management since it allows foreigners to live within hotels without the hustle and bustle. Another motivation to pursue tourism as a career is that many students aspire to become entrepreneurs after completing their schooling. As a result, Omani university graduates will only be a minor part of the tourist and hospitality job market, generating restrictions and hindrances for Omani employees in the industry. An improved salary and benefit package coupled with a better working environment could help in retaining a greater number of local employees in the tourism industry. Most of the students have limited knowledge on the nature of tourism and hospitality jobs and career prospects available in MICE, ecotourism, hotels and resorts, restaurants, catering, travel agencies and tour operations, airlines, cruise, car rentals etc. The research suggests that strengthening the image of tourism and hospitality employment could facilitate in attracting a good quality workforce. More emphasis on providing proper career guidance in the field of tourism and hospitality careers could be a true game changer in minimizing the career impediments.

Further Reading

Bontenbal, M. and Aziz, H. (2013) Oman's tourism industry: student career perceptions and attitudes. *Journal of Arabian Studies* 3(2), 232–248. https://doi.org/10.1080/21534764.2013.864508

NCSI (National Centre for Statistics and Information) (2021) *Oman Statistical Yearbook 2021 – Issue 49*. NCSI, Sultanate of Oman.

Roney, S. and Özdin, P. (2007) Career perceptions of undergraduate tourism students: a case study in Turkey. *Journal of Hospitality, Leisure, Sport & Tourism Education* 6(1), 1–17. DOI: 10.3794/johlste.61.118.

Times of Oman (2019) Omanisation targets for 2020 announced. *Times of Oman*, 10 December. Available at: https://timesofoman.com/article/83064-omanisation-targets-for-2020-announced (accessed 1 November 2021).

19 Leveraging Artificial Intelligence in Hospitality and Tourism for Success

Kathryn Hayat*
University College Birmingham, Birmingham, UK

Abstract
The terms artificial intelligence (AI), machine learning and robotics are no longer deemed futuristic or restricted to the realms of science fiction. Society has embraced and adopted a range of AI technologies. The purpose of this case study is to review established and experimental uses of AI across the hospitality and tourism sector. As a result, students will be able to analyse existing practices and develop innovative ideas for AI implementation. In particular, the case study will evaluate the uses and benefits of AI in providing personalized experiences for tourists through aspects such as service innovation, adding value and empowered AI robots. The case study will introduce and compare narrow and artificial general intelligence and explore applications of big data and machine learning. On completion of this case study, learners will be able to fully understand the relevance of AI in the hospitality and tourism industry and how it can be a significant aspect of business decision making and strategic implementation.

Subjects: artificial intelligence (AI), technology innovation, disruption, tourism, hospitality, personalization

Learning Objectives

At the end of the case study, the learner will be able to:

1. Identify opportunities for artificial intelligence (AI) to disrupt existing practices within the hospitality and tourism sector.
2. Analyse the benefits of AI to consumer decision making.
3. Evaluate uses of AI in a range of sector contexts.

Case

Stephanie and her colleagues from TECHON consulting firm walk into the board room of Venture Hospitality Inc. (VHI) to make a presentation to the Board of Directors. VHI currently owns about 100 luxury hotel properties spread across the UK. Stephanie and her team are invited by the Board of Directors of VHI to suggest strategies for the investment-related decisions on emerging

*k.hayat@ucb.ac.uk

 Teaching Cases in Tourism, Hospitality and Events (eds. S.K. Dixit and S. Piramanayagam)
DOI: 10.1079/9781800621022.0019

customer-driven technologies and incorporating the artificial intelligence (AI)-based services in all the hotel properties owned by the group. TECHON is a technology-based consulting firm that provides solutions to hospitality service firms in identification, budgeting, training and adopting relevant technology to reduce the cost of operation and improve customer experience and satisfaction. Almost one-quarter of UK hospitality companies, ranging from independent to multinational hospitality firms, are major clients of TECHON.

The board of VHI are split over the decision about incorporating AI and other emerging technologies into the business. A set of members supporting the introduction and investment in technology believe that it would help the day-to-day operations of their hotel and with contactless services that have become an essential element of hospitality service after the coronavirus disease 2019 (COVID-19) pandemic. Another set of members is against the introduction of AI and tech-based services as it involves huge investment and the risk of losing the customer touch, the soul of hospitality services. Considering the concerns and fear in the adoption of technology by the board members, prior to the presentation Stephanie circulated a descriptive note (Box 19.1) on the emergence and the use of AI and tech-based products available in the market.

Hope for the best

Before the meeting with the Board of Directors of VHI, Stephanie and her team at TECHON researched VHI's current market size, customer segments, the existing level of adoption of technology, and the current needs and wants of the customers. Stephanie and her team have identified the relevant technology-based products that address the concerns of some of the board members of VHI. Stephanie believed that circulating the descriptive note on AI and tech-based products among the board members would help her pitch the idea of technology adoption.

Discussion Questions

1. List the various technology-based products that help in decision making in the day-to-day operations of various functional departments of a hospitality organization.
2. Briefly discuss the challenges in adopting the new technology-based service in the hospitality industry specific to your country.

Teaching Notes

Tourism and hospitality are among the world's largest and fastest growing economic activities. Its continued growth and success rely on delivering a memorable experience and ensuring customer satisfaction. AI and machine-learning capabilities help an organization work smarter, deliver a quality experience, and provide a seamless customer experience. AI has become a part of hospitality and tourism organizations' everyday operations ranging from marketing to revenue management. The aim of this case study is to familiarize students with the availability of various AI-based products and services and the use of technology-based products adopted in hospitality and tourism. Prior to the classroom sessions, students could read the account of the case about Stephanie and her team from TECHON's pitch to adopt AI and technology-based products. Then the case study can be implemented in the class by assigning the students to visit nearby luxury hotels and list AI-based technological developments used in the day-to-day operations of the hotel. Students may be divided into different groups to identify the AI-based products specific to operations, marketing, human resource, and decision making in the hospitality and tourism organization.

Target audience

This case study is suitable for second-year and third-year undergraduates at bachelor's degree level studying hospitality and tourism. There is no requirement for students to

Box 19.1. Emergence and use of artificial intelligence (AI) and technology-based products.

AI is a branch of computer science that revolves around the concept of machines performing tasks typically requiring human intelligence. It was first mentioned at a seminar in 1956, which broadly centred around the concept of machines operating with human-like intelligence. Within pop culture, AI is often related to notions of robotic or machine world domination, futuristic realities, or a technological advancement to fear. However, AI is no longer deemed futuristic or restricted to the realms of science fiction; it has become a general-purpose technology, established in some practices, with experimental use in others. We live in an era where disruptive technologies such as AI have become widely accessible and used across a range of industries. In some cases, AI has significantly changed the way humans operate and interact, as well as some of the processes within regular day-to-day living. Society has embraced and adopted a range of AI technologies that once were restricted to the domains of movies and computer games. In 2011, Apple Inc.'s Siri was launched for use on a smartphone. Siri, a virtual assistant, is available on a range of Apple's devices and operating systems. Responding to voice, gestures and focus tracking, Siri can answer queries, make recommendations and perform actions for a user. Following Siri's success, other operating systems launched similar virtual assistants with varying success. In 2013, Google launched wearable technology through smart glasses that focused on hands-free AI with a transparent display, using augmented reality to add a virtual layer of AI to the physical view through the glasses.

Wearable technology such as smart glasses are designed to improve efficiency and effectiveness of human intelligence, as well as access a range of digital tools usually found on a smartphone (e.g. calendar, maps, phone book). More recently, AI has infiltrated global brands, where recommended searches on Google or Netflix have become commonplace with machine learning continually improving personalized experiences. Customer service, through automated chatbots within messaging apps and web-based chat services, is in high demand, a convenient access point for customers and a cheaper, standardized way of offering customer service for businesses. Furthermore, affordable smart-speaker assistants in homes and businesses worldwide can, through voice and gesture commands, control lightbulbs at home, turn on a TV or make recommendations in response to human questioning.

The use of AI to make predictions, recommendations or decisions that influence real or virtual environments, has transformed critical business purposes, and infiltrated many aspects of the hospitality and tourism sector, providing personalized experiences for tourists and playing a role in strategic decision making.

Within AI there are two distinct branches:

- narrow *or* weak AI – single tasks performed through a simulation of human intelligence; and
- artificial general intelligence (AGI) *or* strong AI – machines with human-like intelligence performing complex tasks, independently.

Technological advancements in the 21st century mean that industries have access to data on a 'massive' scale: this is referred to as big data. Data comes from a range of sources including customer databases, social networks, Internet traffic and mobile apps. This data is significant enough to potentially outweigh the benefits of survey data or individual website traffic data and big data is a key element of AI functionality. The scale of big data means that AI can be used to analyse and extract data trends that are often too complex for traditional or human-led analysis methods. Through machine learning (using data to identify patterns and make decisions, reinforced by continual activity) the analysis of big data allows AI to make predictions, recommendations, and decisions through a perception of real or virtual environments. Importantly, the machine learning is continuous and as such, the more the analysis happens, the more intelligent the predictions, recommendations and decisions become. The result is that, through the analysis of data rather than human manipulation of data, AI programmes learn and imitate human intelligence and behaviour. As this technology becomes more general purpose, established and adopted by companies and industries, the technology is becoming cheaper and more accurate, resulting in increasing opportunities for the hospitality and tourism sector to capitalize on the value of AI.

Relevance of AI in the hospitality and tourism sector

Historically the hospitality and tourism sector has been good at capitalizing on various evolutions of e-technologies such as GDS (global distribution systems), CRS (computer reservation systems), Web

Continued

Box 19.1. Continued.

2.0 technologies, and more recently mobile technologies to provide personalized services. As a sector where much of the product is a service experienced at the point of delivery, rather than the point of purchase, technologies have provided tools to enhance consumer experience, and distinguish providers and destinations from the competition. Digital technology is a driving force behind a range of marketing and management operations. As a result, AI can: (i) enhance service operations; and (ii) influence product development, marketing campaigns and sales tactics, driving economic growth and transformation. This is increasingly important as the consumer is constantly evolving in tandem with AI, creating a continued need to satisfy customers to remain competitive.

Millennials and Generation Z are generational groups who have only experienced life with the Internet and other digital media, and they depend heavily on the characteristics of AI such as automation, personalization and human-like services in daily life. The next generational group, often referred to as Generation Alpha, are the voice generation who will have only experienced life with AI voice-search functionality. So the need to use AI in hospitality and tourism is key, to meet the continued and increasing demand for such technologies.

Within the tourism and hospitality sector, AI can be cited as a significant aspect of tourist decision making. AI technologies are changing the way in which tourists plan, book and travel. Whether inspiring tourists through recommendations, automating processes and personalizing experiences, AI is firmly establishing a place in the entire tourist experience. It can also influence strategic implementation, becoming a way for an organization to uniquely position a brand in the marketplace and gain a competitive edge. AI has disrupted many ways that consumers behave and as a result, within tourism, hospitality and events there are examples of the adoption of AI reflecting this societal change. Furthermore, AI is by virtue disruptive in nature and has forced change. The COVID-19 global pandemic of 2020 and 2021 became a catalyst for adoption and implementation of AI within the hospitality and tourism industry, reacting to concurrent national lockdowns and widespread travel restrictions.

As part of daily life, AI takes centre stage underpinning aspects such as tailored weather forecasts, Google search predictions, email spam filtering and facial recognition on social media apps. Furthermore, ridesharing apps like Uber, online shopping retailers and virtual assistants all operate on the merits of AI. These are established uses of AI that aim to create the optimal customer experience, through replacing or augmenting human labour. Through the mining of big data, AI can respond to consumer behaviour to target and personalize content through data-driven smart service delivery. From a tourist perspective, AI is empowered through data to know traveller behaviour better and is increasingly evident through all stages of the tourist journey, and importantly AI is increasingly demanded by consumers. Three established ways that tourists can actively benefit from AI include: (i) smart assistants; (ii) self-service; and (iii) wearable technology. The following section identifies uses of AI across the hospitality and tourism sector internationally, in a range of contexts from small- to large-scale use of AI all aiming to provide richer and more-engaging experiences for consumers.

Smart assistants

Smart assistants feature within the technology built into a smartphone, speaker or other digital device. Smart assistants use machine learning and human inputs to influence real or virtual environments through predictions, recommendations or decisions. Examples include Apple Inc.'s Siri, Amazon Alexa and Google Nest. Amazon Alexa has further developed a specific AI function for hospitality called 'Alexa for Hospitality' to manage Alexa devices on scale across a hotel, for example to provide typical room service. Other examples of smart assistants being used within the industry include: (i) Village Hotels, UK; (ii) Marriot International Hotels; and (iii) KLM Royal Dutch Airlines.

Self-service

Although robotics is often considered a further sub-field of AI, robotic machines and other digital devices such as kiosks offer countless self-serve functionality through the automation of tasks, standardized

Continued

Box 19.1. Continued.

responses and performing multiple functions at once. Examples include mobile applications, mobile devices such as smartphones and tablets, interactive kiosks and robots. Bangalore International Airport has implemented a range of self-service devices to support a contactless airport journey for passengers such as using self-check-in kiosks with facial recognition.

Wearable technology

Wearable technology is embedded into a fashion or functional item and worn by the user. Functionality is the key aspect of technology to influence human behaviour. Wearable technology is commonly found in glasses, watches and earpieces, but virtual and augmented reality devices can also be considered as wearable technology, where smartphones or head-mounted devices provide sensory simulation through imagery, videos and sound. Examples of wearable technology include smart glasses and smart watches, pioneered by Google and Apple Inc., respectively. KLM airlines introduced 'flight-beat' wearable technology in 2015 to monitor stressed or sick passengers. More recently, Air New Zealand using PixelBuds trialled live language translation to tackle language barriers that often exist on board flights.

One of the most significant aspects of AI is the benefits it can yield in relation to personalization. As an essential element of user engagement, personalization is increasingly a critical component of successful business and engagement with consumers. Empowered robots have been used in various guises across the industry to automate service and improve quality. For example, the Henn-na Hotel in Sasebo, Japan experimented with robots operating in a variety of roles including front of house, room service, portering and cleaning. Furthermore, personalization can add value to the customer experience and feeds into a co-creation paradigm whereby machine learning and customer interaction creates an emotional connection. Using algorithms, behavioural data is analysed to identify what a consumer was looking for, what was most relevant, to provide the most suitable content or recommendation. Through utilizing data in the most efficient way and identifying patterns, personalization through AI can create a seamless user experience for a consumer. Switzerland Tourism's website www.myswitzerland.com (accessed 15 January 2022) is powered by AI to personalize the user's experience by providing an 'interactive inspiration calendar' with immersive virtual reality videos selected to suit the individual based on their browser history.

AI has the potential to transform the global hospitality and tourism industry, improving efficiencies and service quality, automating services and processes, and as a platform for innovation. These industries driven by consumer demands and economic pressures can use AI as a significant tool to enhance consumer decision making. Therefore, the significance of AI to strategic implementation cannot be overlooked to move beyond current business models, particularly in contributing to a sustainable competitive advantage in a world emerging from COVID-19 disruption.

have a background knowledge of AI and associated technologies.

Teaching methods and equipment

The teaching methods for this case study include hands-on practice, Internet-based activities and brainstorming. This can be enhanced through the use of smartphones, smart speakers such as Amazon Alexa, Google Dot and other robotic equipment, but these are not essential.

Teaching instructions

This case may be used in the classroom using the teaching strategies described in the following teaching instruction plan, which includes two activities (for 90 min each).

Teaching instruction plan

Discussion point	Time required (min)	Teaching strategies
Activity 1		
Opening session of the case	15	Task 1. The tutor instructs the students to share their own uses and experience of AI in their day-to-day lives and as a tourist.
Practical use of AI technology	45	Task 2. Student practical use of AI technology, suggestions are offered in Table 19.1.
Reflective questions	30	Task 3. Students answer the following reflective questions, either individually or through discussion in peer groups, face to face or through use of an interactive discussion board.1. How useful did you find the AI technology? 2. Identify a different use of AI technology in the hospitality and tourism sector 3. Evaluate the value of AI technology to consumer decision making in the hospitality and tourism sector
Activity 2		
Introduction	15	Task 1: Discuss within group – technology-led innovation in the hospitality and tourism sector. (The discussion should be based on students' personal experience and the case about Stephanie and her team from TECHON's pitch to adopt AI and technology-based products, which they will have read prior to class.)
Discussion question 1	45	**List the various technology-based products that help in decision making in the day-to-day operations of various functional departments of a hospitality organization.**To help answer this question students complete Task 2. Task 2: Students will be assigned to visit nearby luxury hotels, where working in groups they identify how AI could be implemented, by listing AI-based products specific to operations, marketing, human resource and decision making.
Discussion question 2	20	**Briefly discuss the challenges in adopting the new technology-based service in the hospitality industry specific to your country.**To help answer this question students complete Task 3. Task 3: Students present a brief outline of their findings from Task 2, including examples of AI to be implemented, their uses and benefits, as well as the challenges faced in adopting the new technology (10 min per presentation).
Closing remarks and discussion	10	The tutor needs to instruct the teams and students to raise questions related to their discussion. The tutor concludes the session by summarizing the facts, application and possible answers for the discussion questions.

Duration

1½ hour per activity or longer depending on group size

Analysis

Discussion questions 1 and 2 will result in diverse solutions. The tutor must ensure that students leave with a solid understanding of established uses of AI in society and within the hospitality and tourism industry. Furthermore, students should have explored experimental uses of AI and generated innovative ideas within the context of increasing competition and the demand for personalization. Students can continue to explore AI outside the classroom, for example pursuing the concepts of machine learning and big data.

Table 19.1. Suggestions of accessible artificial intelligence (AI) for use in the hospitality and tourism sector.

AI equipment	Examples
Facebook autobots in Messenger app	• KLM (airline) • FRAnky (Frankfurt Airport) • Skyscanner (online travel agent) • TripAdvisorBot (travel recommendations)
Smart assistants – speakers such as Amazon Alexa/Echo Show, Google Dot, Google Nest, Apple homepod	Skills include: • Expedia for Alexa (trip details) • My ETA (for travel time) • Stubhub (for theatre tickets) • Flight tracker (live flight information)

Further Reading

Leung, X.Y. (2020) Technology-enabled service evolution in tourism: a perspective article. *Tourism Review* 75(1), 279–282. https://doi.org/10.1108/TR-06-2019-0229

Ruel, H. and Njoku, E. (2021) AI redefining the hospitality industry. *Journal of Tourism Futures* 7(1), 53–66. https://doi.org/10.1108/JTF-03-2020-0032

Samara, D., Magnisalis, I. and Peristeras, V. (2020) Artificial intelligence and big data in tourism: a systematic literature review. *Journal of Hospitality and Tourism Technology* 11(2), 343–367. https://doi.org/10.1108/JHTT-12-2018-0118

20 Solving the Drive-through Crisis with Service Blueprinting

Drew Martin* **and Su Gibson**
University of South Carolina, South Carolina, USA

Abstract
As hospitality services contain high experiential qualities, the service delivery's success depends on customers' assessments of whether the gap between their expectations and perceptions of the service delivery is acceptable. When customers deem the gap is too large, service failure occurs. Identifying the cause of a service failure is not straightforward because delivery systems are complex, and customers are participants in the process. Service blueprinting encourages systems thinking to understand the interconnectedness of processes and people. Mapping out delivery systems helps to identify system fail points. Correcting service-delivery fail points requires a holistic view to understand how controllable (internal) and uncontrollable (external) variables affect the service delivery. This case examines a growing problem in quick-service restaurants' drive-through services to help learners understand the benefits of service blueprinting and systems thinking to improve hospitality services.

Subjects: service blueprint, strategic management, service marketing

Learning Objectives

At the end of the case study, a learner will be able to:

1. Identify the levels of a service blueprint.
2. Create a service blueprint for a restaurant drive-through or takeout service that shows the interconnectedness of the processes and people involved in the service delivery.
3. Identify potential bottlenecks and fail points in the service delivery process.
4. Recommend solutions to the service delivery that consider both controllable and uncontrollable elements in the business environment.

Case

In the USA, quick-service restaurants and drive-through restaurant service emerged independently from each other in 1921. In Wichita, Kansas, White Castle's business model of selling sacks of uniformly produced hamburgers became the first hamburger chain (Kieler, 2015). Around the same time, the Pig Stand opened on the highway between Dallas and Fort Worth, Texas and sent waitpersons to the parking lots to greet customers and receive food orders (Barksdale, 2021). This first drive-through restaurant was a response to growing automobile ownership in the USA

*martin11@mailbox.sc.edu

DOI: 10.1079/9781800621022.0020

(Maze, 2021). The original drive-through concept required the restaurant worker to visit parked automobiles to take orders, deliver food and receive payment. Customers ate the meals in their parked cars. This delivery system gradually became more challenging as more automobiles and customers queued for food. The solution to this problem was the invention of the drive-through window. Early versions of drive-through window service began in 1931 (see Barksdale, 2021), but the full-scale adaptation to drive-through windows started in the 1970s (Witzel, 2013).

Today's quick-service restaurant employees generally stay inside the restaurant, and the drive-through service is like a production line. Customers drive through a series of stations where they order, pay and receive their food. Quick-service restaurant locations match traffic patterns to take advantage of commuting patterns. The convenience and relatively low cost of drive-through service has been popular with customers. For example, McDonald's generated about 70% of sales (pre-coronavirus disease 2019 (COVID-19) pandemic) from drive-through service (Maze, 2021).

The challenges

The pandemic affected both customer and employee behaviour, creating unprecedented challenges for the hospitality industry. For restaurants, inside dining decreased dramatically, pushing service to drive-through and takeout service delivery. In 2020, customers spent US$435 billion on online food orders, and 61% of that spending was at restaurants only offering sit-down dining prior to the pandemic (Guszkowski, 2021). Quick-service restaurant orders have also dramatically increased. McDonald's alone served an estimated 300 million additional cars in their drive-through business during the last 3 months of 2020 (Maze, 2021). Restrictions on inside dining were not the only reason for this shift. A National Restaurant Association (NRA) survey found 61% of quick-service restaurants voluntarily reduced dine-in capacities because they lacked staff (Romeo, 2021).

For restaurant employees, the last couple of years have been difficult. Turnover of quick-service restaurant employees continues to plague the industry. Some estimates peg the turnover rate for quick-service restaurant employees at over 130% (Lucas, 2021). This turnover is not a new problem, but the gap between demand for new workers and the supply of applicants is growing. The NRA (2021) reports restaurant employment remains almost 800,000 jobs below the pre-pandemic peak. To attract applicants, the industry now offers higher starting wages and referral bonuses. These incentives do not seem to be closing the gap, and costs of doing business are rising. Employee turnover is estimated to cost restaurants about US$2000 per worker (Ngo, 2020).

Another industry response to solving the employee shortage is technology adoption. Smart kitchen equipment, digital menu boards and self-ordering kiosks help quick-service restaurant adaptations to streamline services (Gentile, 2019). Unfortunately, a recent study reported that 29% of restaurant operators said their technology use lags behind that of their competitors in the restaurant industry and that identifying the best technology solution is difficult (Riehle and Wilson, 2019). A disconnect exists between restaurant operators and customers (see Table 20.1).

Table 20.1. Customer–restaurant disconnect in technology. Adapted from Riehl and Wilson, 2019.

How customers place orders	Customers' use (%)	Restaurant operators' offer (%)
Delivery orders on restaurant websites	56	45
Delivery orders on a restaurant app	43	18
Ordering via virtual assistant (e.g. Amazon Alexa)	31[a]	12

[a]These are customers who are receptive to ordering via a virtual assistant but may not have actually used one.

The annual drive-through benchmark report results show downward trends in customer service. Ten major brands of quick-service restaurants were visited by mystery shoppers in 2021 (n = 1492 visits). The study reports total time to receive an order increased by over 26 seconds, and inaccurate orders increased by 13% compared with the previous year's study (PR Newswire, 2021). Slower service results in longer queues that discourage customers. An estimated 30% of customers drive away when lines are long (Meisenzahl, 2021). This service decline is more problematic as restaurants' revenues are increasingly dependent on drive-through business.

Recent evidence suggests that strains on service delivery are reducing customer satisfaction. Table 20.2 shows results of the American Customer Satisfaction Index's (ACSI) annual national study on consumer satisfaction for some quick-service restaurants (ACSI, 2022). For example, results suggest consumers were about 4% less satisfied with Chick-fil-A in 2021 compared with 2018 (83% vs 87%). While this study does not show declines for all quick-service restaurants, overall customer satisfaction appears to be declining.

The combination of higher demand, longer wait times and staff shortages may exacerbate customer behaviour issues. Some recent evidence suggests that frustrated customers may create a more hazardous work environment. An analysis of 911 calls (emergency service calls to the police) from 2017 to 2021 found 77,200 violent or threatening customer behaviour incidents in just 643 quick-service restaurants alone (Service Employees International Union, 2021). Solving the drive-through crisis requires a solution that addresses operations, employees and customers.

Table 20.2. Consumer satisfaction (%) for quick-service restaurant service by year. Adapted from ACSI, 2022.

Quick-service restaurant	Year				
	2017	2018	2019	2020	2021
Chick-fil-A	87	87	86	84	83
KFC	78	77	78	79	79
Arby's	80	79	80	79	77
Dunkin'	79	78	78	79	77
Burger King	77	76	76	76	76
Taco Bell	76	74	75	74	74
Wendy's	76	77	77	76	73
McDonald's	69	69	69	70	70

Service blueprinting – mapping systems

Improving an existing service is challenging due to the difficulty of describing and depicting the service concept. Service blueprints are maps that portray service systems (see Shostack, 1984). In addition to providing an overall visual of the service system, a service blueprint helps managers to identify possible system fail points. A service blueprint has several components identified in Fig. 20.1.

At the very top of the blueprint, is **physical evidence**. Typically, above each point of contact, the actual physical evidence of the face-to-face meeting includes tangibles like the decor, clothing of the service provider, and even other customers. Other less tangible elements of the servicescape, including sounds, scents and lighting, also affect interpretations of the physical evidence.

Customer actions are the steps taken when the customer performs the process of purchasing, consuming and evaluating the service. The steps may include evaluating choices, performing an activity and interacting with service providers as well as other guests. The example in Fig. 20.1 shows a series of steps necessary for the hotel guest to complete upon arrival, from making a reservation, arriving at the property and entering the hotel room. In addition to interpreting the physical evidence, the customer also interacts directly with **onstage contact employee actions** throughout the process. Onstage contact employees perform service delivery activities that are visible to the customer. In this example, an employee who greets the guest and loads the luggage onto a cart is an example of a front-line or onstage contact. The *line of interaction* separates the guest from the employee, but these two actors co-create the service delivery.

Hospitality services typically require the coordinated efforts of a team. Many components of service delivery occur behind the scenes or beyond the customer's direct view. **Backstage contact employee actions** are the steps and activities not visible to the customer. These activities may be performed by the contact employees (e.g. the delivery of luggage to the guest's room) or people who primarily work backstage, but they provide critical support to service delivery. At the hotel, housekeeping services are critical for keeping the physical appearance of the property clean. Usually, these backstage contact people enter a guest's room when they are gone and perform cleaning, replace linens and replenish toiletries as needed. The *line of visibility* shows the separation from the customer's view.

Finally, **support process actions** are internal services, steps or interactions that support the contact employees. In this example, the people who maintain the hotel's website and online reservation system are critical for sorting guests by their arrival days and helping revenue managers to adjust pricing based on room inventory. Other support actions include procurement of supplies, human resources activities like hiring and training new employees, and accounting services to monitor revenues and expenses.

A service blueprint showing a complete service delivery process is likely to be more complex than this example. After the process is completed, managers can take a close look at the service holistically and identify potential bottlenecks. These bottlenecks can slow or even stop a service delivery, leading to a potential service failure. These bottlenecks are fail points (i.e. the shaded letter 'F' inside a circle). Figure 20.1 shows two possible fail points for the hotel check-in service. In this case, fail points are identified during the reservation process (e.g. unable to match inventory with guest preferences) and during the check-in process (e.g. many guests arriving simultaneously slows the process).

Project or Discussion

The current business environment suggests that the industry needs some guidance. To address these issues, a short study is needed. Choose a local quick-service restaurant to study. If local restaurants do not offer a drive-through service delivery, study the restaurant's takeout service. An actual visit to the restaurant helps to understand the service delivery process.

1. Create a complete service blueprint. Each blueprint should include all the components listed and discussed in class. All lines should be labelled clearly, and arrows should be included to indicate the order of activities. The blueprint should include at least four 'boxes' of support processes. Potential 'bottlenecks' or 'fail points' in the service delivery process also should be identified.

2. What operational improvements address the problems? Suggest possible solutions/alternatives to address these potential problem areas. Solutions should come from best practices in industry (include references and citations from quality outside sources).

3. What technological improvements address the problems? Suggest possible technology-based solutions/alternatives to address these potential problem areas. Solutions should come from best practices in industry (include references and citations from quality outside sources).

4. The deliverable to company management is a four- to six-page professional report (double-spaced, plus references and appendices). Write as an outside consultant reporting to that company's management. Do not write from the perspective of personal experience as a customer or an employee. The following outline provides possible titles and suggestions for each section of the report.

a. Introduction: Briefly describe the business that you are studying and the purpose of the report. The business history or background is not needed because the audience is the company's management team. Four or fewer sentences should suffice.

b. Turnover challenges: How does employee turnover affect service delivery? Some sources listed may be cited, but management expects further investigation as well.

c. Service design: Describe the service process. Refer to your blueprint, which should be in an appendix. What key elements of physical evidence influence the customer's experience? Identify and describe at least three potential bottlenecks (points in the process where backups or slow delivery may occur) and/or fail points (points in the process where problems may occur).
d. Service modifications: Suggest possible solutions/alternatives to address these potential problem areas. Solutions should come from best practices in industry (include references and citations from three or more quality outside sources).
e. Conclusion: Summarize the key points in a short paragraph.
f. References (on a separate page).
g. Appendices: A separate page with the service blueprint (see Fig. 20.1) and other tables or figures to support findings.

Teaching Notes

This case introduces students to service blueprinting as described by Shostack (1984). The goal is to illustrate the importance of systems thinking when designing services and allow students the opportunity to practise planning out service delivery in detail. Since there are myriad ways to design quick service food delivery and an even greater number of possible fail points, this exercise is excellent for group work and class discussion. The scenario should be familiar to many students, depending on the market in which the class is based. New technologies, evolving market conditions and post-pandemic customer demands make the case subject relevant and interesting. An added benefit to learning this technique is a fundamental shift in viewing business activities as a dynamic system. The industry is evolving at a fast pace, and time to recover from mistakes is shrinking.

An understanding of the typical guest journey, guest–staff interactions and supporting processes during a quick service drive-through or takeout experience is required. An in-class review of restaurant operations might be helpful, depending on the prevalence of fast food takeout and drive-through restaurants in the local market. Exposure to the basic format and elements of service blueprints is also necessary for students to complete this exercise.

Benefits of blueprinting

Service blueprinting offers many benefits for hospitality managers. First, mapping the

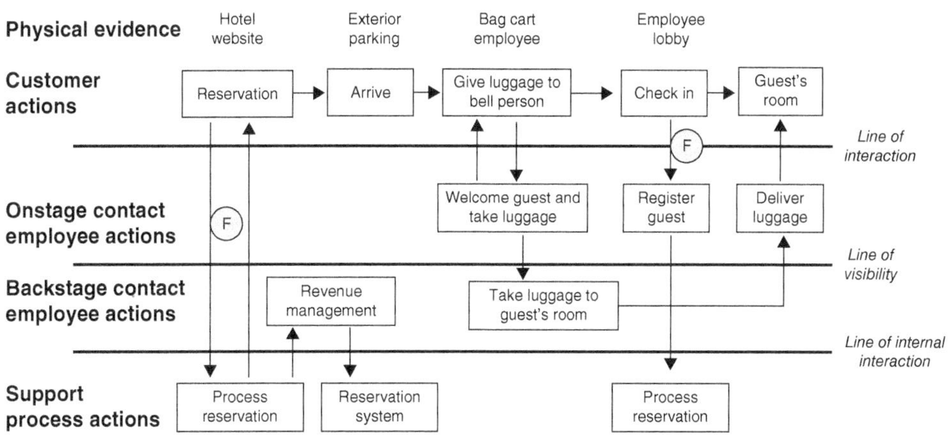

Fig. 20.1. Service blueprint elements for a hotel reservation and check-in service. The shaded letter 'F' indicates a potential fail point (i.e. a bottleneck that can slow or even stop service delivery leading to service failure). Authors' own figure.

service process shows employees how the different functions fit together. For larger systems, seeing how an individual's contribution supports the overall service delivery often is difficult for employees to understand. Second, a closer look at the chain of service activities helps to identify potential fail points and weak links in the chain of service activities. This process encourages continuous quality improvement. Third, managers can use the blueprint to assist with resource allocations. To maintain or improve service delivery, blueprints help to identify and assess costs, revenues and capital investments for various activities. Finally, service blueprinting provides rational for internal and external marketing. Employees need to understand their roles in the service delivery, and customers' expectations need to be managed.

Target audience

Undergraduates in their third or fourth year of study or graduate students in a business, tourism or hospitality management degree programme would benefit from completing this case. The deliverables could be adjusted for different levels of education; however, a basic understanding of the hospitality industry, management and foodservice operations is necessary for successful completion of the tasks. The case is designed to be completed individually or in small groups. Groups should not exceed four members to allow for the coordination of field visits and provide each group member the opportunity to creatively contribute to the service blueprint.

Teaching methods and equipment

Students are encouraged to make a field visit to a local quick-service restaurant to observe service delivery in action. There, they should take written field notes about the sequence of a guest's journey, observations about interactions between staff members and guests, apparent service delivery failures, visible operational processes, and support processes if possible. This information can then be used to guide their creation of a service blueprint and discussion about ways to optimize the service delivery system.

If students have ready access to computers and software, such as Microsoft PowerPoint and Microsoft Word, they can use them to create the service blueprint and corresponding management report.

Alternatively, students could be provided with whiteboards or large paper to draw the service blueprints. In-class presentations are another option. Students can address all questions listed in the case instructions while showcasing their service blueprints.

Teaching instructions

Instruction for this case could be divided between two to three class sessions of roughly 1 hour each. Students should plan to visit a local quick-service restaurant for observation in between the first and second session. In the first session, the instructor should familiarize students with basic quick-service operations and drive-through processes. They should also provide an overview of service blueprinting and explain the deliverables for the case study. Students may be organized into groups at this time, if applicable, so they can share contact information, choose a local restaurant to observe, and coordinate their visit for the fieldwork. Students may also discuss roles within the group to ensure expected contributions to the project are understood by all members. A group contract or pledge that group members sign which itemizes these expectations is recommended.

During the visits to the local quick-service restaurant(s), students will observe the service delivery and write field notes to help them develop their service blueprint and report or presentation. These field notes may or may not be graded.

The second class session is dedicated to the development of the service blueprints and discussion of the questions posed in the case. Service blueprints could be developed following the steps provided by Shostack (1984):

1. Identify processes.
2. Isolate fail points.
3. Establish time frame.
4. Analyse profitability.

A third class session may be required for group presentations if assigned as an alternative to the written report, or for debriefing and further discussion once the student projects are complete.

Teaching instruction plan

Steps to take	Time required (min)	Teaching strategies
Reading	30	Students should receive the case study and any supporting literature with which the instructor would like them to be familiar for the exercises. Reading could be done outside of class time.
Introduction	25–30	The instructor should introduce the case study and ask students for examples to share from their own experiences in quick-service environments. Examples from both customer and employee perspectives should be solicited, if possible. The instructor should then explain the process of service blueprinting and present a visual example, such as the one provided in the case study. Key features, including the directional connecting arrows, line of interaction, line of visibility, line of internal interaction, and fail points should be included when examining the blueprint example with students. The benefits of service blueprinting and careful service delivery system planning should be emphasized.
Group assignments (optional)	5	If appropriate, students could be sectioned into small groups, with a suggested maximum of four per group. They could be assigned into groups by the instructor or self-select. Students should then gather in their groups for organization.
Group organization (optional)	15–25	If a group project, students should be given time to decide on a local restaurant to visit for the field work portion of the case. They will need to decide on a time, exchange contact information, and arrange transportation to and from the restaurant.
Field work	30	Students visit a local quick-service restaurant to observe the takeout and/or drive-through process. They should write notes about their observations, including all visible steps in the process, which steps include interactions with staff members, any apparent support processes happening outside of customer view. Any evidence of failure points, bottlenecks and customer dissatisfaction should also be noted.
Service blueprinting	20	Using their field notes, students create a service blueprint of the drive-through or takeout service they observed. All elements should be included, such as: (i) the steps in the customer journey; (ii) physical evidence; (iii) interactions with staff members; (iv) supporting processes; (v) directional connecting arrows; (vi) line of interaction; (vii) line of visibility; (viii) line of internal interaction; and (ix) fail points. Students may draw these service blueprints by hand on whiteboards or paper, and/or use software such as Microsoft PowerPoint to create them digitally. If students want to use a computer to create a digital image, they may benefit by drafting the blueprint by hand first while they work out all of the elements that should be included. Instructors should circulate through the classroom during this time to provide guidance and assistance, since students may have questions about the blueprinting format.

Continued

Teaching instruction plan Continued.

Steps to take	Time required (min)	Teaching strategies
Identification of fail points	10	Once the main elements of the service blueprint are in place, students can identify and mark possible bottlenecks and fail points in the service delivery process.
Brainstorming solutions	20	Students should use the bottlenecks and fail points identified in the above step to brainstorm solutions. These solutions should be based on techniques and strategies used by other quick-service restaurants. If there are no suitable solutions currently deployed in the field, students may seek solutions from other industries that could be adapted for their purposes. This step is critical. Students should investigate best industry practices to support recommendations based on evidence.
Finalizing paper/ presentation	10	Students should be alerted that the exercise is almost over and given some time to review the instructions for the assignment, their service blueprints and their paper/presentations.
Group presentations (optional)	50–60	If the course schedule allows for a third class session dedicated to the case study, much value and educational enrichment can be gleaned from having the students share their experiences during the field work, service blueprints and solutions. This session could be done with recorded video presentations and/or posting of graphic files of the service blueprints to an online course page, if the course is delivered in a non-traditional format.

Duration

2.5–4 hours, including field observations

Analysis

1. Create a complete service blueprint.

Students should use the service blueprint example (Fig. 20.1) as a guide. Service blueprints should include physical evidence, the steps of the customer journey, the line of interaction between staff members and guests, staff members and the activities in which they interact with guests or support the process, classification of these activities as onstage or backstage separated by the line of visibility, and support processes separated from the other activities by the line of internal interaction. All items should be connected by directional lines and arrows to show the effects and how processes proceed over time, as illustrated by the sample blueprint in Fig. 20.2. Figure 20.2 shows an example of a drive-through service blueprint.

2. What operational improvements address the problems? Suggest possible solutions/alternatives to address these potential problem areas.

Answers will depend on what bottlenecks and fail points the students identify in their service blueprints. For example, if a student or group highlights a fail point during the ordering process, they may suggest that the staff member verbally summarize the order at the end of the process so the customer can confirm the accuracy.

3. What technological improvements address the problems? Suggest possible technology-based solutions/alternatives to address these potential problem areas.

Answers will depend on what bottlenecks and fail points the students identify in their service blueprints. Using the example of a fail point at the ordering process, the students may suggest having the ordered items displayed on a digital screen at the ordering stage, as some fast food chains have implemented, so that guests can visually check for errors in the order before moving on to the next stage of the process. In this case, students should find an

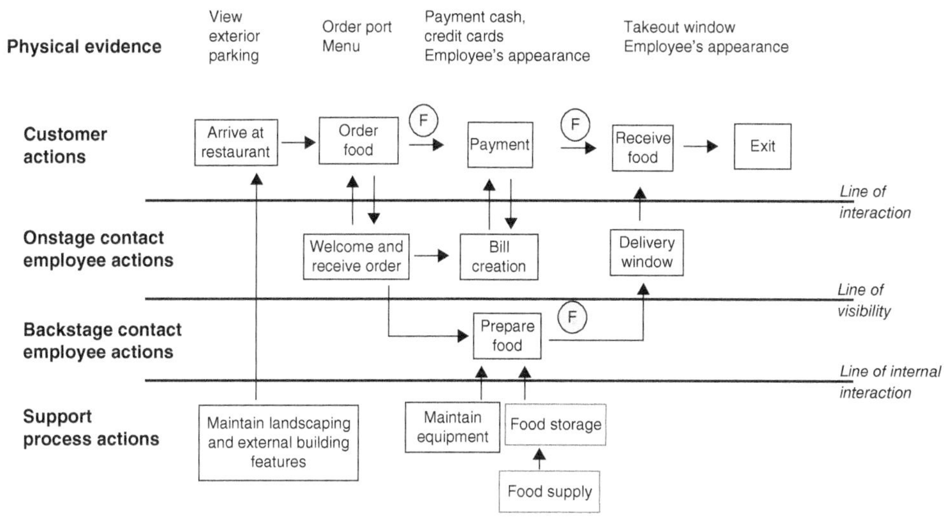

Fig 20.2. Example of a service blueprint for a drive-through restaurant. The shaded letter 'F' indicates a potential fail point in service delivery. Authors' own figure.

example of this feature in action and provide a reference, such as to an online article showcasing the technology (e.g. https://kioskindustry.org/mcdonalds-drive-thru-designs-deployed/ (accessed 24 April 2022)).

4. The deliverable to company management is a four- to six-page professional report (double-spaced, plus references and appendices). Write as an outside consultant reporting to that company's management.

Students are provided with an outline for the written paper in the instructions for the case. They should write in full sentences using proper citations and professional language. Grading should include emphasis on error-free, level-appropriate writing in a university environment. This part of the case could be converted into in-class group presentations or recorded presentations for online and web-based courses.

References

ACSI (American Customer Satisfaction Index) (2022) Benchmarks by industry: limited-service restaurants. Available at: https://www.theacsi.org/index.php?option=com_content&view=article&id=149&catid=&Itemid=212&i=Limited-Service+Restaurants&sort=Y2017 (accessed 24 April 2022).

Barksdale, N. (2021) How Drive-thru Dining Changed Fast Food. Available at: https://www.history.com/news/drive-thru-dining-history-in-n-out-burger (accessed 24 April 2022).

Gentile, M. (2019) 4 Game-changing innovations every restaurant should adopt. *QSR*, May. Available at: https://www.qsrmagazine.com/outside-insights/4-game-changing-innovations-every-restaurant-should-adopt (accessed 24 April 2022).

Guszkowski, J. (2021) The pandemic swept restaurants into the digital age. *Restaurant Business*, 16 March. Available at: https://www.restaurantbusinessonline.com/technology/pandemic-swept-restaurants-digital-age (accessed 24 April 2022).

Kieler, A. (2015) The White Castle Story: the birth of fast food & the burger revolution. *Consumerist*, 14 July. Available at: https://consumerist.com/2015/07/14/the-white-castle-story-the-birth-of-fast-food-the-burger-revolution/ (accessed 24 April 2022).

Lucas, A. (2021) Fast-food wages climbed 10% in latest quarter, the largest jump in years, report says. *CNBC*, 23 August. Available at: https://www.cnbc.com/2021/08/23/fast-food-wages-climbed-10percent-in-latest-quarter-the-largest-jump-in-years-report-says.html (accessed 24 April 2022).

Maze, J. (2021) The year of the drive-thru. *Restaurant Business*, 17 March. Available at: https://www.restaurantbusinessonline.com/operations/year-drive-thru (accessed 24 April 2022).

Meisenzahl, M. (2021) The fast-food industry had a uniquely tumultuous 2021, *from staff shortages to supply chain issues. Business Insider*, 31 December. Available at: https://www.businessinsider.com/the-fast-food-industry-had-a-tumultuous-2021-staff-shortages-supply-chain-issues-2021-12 (accessed 24 April 2022).

Ngo, H. (2020) The fast food industry crisis that's only getting worse. *Mashed*, 20 January. Available at: https://www.mashed.com/184286/the-fast-food-industry-crisis-thats-only-getting-worse/ (accessed 24 April 2022).

NRA (2021) Restaurant job growth slowed in recent months. 16 December. Available at: https://restaurant.org/research-and-media/research/economists-notebook/analysis-commentary/restaurant-job-growth-slowed-in-recent-months/ (accessed 24 April 2022).

PR Newswire (2021) SeeLevel HX 21st annual drive-thru study uncovers delays and inaccuracy as QSRs struggle with labor shortage. 23 September. Available at: https://www.prnewswire.com/news-releases/seelevel-hx-21st-annual-drive-thru-study-uncovers-delays-and-inaccuracy-as-qsrs-struggle-with-labor-shortage-301383881.html (accessed 24 April 2022).

Riehle, H. and Wilson, M. (2019) Harnessing technology to drive off-premises sales. Available at: https://restaurant.org/getmedia/dcc5cb1b-a4da-4140-9658-939b07f792b5/research_offpremises_201910.pdf (accessed 24 April 2022).

Romeo, P. (2021) New snapshot shows a backslide in restaurants' recovery. *Restaurant Business*, 29 September. Available at: https://www.restaurantbusinessonline.com/financing/new-snapshot-shows-backslide-restaurants-recovery (accessed 24 April 2022).

Service Employees International Union (2021) Beaten, stabbed, silenced: violence in California's fast-food industry and workers' fight for voice. Available at: http://fightfor15.org/wordpress/wp-content/uploads/2021/11/Beaten-stabbed-silenced-violence-in-California%E2%80%99s-fast-food-industry-and-workers-fight-for-a-voice.pdf (accessed 24 April 2022).

Shostack, L. (1984) Designing services that deliver. *Harvard Business Review* 62(1), 133–139.

Witzel, M.K. (2013) Drive-ins. In: Smith, A. (ed.) *The Oxford Encyclopedia of Food and Drink in America*, Vol. 1, 2nd edn. Oxford University Press, New York, pp. 654–655.

Further Reading

Bitner, M.J., Ostrom, A.L. and Morgan, F.N. (2008) Service blueprinting: a practical technique for service innovation. *California Management Review* 50(3) 66–94.

Hossain, M.Z., Enam, F. and Farhana, S. (2017) Service blueprint a tool for enhancing service quality in restaurant business. *American Journal of Industrial and Business Management* 7(7), 919–926.

Shostack, G.L. (1992) Understanding services through blueprinting. In: Swartz, T.A., Bowen, D.E. and Brown, S.W. (eds) *Advances in Services Marketing and Management*, Vol. 1. JAI Press, Greenwich, Connecticut, pp. 75–90.

21 Use of Customers' Satisfaction Survey to Segment the Tourism Market

Noelia Sánchez-Casado,* José Miguel Navarro-Azorín, María Pilar Jiménez-Medina and Clara Cubillas-Para
Universidad Politécnica de Cartagena, Spain

Abstract
Nowadays, in the tourism sector firms are focusing more and more on satisfying customers. Nevertheless, the special restrictions applied to reduce the risk and level of contagion of coronavirus 2019 (COVID-19) in some establishments (e.g. hotels, restaurants, coffee shops) could modify the customers' level of satisfaction. For this reason, this case study proposes using a survey of customers' satisfaction to identify potential differences that may serve as segmentation criteria. Then, this case study helps students to better understand the importance of market segmentation and introduces them to the statistical methods and tools necessary to implement the corresponding business' marketing strategies.

Subjects: segmentation, customer satisfaction, marketing, management

Learning Objectives

At the end of the case study, a learner will be able to:

1. Explain the concept of market segmentation and how this can help firms to identify different types of customers.
2. Identify the usefulness of surveys as a tool to retrieve market information.
3. Evaluate why market research is necessary in order to consider all ideas and not take some ideas for granted.
4. Apply basic statistical methods.

Case

From 2020 onwards, the world has been affected by the coronavirus 2019 (COVID-19) pandemic. This has caused an unprecedented crisis with some economic sectors such as tourism being more severely affected than others. During the first months of 2020, travel restrictions led to a fall of 97% in international tourist arrivals and, despite some recovery, it was the worst year on record for tourism with a decrease of international arrivals of about 73%.

*noelia.sanchez@upct.es

(eds. S.K. Dixit and S. Piramanayagam)
DOI: 10.1079/9781800621022.0021

Given that tourism activity involves human traffic, governments of each country imposed harsh measures and restrictions affecting the entire tourism value chain in order to prevent the spread of the virus. These measures, which were different and dynamic in each destination, included, among others, the prohibition to travel, the obligation to show the COVID-19 vaccination certificate (also known as a COVID-19 passport which was needed for both travelling and/or attending establishments indoors), and the imposition of quarantines upon arrival at destinations.

In Spain, some examples of these measures were the use of masks, disinfection of facilities, social distancing, or capacity restrictions. Although all the economic sectors were negatively affected by the situation, hotels, restaurants and coffee shops were hit the hardest. Actually, because of the nature of the service they offer, governments applied special measures to these types of establishment, such as limiting their opening hours, imposing strict capacity restrictions, limiting the number of people at a table, prohibiting consumption at the bar and requiring use of a COVID-19 passport to enable entry indoors.

Within this general context, the impact of the COVID-19 pandemic is not constrained to a fall in the number of tourist arrivals. Indeed, it has brought outstanding changes in the behaviour of the travellers. Along these lines, several new trends in the tourism markets have emerged, such as a stronger preference for open-air activities, nature-based products and rural tourism, that mirror a greater concern for health and security issues in the consumption of hospitality services.

In light of the above, many companies and tourist groups thought a lot about what to do to face such a different and complex scenario that posed numerous challenges that had to be overcome in order for the companies to survive. The fear of contact between people encouraged the development of new technologies to minimize contact such as: (i) mobile payment; (ii) mobile access control systems; (iii) online video conferencing tools (e.g. Zoom, Skype, Meet, Microsoft Teams); and (iv) guest management systems. Moreover, firms have also adopted changes involving distribution through online delivery services and reinforcement of their online sales platforms. In short, all companies have struggled to reinvent themselves to ensure their continuity.

In this case study we propose the initiative of the sectoral association HOSP&MUR (hospitality and tourism) located in a region in the Spanish Mediterranean fringe. This association gathers small sub-sector associations (hotels, restaurants, coffee shops) with the aim of facing together the crisis caused by the pandemic that was obviously endangering their survival. After debating a lot and saving the different points of view that all of them could have had during this process of transformation and adaptation, the members of HOSP&MUR reached a consensus and different decisions were made in different areas within the association. Specifically, this case refers to one of the decisions in the field of marketing as the HOSP&MUR concluded that it was essential to have a better understanding of their customers' behaviour. It was evident that the degree of customer satisfaction is of the utmost importance in the tourism sector, since it can affect future behaviour, such as the intention to recommend or repeat a visit to a hotel, restaurant or coffee shop in the future. Furthermore, since tourism services are intangible by nature, it is vital that the tourism sector understands what factors lead to high satisfaction with the service provided. Moreover, as health and safety is now one of the main priorities of consumers, having information on customer satisfaction with how each establishment deals with these issues may help improve business management.

It is also important to bear in mind that satisfaction may vary according to: (i) the profile of the customer defined by variables such as socio-demographic characteristics (e.g. gender, age, origin, level of studies); or (ii) the type of establishment where the service is provided (hotels, restaurants, coffee shops). Therefore, special conditions and the context in which the establishment is visited can modify the customers' perception, and their opinion about the product or service they are consuming can vary from one to another. For this reason, marketing research aims to classify customers according to their particular characteristics, preferences and behaviour. This process, called segmentation, seeks to gather similar customers in order to develop an effective marketing tool to accomplish the business strategy.

Identification of segments in tourism is mainly made based on their geographical, demographical, socio-economic, psychographic, behaviour characteristics or a combination of them. Moreover, according to Ahani *et al.* (2019), segmentation refers to identifying and matching groups to distinct product and/or service offerings. Sometimes customers' needs can be difficult to manage or satisfy, and segmentation may help firms to group similar customers, identify preferred target markets and inform marketing mix elements to ensure the effective management of marketing expenditures. All the previously exposed factors finally lead the different associations integrated in HOSP&MUR to ask their managers for a study to resolve key aspects that provide information on client segmentation. In this way, the manager of HOSP&MUR decides to entrust the newly recruited management trainee in the administration department of the recently created association to carry out a survey and its subsequent analysis to resolve customer segmentation.

Data analysis

To analyse the overall degree of satisfaction of customers during their visit to tourist establishments such as hotels, restaurants and coffee shops, some data were collected through a survey. This survey focused on assessing the levels of satisfaction obtained by users of these facilities, based on the correct application of the criteria imposed by state and regional public institutions to minimize COVID-19 infections. Data were collected during 2021 through personal interviews with customers of tourist establishments (hotels, restaurants and coffee shops) in the region of Murcia (Spain).

During this period, some of the COVID-19 measures applied to the hospitality sector included the following:

- use of masks by employees;
- use of masks by customers while they are not eating or drinking;
- disinfection of facilities after each use;
- avoiding the use of a print menu, prioritizing a digital version (e.g. accessed using a quick response (QR) code);
- avoiding the use of shared products such as napkin holders/oil bottles, prioritizing the use of disposable or individual napkins/oil bottles;
- social distancing (2 m) between employees and customers but also among customers (when they were not housemates);
- capacity restrictions;
- limiting opening hours (closing before midnight);
- limiting the number of people (ten) at a table (when they were not housemates);
- prohibiting consumption at the bar; and
- requiring customers to present a COVID-19 passport before allowing them entry to indoor premises.

To survey customers' satisfaction level a questionnaire was used. A total of 466 customers answered the whole questionnaire. The resulting dataset included information regarding the profile of the customer (i.e. whether they were a resident or tourist, and their gender) along with some specific questions regarding the type of establishment they visited and their overall level of satisfaction. Table 21.1 shows the exact wording of the corresponding questions in the survey.

The question relating to satisfaction was posed as a Likert item with five response options ranging from 'not at all satisfied' to 'very satisfied'. To analyse data from this question, a value is assigned to each option. In this case, 'very satisfied' scored 5 points and 'not at all satisfied' scored 1 point; these numbers were used to compute several statistical descriptive measures such as the mean ($\bar{x}$) and the standard deviation (SD) to make the results easier to interpret. The data from these measures are summarized in Table 21.2 according to the type of

Table 21.1. Questions included in the survey and respondents' possible answers. Authors' own elaboration.

Question	Options for the answer
Please think about any establishment that you are visiting during your stay and indicate which type it belongs to ...	Hotel Restaurant Coffee shop
Indicate your level of satisfaction with the establishment ...	1 = Not at all satisfied 5 = Very satisfied
Type of customer	Resident Tourist
Gender	Male Female

Table 21.2. Customer satisfaction by type of establishment visited, type of customer (resident or tourist) and gender. Authors' own elaboration.

	n	$\bar{X}$[a]	SD[a]
Hotels	206	4.092	0.763
Resident	10	4.60	0.52
Tourist	196	4.07	0.77
Restaurants	161	3.882	0.876
Resident	74	4.07	0.91
Tourist	87	3.72	0.82
Coffee shops	99	3.737	0.954
Resident	76	3.93	0.75
Tourist	23	3.09	1.24
Gender			
Female	273	3.98	0.81
Male	193	3.90	0.92
Total	466	3.944	0.856

[a]$\bar{X}$, mean; SD, standard deviation.

establishment visited by the customers. Moreover, a deeper analysis considering type of customer (resident or tourist) and gender as segmentation criteria was conducted. That is, it may be the case that differences in the level of customer satisfaction arise between residents and tourists as well as between males and females.

A more complete picture of the survey data can be obtained from the boxplot of the responses gathered in each type of establishment shown in Fig. 21.1. From the latter, it seems that customers of hotels tend to concentrate their responses on the higher levels of satisfaction (i.e. those scored with 4 or 5 points), whereas customers from coffee shops and restaurants appeared to exhibit a somewhat lower degree of satisfaction (i.e. they tended to select options scored with 3 or 4 points more frequently).

Discussion Questions

1. Regarding segmentation and according to the data shown, do you consider the segmentation process is advisable as a business strategy for tourist establishments?
2. From a statistical point of view, do you consider it would be appropriate to develop any segmentation processes for this case? Justify your answer in detail.

Teaching Notes

Summary

This case study demonstrates the use of a customer satisfaction survey to identify potential differences that may serve as segmentation criteria. The case intends to familiarize students with the concepts of segmentation and statistical methods that may be used in designing more effective segmentation strategies.

Target audience

The target audience is undergraduate and postgraduate students, who are required to possess a basic knowledge of descriptive statistical analysis, its interpretation, and the standard test of equality of means between samples. The case study is suitable for individual or group work.

Teaching methods and equipment

The case study uses hands-on practice and brainstorming as its teaching methods. There is no need for equipment, apart from

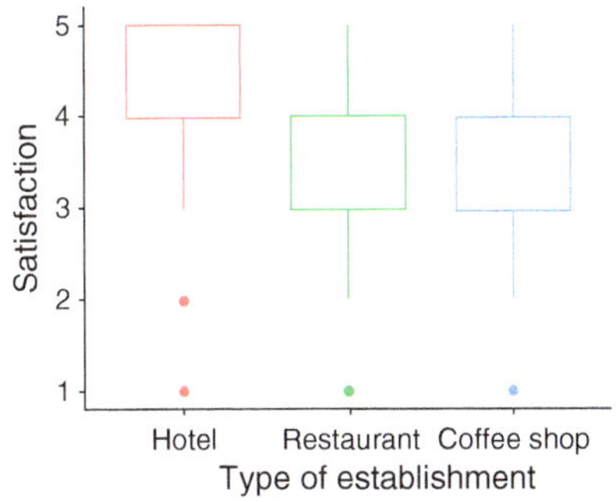

Fig. 21.1. Customer satisfaction of visits to tourist establishments (hotels, restaurants and coffee shops) based on criteria imposed to minimize COVID-19 infections. Satisfaction was measured using a five-point Likert scale whereby a score of 1 indicated the customer was 'not at all satisfied' while a score of 5 showed the customer was 'very satisfied'. Authors' own elaboration.

the document with the information about the case study that students may read to then answer the discussion questions.

Teaching instructions

The students should read the text of the case study before the class session. Then they can elaborate reasoned answers to the discussion questions included at the end. It would also be interesting to encourage discussion during the class, and for students to hold a brainstorming session. Students can work both individually or in groups of two to three people.

At the end of the exercise, teachers will help to summarize and form a final block of learning results and conclusions emerging from the case study.

During the development of the case study teachers should follow steps 1–3. The teaching instruction plan that follows steps 1–3 shows how the case may be taught in class.

Step 1

Teachers can help students identify the usefulness of surveys to retrieve market information, especially in the tourist sector, where the intangibility of the service provided makes it more difficult for customers to perceive value. In order for firms to find out what their customers' needs are, it is usually necessary to develop market research.

Teachers can refer to the survey developed in the case study as an example: it includes some customer characteristics (e.g. whether customers were residents or tourists, their gender) and indicates the sort of questions that customers may be asked. A key issue in satisfaction surveys is related to the way in which people are asked about their level of satisfaction. Teachers can explain the convenience of using a five-point Likert scale in order to score satisfaction as demonstrated in this case.

Step 2

Once students understand the importance of using surveys for market research purposes, teachers can focus on analysing the retrieved data, that is the average customer satisfaction according to the type of establishment visited and origin of customers (included in Table 21.2).

Interpretation of the data can lead one to think that some tourist establishments such as hotels can generate greater customer satisfaction (4.09) than others, for example coffee shops (3.74) or restaurants (3.88). Similarly, the boxplot with the distribution of the responses shown in Fig. 21.1 also suggests the existence of differences in the responses given by users of each type of establishment and, specifically, between visitors lodged in hotels and users of restaurants and coffee shops.

Step 3

At this point, the teacher may stress the importance of rigorously testing differences in the responses from different groups of consumers (i.e. to test if such differences are statistically significant) and introduce the basics of a difference of means statistical test which can be applied straightforwardly in this case using the available information. The teacher may also highlight the distinction between statistical and practical significance and introduce an effect size test (for further explanation see 'Analysis' section, '2. Developing segmentation from a statistical point of view').

Teaching instruction plan

Discussion point	Time required (min)	Teaching strategies
Introduction	10	The instructor briefly explains the general context created by the COVID-19 pandemic and how it affected the hospitality sector. The instructor randomly asks a few students to narrate their own perception of the reaction of the hospitality firms to the pandemic. The instructor writes the list of measures highlighted by the students along with the effect and usefulness they have had in their view. If responses by the students are heterogeneous, the instructor may take these as an example to remark on the fact that the same measures may have different effects on different groups of customers. In case all the students share the same perception (i.e. they respond in the same way), the instructor may remark that this is not always the result in reality. Then, the instructor introduces the concept of market segmentation as a tool for dealing with a situation where not all consumers behave in the same way.
Summary of the case	10	The instructor briefly discusses the background information with special emphasis on the availability of survey data. The instructor briefly comments on the structure of the data and explains the responses obtained by each group of customers by type of establishment (hotel, restaurant, coffee shop). The instructor also remarks on the potential usefulness of this type of data for obtaining a better understanding of consumers and implementing segmentation strategies. After summarizing the background information, the instructor may create multiple teams of two to three people and raises the discussion questions 1 and 2.
Discussion question 1	15	The instructor allows the teams to discuss the convenience of adopting a segmentation strategy by firms in the hospitality sector. The instructor may encourage students to base their discussion on the survey data. The instructor asks the teams to provide their conclusion of the analysis on a sheet of paper; then each team should present their conclusions to the class and the instructor asks the teams to debate their conclusions with their rationale. Some general debate is allowed.
Discussion question 2	30	The instructor introduces the basis of the statistical test for difference of means and for effect size. The instructor also discusses the steps to compute these statistics in a standard spreadsheet (MS-Excel). The instructor now asks the teams to revise their previous conclusions in the light of rigorous statistical tools. Each team is asked to write their revised conclusions on a sheet of paper and present them to the class. Some general debate is allowed.
Closing remarks and discussion	5	The instructor concludes the case analysis session with his/her analysis using statistical tools. The instructor emphasizes the usefulness of survey data and statistical rigorous analysis to achieve management objectives.

Duration

70 mins

Analysis

1. Segmentation process as a business strategy

As some authors propose (e.g. Ahani *et al.*, 2019), market segmentation or customer segmentation is a process that provides a solution for firms as it can help managers in observing several markets and associated customer segments. It can be considered a strategy that includes the separation or classification of a marketplace into segments of customers with dissimilar needs, requirements, features or behaviours which might require distinct marketing strategies.

Different levels of satisfaction for different types of establishment point towards the need for each establishment to accomplish different strategies in order to improve their customers' satisfaction. In this case, the analysed data identifies different levels of satisfaction in customers who visited different tourist establishments in a specific destination. That is, specifically, customers who visited hotels were more satisfied than customers who visited restaurants or coffee shops.

Although several reasons can influence their level of satisfaction, the results suggest that destination managers should be more aware of restaurants and coffee shops than of hotels. Therefore, from the perspective of destination management, efforts should be focused on these establishments that exhibit lower levels of customer satisfaction in order to encourage a thorough revision of the products and services offered to customers. In particular in this case study, the focus is on customer satisfaction in relation to the application of procedures adopted to minimize the spread of COVID-19 infection, in order to make customers feel safe and comfortable while not having their enjoyment compromised.

Apart from that, deeper research can be conducted to understand how to increase the level of satisfaction in customers of coffee shops and restaurants. One reason can be their needs are not properly fulfilled and managers might identify why this might be the case to try to solve this.

Segmentation as a business strategy is a useful tool as it enables firms to understand the character and needs of the different groups of customers and perhaps target their marketing activities towards specific customer types. For example, in this case study, such a target group may be the recently retired, who may be comfortably off and have more leisure time than when they were working. A marketing strategy to encourage them to visit and spend their money in coffee shops and restaurants may increase customer numbers (as long as the current issues about customer satisfaction discussed above have been addressed first).

2. Developing segmentation from a statistical point of view

Although a casual inspection of the data may suggest there is a difference in the responses of customers depending on the type of establishment visited, it is important to confirm this preliminary conclusion through an appropriate statistical test. In this case, both the estimated means and the boxplot are indicative of a difference in the satisfaction of customers of hotels compared with those of restaurants and coffee shops. Moreover, there seems to be no relevant difference between these two last types of establishment. Further investigation on this result may be conducted by applying a standard test for difference of the means.

The departure point in this case is that even though the mean satisfaction for hotel customers is clearly above the mean estimated for restaurant and coffee shop customers, since there is variability in the measurements (the standard deviation is not a null value), differences of means can be explained by chance due to sampling. This means that we need to take that into account when comparing the sets of samples and such a task may be accomplished by a 'hypothesis testing' procedure. In short, the aim is using statistics to determine the probability that a given hypothesis is sufficiently supported by the data.

To test for the difference of means we can use the *t*-test. When we assume equal variances between different population groups, we calculate the following quantity:

$$t = \frac{\bar{X}_1 - \bar{X}_2}{SD_{12}\sqrt{\frac{1}{n_1} + \frac{1}{n_2}}}, \qquad \text{(Eqn 21.1)}$$

where SD_{12} is the pooled standard deviation:

$$SD_{12} = \sqrt{\frac{(n_1 - 1)SD_1^2 + (n_2 - 1)SD_2^2}{n_1 + n_2 - 2}}. \qquad \text{(Eqn 21.2)}$$

The *t*-statistic (t) is distributed as a student's t with $n_1 + n_2 - 2$ degrees of freedom,[1] so that by comparing the quantity t with the appropriate percentile of the t distribution we can conclude whether the null hypothesis should be rejected or not at a given degree of significance.

We now apply the *t*-test for difference of the means to compare the overall level of satisfaction for hotels and coffee shops. The *t*-statistic is:

$$t_{(\text{Hotel}\,|\,\text{Coffee shop})} = \frac{4.092 - 3.737}{0.830\sqrt{\frac{1}{99} + \frac{1}{206}}} = 3.498$$

which is distributed as a student's t with 303 degrees of freedom. Since the p-value of < 0.001 is not less than the usual level of 0.05, we reject the null hypothesis.[2] Consequently, there is sufficient evidence to say that the level of satisfaction with anti-COVID-19 measures is not the same for hotels and coffee shops. Similarly, the *t*-test for difference of means for customers of hotels and restaurants suggests that their satisfaction differs significantly (the *t*-statistic is 2.454, with a p-value < 0.05).

Although we have just concluded that the observed differences in the level of satisfaction of customers of different hospitality establishments are statistically significant, it is advisable to make the distinction between statistical significance and practical significance. In the present case, the follow-up question *is the difference between means really important?* must be addressed separately. To find an answer we can use an effect size test (an effect size is a value measuring the strength of the relationship between two variables in a population, or a sample-based estimate of that quantity), such as the Cohen's d to put the differences in context. The d-statistic is:

$$d = \frac{\bar{X}_1 - \bar{X}_2}{SD_{12}}, \qquad \text{(Eqn 21.3)}$$

A high d indicates that there is a big effect relative to the internal variability of the data. A commonly used interpretation is to refer to effect sizes as small ($d = 0.2$), medium ($d = 0.5$) and large ($d = 0.8$).

In the comparison between users of hotels and those of coffee shops we obtain:

$$d_{(\text{Hotel}\,|\,\text{Coffee shop})} = \frac{4.092 - 3.737}{0.830} = 0.428$$

which is considered a moderate effect. The effect size test when comparing customers visiting hotels and restaurants is 0.258, so that in this case the effect is small and of a rather reduced practical relevance.

In light of these findings, managers should decide whether to concentrate efforts and resources in developing strategies that focus on restaurants and coffee shops or, alternatively, only on coffee shops. For example, for coffee shops the available data from the survey seem to support a segmentation strategy given the statistically significant difference between residents and tourists (the *t*-statistic is 3.975 with a p-value < 0.001) and the size of the effect is large (the d-statistic is 0.946). The values of the *t*-statistic for difference of means and d-statistic for effect size are shown in Table 21.3.

Regarding customer's gender, the *t*-statistic clearly indicates that the level of satisfaction attained by males and females is indistinguishable from a statistical point of view. More specifically, the p-value obtained is 0.3207, well above the usual levels considered in the literature, so that we cannot reject the null hypothesis of the test (i.e. the means are the same). Thus, a segmentation

Table 21.3. Statistical tests for difference of means and size effects for residents vs tourists and women vs men. Authors' own elaboration.

	Resident/tourist			
	Coffee shops	Restaurants	Hotels	Female/male
$\bar{X}_1 - \bar{X}_2$	0.840	0.350	0.530	0.080
SD_{12}	0.888	0.862	0.756	0.856
t-statistic	3.975	2.568	2.163	0.994
p-value	0.0001	0.0112	0.0317	0.3207
d-statistic	0.946	0.406	0.701	0.093

strategy based on gender would be of little interest in this market.

For completion, it must be emphasized that in this case both the availability of survey data and the statistical analysis has proven to be a powerful tool that helps develop more effective business strategies, especially when scarce resources have to be applied to achieve alternative objectives. Moreover, collecting additional data from customers may also be of great importance to gain a deeper insight of the specific aspects that generate a lower level of satisfaction with the service.

Notes

[1] When we cannot assume equal variances, we should use a slightly modified version of the *t*-test known as Welch's *t*-test.

[2] Finding the *p*-value for the *t*-test is straightforward in virtually any statistical package, as well as in the popular spreadsheet Microsoft Excel (the formula '=DISTR.T.2C(ABS(3.498);303)' will do the task for a two-tail test).

Reference

Ahani, A., Nilashi, M., Ibrahim, O., Sanzogni, L. and Weaven, S. (2019) Market segmentation and travel choice prediction in spa hotels through TripAdvisor's online reviews. *International Journal of Hospitality Management* 80, 52–77. https://doi.org/10.1016/j.ijhm.2019.01.003

22 Present Your Choice of Management Information System (MIS) Mix

Miguel Mayol-Tur* **and Andres Artal-Tur**
Universidad Politécnica de Cartagena, Spain

Abstract
To find the management information system (MIS) software that best suits a business is a hard task, given the growing complexity of suppliers and offers existing worldwide. For this task we usually rely on reviews, benchmarks, market share and advisors. This case follows an example of how to choose the MIS that best suits a hotel business. In doing so, we will use a spreadsheet with prices and hotel features after searching the data on the Internet. Such an activity will improve students' knowledge of MIS features, while they learn what property management systems (PMS) and other MIS can do for designing and managing some business operations.

Subjects: case studies in general management, management information system (MIS), MIS software, databases, software research, software evaluation, applications, strategic management

Learning Objectives

At the end of the case study, a learner will be able to:

1. Acquire knowledge of management information system (MIS) features.
2. Identify MIS categories.
3. Analyse, evaluate and compare MIS features, public benchmarks and prices using information gained on the Web.
4. Develop a presentation using free and open source software (FOSS) fonts, spreadsheets and PDF showing it with Okular (a multiplatform software that allows you to view and sign portable documents files and with some other features) in presentation mode.
5. Add knowledge of each MIS software and its features to their curriculum vitae (CV).

Case

Management information systems (MIS) are a set of software tools used for decision making, and for the coordination, control, analysis and visualization of information in an organization. Most of the software categories found under the MIS umbrella are presented in Table 22.1.

*mitcoes@gmail.com

 Teaching Cases in Tourism, Hospitality and Events (eds. S.K. Dixit and S. Piramanayagam)
DOI: 10.1079/9781800621022.0022

Table 22.1. Software categories that come under the umbrella of management information systems (MIS).

Category	Description
DSS	Decision support system
EIS	Executive information system
Marketing IS	Marketing information systems
Accounting IS	Accounting information systems
Human resource MS	Human resource management systems
OAS	Office automation systems
ERP	Enterprise resource planning
PMS/HMS	Property management systems/hotel management software
CMS (also called ECM or WCM)	Content management systems (also called enterprise content management or web content management)
Databases	Organized collection of data stored that can be accessed electronically
BI	Business intelligence (combines business analytics, data mining, data visualization, data tools and infrastructure so business can make more data-driven decisions)
Hosting services (may be on site or managed by the business using others' cloud infrastructure)	Software management systems can be hosted on public clouds or a business can use a hosting platform, for example with Nextcloud, OMV Open Media Vault, TrueNAS (NAS Network attached storage) and Podman, Docker and Kubernetes (containers)[a]

[a]Cloud containers are packages of software that contain all of the necessary elements to run in any environment (from a private data centre to the public cloud).

Scenario

Hotel company ABC have decided that it was not happy with the MIS mix they are currently using and they want to know about: (i) the alternatives available; (ii) the cost in monetary terms; (iii) computer software that may need to be replaced; and (iv) ease of use including courses needed to familiarize staff with the new system. They ask you to prepare a presentation to make to the company's Board of Directors in 1 week's time. You must explain the reasons for your choices over at least three other alternatives so with the information you provide they can decide about the MIS mix they would like to use.

The suggested MIS mix must be cheaper (taking human resources inside cost considerations), better and sustainable in the long term (bearing in mind that information and communications technology (ICT) changes fast). Their goal is to continuously improve their processes and that means their productivity, and their staff and customer satisfaction.

The task

Search the Internet for property management systems (PMS), hotel management software (HMS) and other MIS for the hotel company (learners can use a hotel they are familiar with to represent the hotel company), and make a presentation explaining your recommendations using spreadsheets that will be embedded in PDF – editable – for your MIS mix. You must make a presentation defending an MIS mix for the hotel company with special dedication to HMS/PMS as it is essential. The aim is to achieve the best performance with the lowest budget, without lacking features, using FOSS software if possible. Explain the alternatives, and give reasons for the choices made for your mix, showing features, prices, development stats, if it is open source, and other considerations. In particular, explain why you choose FOSS alternatives or not for each category. You must indicate where to host the software: public clouds, or using the hotel company's own cloud with Nextcloud and OMV Open Media Vault or TrueNAS

(Network Attached Storage) and NFS (Network File System) and Podman containers, or other long-term cost-effective solutions. (*Note:* There is not a FOSS PMS/HMS for hotel chains with big market share (PMS/HMS is the category of MIS needed for hotels), but there are PMS/HMS for GNU/Linux servers that improve performance and save costs.)

The choices you make are less important than the way you explain *why* you think that a particular MIS mix is better for the hotel company. You must explain *why* you think some software is more suitable while other options are not. You must evaluate each software category, if the hotel brand needs to use it or not (e.g. it may not be needed but it may improve the hotel company's performance) and indicate whether it is sustainable (i.e. are the features just useful today or both today and in the near future).

Assumptions

The most sustainable software purchase strategy is, when possible, FOSS (read Appendix 1 in 'Teaching Notes').

Instructions

1. Read the following article, not necessarily as a template, about how something similar was done for BI Business Intelligence on a web-book (take into consideration they are advertising Holistics BI software): https://www.holistics.io/books/setup-analytics/navigating-the-business-intelligence-tool-space/ (accessed 5 May 2022). If you are not a fast reader just read the tables today.

2. Download Montserrat GNU font (available at: https://www.1001freefonts.com/d/5711/montserrat.zip (accessed 5 May 2022)), and other free fonts if you like to use more than one font. The following are good FOSS fonts: https://www.fontsquirrel.com/fonts/ and https://fonts.google.com/ (both accessed 5 May 2022). *You cannot use proprietary fonts* in public or, if in business, outside your business *without paying extra* (read Appendix 2 in the 'Teaching Notes' with regard to Monotype, Microsoft or Adobe fonts End User License Agreement (EULA)).

3. Download and install LibreOffice (Collabora Office for iOS (a mobile operating system created and developed by Apple Inc.) or Android or Chrome OS (operating system) – with its android compatibility layer service).

4. Download and install Okular (FOSS PDF software; this was originally awarded Blue Angel, the world's earliest established environmental label introduced in 1978, now certified as ecological software, obtained in February 2022).

5. Make a spreadsheet in LibreOffice Calc, save it as 'MISmixyourname ...' with all the MIS, beginning with the PMS/HMS software you found interesting, their features and prices.

- Classify all the MIS that you find worth presenting into four or less groups. Inside each group you must put a maximum of four applications. Select the best applications for each of the groups that matches your criteria.
- Add sheets for each group with four or fewer items, the best ones following your criteria, as more than 16 similar products can not be easily processed in one presentation by non-experts.
- Use different font colours to make the spreadsheets easy to understand at first sight. Blues, greens, reds and similar close colours work well on screens and projectors.
- Use A3 landscape – Format/Page Style/Page.
- When finished: Edit/export as a PDF/Hybrid PDF (embed the ODF (open document format) file) that would make your PDF editable with LibreOffice.
- Adjust spreadsheets to the page size with the Edit/Print preview/Margins slidebar.
- Read it with Okular, test/view/presentation.
- Add cases and customers and note if they have a trial or community free edition from the visited MIS web and if they share the information.

- Classify each software option according to the features and price criteria that you have identified match the hotel company's needs and budget.
- Create a way of keeping youreself informed about new MIS features, prices and software.
- Select one PMS/HMS as your favourite and explain why you prefer it over the other possible choices.
- Add one or several pages to your spreadsheet about the MIS mix you would choose to add to your PMS/HMS of choice and explain what your decision mix would be and *why*.

Exercise 1 – Alfresco versus Twiki dilemma

Begin by making a comparison between two enterprise content management (ECM) options, Alfresco (available at: https://www.alfresco.com/customers (accessed 5 May 2022)) and Twiki (available at: https://twiki.org/cgi-bin/view/Main/TWikiSuccessStories (accessed 23 September 2022)). You should make the comparison and present your preferred choice based on customer experiences with special attention to hotel cases (you can use Disney as hotel case).

You should not only use several customers' experiences, but also include the prices and features that customers report. Both Afresco and Twiki can fit any business needs, but you must explain why you would choose one of them and not the other one and explain it well enough that the Board of Directors can reject your choice because they have a different view about their business as they have more information about it.

Exercise 2 – PMS/HMS, property management systems/hotel management software

Design a benchmark spreadsheet for PMS/HMS based on the links provided (see PMS/HMS in Appendix 3 in the 'Teaching Notes'). Using these links, search for and collect information about each type of PMS/HMS software, focusing on: (i) prices; (ii) features offered; (iii) web-perceived customer satisfaction; and (iv) which important customers are using each brand. Divide the tasks between different members of the team. Fill in one spreadsheet for each benchmarked software.

Exercise 3 – choosing the MIS mix that you will recommend

Divide the 12 MIS software categories shown in Table 22.1 as follows: (i) rows 1–3 (i.e. decision support system (DSS), executive information system (EIS) and marketing information systems (IS)); (ii) rows 4–6; (iii) rows 7–9; and (iv) rows 10–12. Now divide your team into four groups and assign each group three MIS software categories, so for example learner group one will be assigned MIS software categories in rows 1–3, learner group two rows 4–6, and so on. Each group should design the benchmark spreadsheet for each MIS category they have been allocated. As in Exercise 2, using the links provided (see Appendix 3 in the 'Teaching Notes'), search for and collect information about each type of software category, focusing on: (i) prices; (ii) features offered; (iii) web-perceived customer satisfaction; and (iv) which important customers are using each brand. Each group of learners should adjust the extent of this exercise to the time you have available to accomplish the task, and fill in the spreadsheet just the basics of what is needed to remember. Finally, each group should decide which MIS mix you would choose with reasons for your choice and share it with your class.

Theories on which this case is based

This case is based on the following theories:

- knowledge management systems as a knowledge hierarchy in technology that uses a data, information, knowledge, wisdom (DIKW) pyramid or a vector where the route is DIKW (Bellinger *et al.*, 2011);

- Teece *et al.* (1997)'s framework about dynamic capabilities and strategic management;
- a firm's open innovation performance (OIP) (Chesbrough, 2003);
- innovation translation as originally defined by Schumpeter (1934); and
- other theories ranging from innovation diffusion (Sharif and Ramanathan, 1981; Jaakkola, 1996) to compilation (Geroski, 2000 and Guardiola *et al.*, 2002).

Discussion Questions

1. How can periodic MIS benchmarks help hotels to improve their productivity and staff satisfaction?
2. How can this study help hotel managers to attract more tourists and raise their satisfaction rates?
3. Do you think hotel chains need ECM software? Do independent hotels (i.e. those not belonging to a chain) also need ECM software?
4. Give your opinion on how a better MIS mix, adjusted in price and performance to the needs of the hotel or hotel chain, would help make hotel business improvements after the pandemic.
5. Could you identify any of these potential improvements that could result from switching to FOSS when possible, even if it is only for servers, containers, Nextcloud, independent FOSS-powered clouds and other backbone solutions?

Teaching Notes

Summary

This case will introduce learners to MIS and how to find their required MIS, by following an example of how to choose the MIS that best suits their hotel business.

Presenting the case

The outline described here is our suggestion, but as teachers know the characteristics of their learners best a teacher can change it to suit their particular learners. Split the class into two groups, so that there are two final presentations, and each group should choose a speaker who will present their case. To make the choice of speakers use any game or random method. Each group needs to prepare their presentation making a list of reasons that explain why a particular MIS mix should be chosen by the hotel chain.

Target audience

The target audience is computer and Internet users studying MIS for hotels.

Teaching methods and equipment

The equipment needed includes a reliable Internet connection and computers with screens (even pocket computers such as mobile phones or tablets can be used). The teacher can decide whether or not to use the hints and links provided in Appendix 3.

Shrinking the case

You can shrink the case to the bare minimum by just asking learners to defend the Alfresco versus Twiki dilemma (i.e. just using Exercise 1).

Teaching instructions

The following teaching instruction plan is suggested. The discussion questions are not included in the plan and for this case study are designed for an additional follow-up session based on a class discussion to test what the students have learnt through the practical sessions of the case study.

Teaching instruction plan

Discussion point	Time required in min (cumulative time in parentheses)	Teaching strategies
Introduction	7 (7)	The instructor begins the case analysis by asking questions about the learners' recent visit to MIS web pages. The instructor randomly ask a few students to volunteer themselves to narrate the good and bad elements that impacted their benchmarking experience about Alfresco and Twiki ECM pages.
Briefing on the concept of MIS in relation to tourism	6 (13)	The instructor introduces the concept of MIS in relation to tourism and explains the software categories found in MIS (as per Table 22.1). The instructor asks randomly selected students to provide a software example for each MIS category of the 12 in Table 22.1 and asks students which they actually know.
Summary of the case	7 (20)	The instructor briefly discusses the background information, the scenario, the task and how to make the presentation to the Board of Directors. The instructor explains the adaption of the spreadsheet to use as a presentation tool for benchmarking. After summarizing the case, the instructor creates two teams (or groups) to make the presentation tables.
Exercise 1	15 to read and debate 6 to explain (41)	**Alfresco versus Twiki dilemma** The instructor now facilitates the teams to compare Alfresco and Twiki and then debate their preferred choice with their rationale. Each team must defend their choice and write part of the final presentation using a spreadsheet.
Exercise 2	15 to read and debate 10 to explain (66)	**PMS/HMS, property management systems/hotel management software** The instructor now facilitates the teams to work through Exercise 2. The teams must split the tasks, collecting information from each type of software, focusing on prices, features and web-perceived customer satisfaction and which important customers are using each brand. The teams must fill in a spreadsheet for each benchmarked software so they can present their reasons for selecting a particular software package to recommend to the Board of Directors.
Exercise 3	28 to read and debate 12 (3 per team) to explain (106)	**Choosing the MIS mix that you will recommend** The instructor divides the 12 MIS software categories in Table 22.1 as follows: (i) rows 1–3; (ii) rows 4–6; (iii) rows 7–9; and (iv) rows 10–12. Each team of learners is divided into four groups and each group is assigned three MIS software categories (so e.g. learner group one will be assigned rows 1–3). Each group should design the benchmark spreadsheet for each MIS category they have been allocated. Using the provided links they should search for and collect information about each type of software category, and complete the MIS benchmark spreadsheet for each category for prices, features offered and customer satisfaction. Finally, each group must decide their choice of MIS mix and present this to their classmates and instructor, giving reasons for their choice.
Closing remarks and discussion	14 (120)	The instructor concludes the case analysis session sharing all the MIS mix with Okular in presentation mode and explaining how benchmark tables can be improved and updated over time to improve MIS mix choices.

Duration

120 min

Appendix 1

FOSS is used a good deal, and it is more reliable and secure to use FOSS than any other software that can be cancelled for any reason at any time. The reason is that FOSS exists beyond their developers or companies developing it, because it is free and open. When software is not free, even if it is open, the software can 'die' when the company wants it to (e.g. Google closed G+, G reader, Orkut, Picasa, and more, Microsoft, Activesync, BackOffice Server, Explorer, and more) or the software disappears should the company become bankrupt.

Examples of FOSS software include LibreOffice and GNU/Linux. LibreOffice is a GPL (General Public License) fork of Apache Software Foundation's Open Office with MIT licence (a permissive free software licence originating from the Massachusetts Institute of Technology), that was open sourced by Sun Microsystems, as Sun Open Office after being Sun Star Office with a private licence. (Forking is to take the source code from an open source software program and develop an entirely new program.) It was also forked by IBM Lotus Symphony, and inspired others such as Google Documents. Sun was purchased by Oracle, but as they open sourced LibreOffice (and Java and mySQL) other companies and foundations continued some of their open source software business. MariaDB or Android are based on this FOSS created by Sun.

FOSS does not mind free of charge, for instance Alfresco prices are usually not affordable for small and medium enterprises (SME), (some of their prices are 150€/y/user in Amazon Web Services (AWS) or 48.000€ on premise for 100 users). If you like the software it is better to use the Community Edition. However, there is a big cost in learning how to use, install and configure it if you are not a computer expert, or you will have to hire an expert.

GNU/Linux Operating Systems – lignux or just linux - servers for self-host your MIS, can be more expensive in manpower to install configure and maintain than hire more expensive cloud developments, but cheaper in human resources if you are a SME, size matter when you must choose your MIS mix. Or IBM's Red Hat Enterprise Linux RHEL fees instead of Debian linux – free of fees – can be better for your organization because the RHEL fees are cheaper than learning or hiring experts in your zone.

Appendix 2

The font Times New Roman is used as a default font a lot without paying Monotype its rights. Monotype is owner of ***Times New Roman***, and the End User License Agreement (EULA) states:

> 2.- **License Grant.** You are hereby granted a non-exclusive, non-assignable, non-transferable (except as expressly permitted herein) license to access the Font Software (i) only in a Licensed Unit, (ii) ***only for your Personal or Internal Business Use***, and (iii) only subject to all of the terms and conditions of the Agreement.
>
> (Monotype Imaging Inc., 2022)

Every company should select a group of free fonts for its use in public, or it should pay the fees for some proprietary ones. They should always read the font EULA – as they should with every contract. This is because managers must know the laws that apply to them, and make sure their companies follow them, even if they are not enforced currently.

Appendix 3

Optional hints before you start include:

- Keep in mind that even the more expensive software, such as GNU/Linux servers, and the Open Travel Alliance intercommunication gateway, need FOSS to work.
- Watch this short video: Open Travel Alliance available at: https://www.youtube.com/watch?v=L-S4_mI66Mc (accessed 19 September 2022).

- Now begin your search. You can search the partners in this web page to collect features advertised and prices: Open Payment Alliance available at: https://bavelnetwork.voxelgroup.net/en/travel/partners.html (accessed 19 September 2022).

HMS software brands include Alexandros, AmonRa, Arab Soft, Protel, Host, Fidelio Suit 8, Noray, Sihot, Silverbyte, Prestige, Tesipro, Indra, Navision, Eurowin, Winhotel and Oracle Opera. The list is so long because it is the most important MIS for any hotel, and it is mandatory to use one of them these days.

Try to search in Bing, Google, Duckduckgo and other search engines. However, should you fail to find some good links here are some that can improve your search:

The main **MIS software** in hotels are PMS/HMS (property management systems hotel management software) as hospital ones are HIS (hospital information software). However, sometimes you need more than one software package, such as one PMS-HMS and a content management system (CMS) such as Alfresco or Twiki (Alfresco and Twiki are both CMS and ECM) to manage information, or custom databases, to add to the MIS mix you will choose. MIS are used to create reports, extract data and assist in the decision-making processes of managers.

Enterprise resource planning (ERP) manages the information between all business functions. Links are available at:

- https://www.thirdstage-consulting.com/top-10-erp-systems-2022-top-software-vendors/ (accessed 19 September 2022); and
- https://dynamics.folio3.com/blog/open-source-enterprise-resource-planning-software/ (accessed 19 September 2022).

These links are just for your information, as hotels use PMS-HMS or PMS-HMS modules and not generic ERP. Use these links only for your own general knowledge or perhaps to find a module that is not available when we wrote this case.

PMS/HMS are available at the following sites:

- https://www.techjockey.com/blog/free-open-source-hotel-management-software (accessed 19 September 2022);
- https://blog.capterra.com/top-free-and-open-source-hotel-management-software-solutions-compared/ (accessed 19 September 2022); and
- https://digital.com/best-hotel-management-software/ (accessed 19 September 2022).

At KW*H*otel Free (available at: https://kwhotel.com/ (accessed 19 September 2022)) there is a free version of HMS for trial. You can use this if you want to learn about the common features available on them. Alexandros, AmonRa, Arab Soft, Protel, Host, Fidelio Suit 8, Noray, Sihot, Silverbyte, Prestige, Tesipro, Indra, Navision, Eurowin, Winhotel, Oracle Opera Cloudbeds are great contenders in the PMS/HMS space too.

Marketing information systems (marketing IS) focus on marketing (available at: https://financesonline.com/top-20-marketing-analytics-software/ (accessed 19 September 2022).

Accounting information systems (accounting IS) focus on accounting analysis and are available at:

- https://www.selecthub.com/enterprise-resource-planning/accounting/open-source-accounting-software/ (accessed 19 September 2022);
- https://www.investopedia.com/best-accounting-software-for-small-business-5069679 (accessed 19 September 2022); and
- https://www.businessnewsdaily.com/7543-best-accounting-software-v2.html (accessed 19 September 2022).

Human resource management systems (human resource MS) are used for personnel aspects and are available at:

- https://www.predictiveanalyticstoday.com/top-free-open-source-human-resource-hr-software/ (accessed 19 September 2022); and

- https://www.softwaretestinghelp.com/best-hr-software/ (accessed 19 September 2022).

Office automation systems (OAS) are used for communication and productivity by automating workflow and eliminating bottlenecks (available at: https://www.jotform.com/blog/office-automation-software/ (accessed 19 September 2022)).

The **executive information system** (EIS) provides summarized reports coming from departments such as accounting, human resources and operations (available at: https://www.g2.com/software/executive-information-system (accessed 19 September 2022)).

Decision support systems (DSS) compile information from many sources to support problem solving and decision making. A DSS is used mostly for non-structured decision problems. Links to DSS are available at:

- https://www.capterra.com/decision-support-software/s/free/ (accessed 19 September 2022); and
- https://www.itbusinessedge.com/business-intelligence/decision-making-tools/ (accessed 19 September 2022).

Content management systems (CMS) also known as enterprise content management (ECM) and web content management (WCM) are available at:

- https://www.softwaresuggest.com/alfresco-dms/alternatives (accessed 19 September 2022); and
- https://www.wpbeginner.com/showcase/best-cms-platforms-compared/ (accessed 19 September 2022).

Examples include Alfresco, Twiki, Opentext, Nuxeo, Magnolia, KnowledgeTree, OpenKM, Athento, Documentum and SharePoint.

Databases include PostgreSQL, MariaDB, CockroachDB. Clickhouse, Neo4j, MongoDB, RethinkDB, Redis, SQLite, Cassandra, Timescale, CouchDB, Apache Ignite and Oracle. The following are links to databases:

- https://geekflare.com/open-source-database/ (accessed 19 September 2022); and
- https://appinventiv.com/blog/top-web-app-database-list/ (accessed 19 September 2022).

Tips

- You must search for links and similar web pages as these can help you in your presentation.
- All MIS are databases, and/or use databases, with the fields and functions needed for their specific tasks.
- If you know the database software used by an MIS application, you can add their benchmarks to the information mix.
- You will have to choose (or advise) which MIS mix to use, or if may be the case that there are several compatibles to avoid dependence on one brand and be up to date regarding all the features available.
- Finally, the best MIS books – in their opinion – are available at: https://www.findthisbest.com/best-management-information-systems-books (accessed 19 September 2022).

Analysis of discussion questions

In what follows we add new material to the case study in order to test if students have learnt the content well through doing the exercises in the class sessions. In this case, this additional part can be thought of as a series of questions to settle the main outcomes and take-away messages of the case study. In this way, we provide a number of guidelines which the teacher could develop and use to conduct the in-class discussion with students.

1. How can periodic MIS benchmarks help hotels to improve their productivity and staff satisfaction?

New software developments, new features and new brands can inspire management to improve some processes implementing these novelties. Sometimes management asks their MIS providers for specific add ons that other brands will enjoy too.

2. How can this study help hotel managers to attract more tourists and raise their satisfaction rates?

If hotel managers *just know* all the features that the MIS software installed on their system can do, even if these features are not used on a daily basis, the hotel managers may find that they can use some of these features periodically to improve their processes. Using the features that would help them is likely to improve the hotel business's performance, including increasing the number of tourist visits because of good comments and reviews and thus increasing returns.

3. Do you think hotel chains need ECM software? Do independent hotels (i.e. those not belonging to a chain) also need ECM software?

ECM saves time, if your know-how and paperwork is small, you can handle it with templates and sharing directories over a personal network. But when it grows in complexity and quantity, having it organized and shared in a NAS (network attached storage) with any ECM will save you time, even if it is just filing, and this will result in performance improvement, even if it is just staff satisfaction, because of the time saved. (NAS is basically private hard drives connected to the Internet thanks to specialized operating systems.)

4. Give your opinion on how a better MIS mix, adjusted in price and performance to the needs of the hotel or hotel chain, would help make hotel business improvements after the pandemic.

Of course having better tools for your needs will improve your performance, and how ever many staff a hotel may have, even saving on organizational tasks will enable hotel managers to improve other services with those savings, or just improve the return of investment (RoI).

5. Could you identify any of these potential improvements that could result from switching to FOSS when possible, even if it is only for servers, containers, Nextcloud, independent FOSS-powered clouds, and other backbone solutions?

Just by not using major brands that almost force you to use their expensive solutions that depend on each other, you save in cloud storage and servers. If you add your own servers as the main source and cloud containers as backups, savings will increase. Also if part of your MIS mix is open source, you are safe from future code problems or business transactions, and if it is free, it is the most cost-efficient software solution you can find.

References

Bellinger, G., Castro, D. and Mills, A. (2011) Data, Information, Knowledge, and Wisdom. Available at: https://www.semanticscholar.org/paper/Data%2C-Information%2C-Knowledge%2C-and-Wisdom-Bellinger-Castro/b553609347b9b8bc5698ccaef823b3acc1128dd7 (accessed 5 May 2022).

Chesbrough, H.W. (2003) *Open Innovation: the New Imperative for Creating and Profiting from Technology*. Harvard Business School Press, Brighton, Massachusetts. Available at : https://www.academia.edu/2485514/Open_innovation_The_new_imperative_for_creating_and_profiting_from_technology (accessed 5 May 2022).

Geroski, P.A. (2000) Models of technology diffusion. *Research Policy* 29(4–5), 603–625. https://doi.org/10.1016/s0048-7333(99)00092-x

Guardiola, X., Díaz-Guilera, A., Pérez, C.J., Arenas, A. and Llas, M. (2002) Modeling diffusion of innovations in a social network. *Physical Review E* 66(2). https://doi.org/10.1103/physreve.66.026121

Jaakkola, H. (1996) Comparison and analysis of diffusion models. In: Kautz, K. and Pries-Heje, J. (eds) *Diffusion and Adoption of Information Technology*. IFIP — The International Federation for Information Processing. Springer, Boston, Massachusetts, pp. 65–82. https://doi.org/10.1007/978-0-387-34982-4_6

Monotype Imaging Inc. (2022) Font Software End User License Agreement. Available at: https://www.monotype.com/legal/eula (accessed 19 September 2022).

Schumpeter, J.A. (1934) *The Theory of Economic Development: an Inquiry into Profits, Capital, Credit, Interest, and the Business Cycle*. Harvard University Press, Cambridge, Massachusetts.
Sharif, M.N. and Ramanathan, K. (1981) Binomial innovation diffusion models with dynamic potential adopter population. *Technological Forecasting and Social Change* 20, 63–87. https://doi.org/10.1016/0040-1625(81)90041-X
Teece, D.J., Pisano, G. and Shuen, A. (1997) Dynamic capabilities and strategic management. *Strategic Management Journal* 18(7), 509–533. http://www.jstor.org/stable/3088148

23 Small Events Snowballing into Large Effects – Using Economic Impact Analysis with a Winter Sports Non-profit Organization

Cass Morgan* **and Matt Gnagey**
Weber State University, Ogden, Utah, USA

Abstract

Leveraged as a means for economic development, event tourism is positioned as a powerful way to drive local, regional and national economies. For small-to-medium service providers disentangling the value of an industry or large-scale events (e.g. professional sports, festivals) to smaller-scale events can be difficult to determine without conducting a self-study. The purpose of this case study is to illustrate the economic impact that a small organization can have on a community using a non-profit youth ski club as an example. Equipped with economic impact data, event managers can demonstrate the 'micro-benefits' of small events to stakeholder groups.

Subjects: strategic management, data-driven decision making, positioning

Learning Objectives

At the end of the case study, the learner will be able to:

1. Comprehend the value of an economic impact study for event and tourism providers.
2. Calculate and analyse visitor spending data from a series of ski racing events to determine their economic impact on the community.
3. Discuss how to use economic impact data to position a service provider as an asset to local businesses and the broader community.

Case

In the USA, there are over 50 US Ski & Snowboard Association-certified clubs that offer developmental and competitive ski programming for youth (US Ski & Snowboard Association, 2019). These programmes vary across a number of factors including size, scope of programming, budget, staffing and the number of sporting events held. To cover operational costs, clubs rely on membership fees, fundraising and hosting competitive races, which can draw athletes from across the nation. For small clubs, hosting a competitive event is an important source of revenue, but

*cassmorgan@weber.edu

 Teaching Cases in Tourism, Hospitality and Events (eds. S.K. Dixit and S. Piramanayagam)
DOI: 10.1079/9781800621022.0023

can also incur substantial overhead costs. One method to evaluate the cost–benefit of hosting ski racing events is to conduct an economic impact study.

Always Winter Ski Club is a small non-profit youth ski club in the Western United States. For over 40 years, the ski club has offered competitive ski programming with a focus on alpine ski racing for youth ages 8–20. Alpine ski racing consists of several disciplines including Giant Slalom, Super G, Slalom and Downhill, and combined disciplines. Annually, the organization directly serves 75–100 youth who participate in competitive ski race events. Depending on age and skill, athletes compete in anywhere from 10 to 35 races in a season that are hosted in specific regions and across the nation. As a nationally certified club, the Always Winter Ski Club serves as a site to host ski racing events. Hundreds of athletes across the region, the USA and internationally regularly participate in these events.

Last winter season, the ski club found out they won a bid to host six different races that will last approximately 17 days over the course of the winter season. By the end of the winter season, the club expected to host nearly 100 teams with an estimated 1000 out-of-town athletes, plus spectators coming to the area. The six races that the club planned to host varied in the alpine racing events offered. Throughout the season, the club estimated it would host 94 teams of racers. Below is a brief description of each racing event.

- *University Invitational Race* (Race 1) – The first event of the season was a college race and consisted of 22 different teams from across the USA. University-aged athletes were eligible to compete.
- *Western Regional Race* (Race 2) – The second event was a Western Regional race that included Federation of International Skiing (FIS) and Giant Slalom races. This race was open to athletes in the region who are under 19 years old (U19) (ages 16–18). Athletes representing 18 teams participated.
- *Intermountain Racing Cup U14* (Race 3) – The third event was a Super G racing event open to teams and athletes within the division with athletes in the U14 category (ages 12–13). Approximately 14 teams participated.
- *Intermountain Racing Cup U16* (Race 4) – The fourth event was a Super G racing event open to teams and athletes in the division with athletes in the U16 category (ages 14–15). Approximately 14 teams participated.
- *South Series* (Race 5) – The fifth event was a Slalom and Giant Slalom racing event for athletes in the Intermountain division with eligible aged athletes in the U10, U12 and U14 categories. Approximately nine teams participated in the race.
- *Divisional Championship* (Race 6) – The final event of the season was a tri-divisional championship including Super G, Giant Slalom and Slalom alpine racing events. Seventeen teams across three divisions participated.

Prior to the ski season, the Board of Directors of the Always Winter Ski Club met with the executive director to discuss the potential economic benefit of hosting the six races. One board member shared a recent newspaper article from another ski area that described how ski racing events boosted the local economy. The board member suggested that it could be worthwhile to conduct an economic impact study to determine how hosting ski races benefit the broader community, like retailers, grocery stores, restaurants, fuel stations and ski resorts. The board member reasoned that if the club could show how companies within the local community benefit from the ski racing events by bringing visitors into local establishments, local companies might be more willing to sponsor or donate to the club. The board reached out to their contacts at the local university to assess the feasibility of conducting an economic impact study. A research team was assembled as a part of a sustainable tourism class.

The main tasks for the team were:

- to collect data on the number of visitors at each event and length of the visit;
- to collect data on visitor spending during the event, broken down into categories;

- to analyse the data by identifying local spending that would not have occurred without the events; and
- to apply the appropriate multipliers.

Data collection

The research team put together a nine-item survey that was adapted from a version of Crompton's questionnaire (Crompton, 2010, p. 36). Items included the following: (i) number of days and nights spent in the area; (ii) how many individuals were in the respondent's group; (iii) expenditure during the trip; (iv) type of visitor (athlete, spectator, coach, etc.); (v) the reason for the visit; and (vi) their home zip code. During each of the races throughout the winter season, the team was on site at the ski resort to determine the number of visitors, and interview individuals participating and/or attending the competitive events.

Step 1: Determine the visitor count

The accuracy of an economic impact study is reliant on having a good visitor count. When counting event visitors, the team made sure to distinguish local residents from out-of-town visitors, as resident spending reflects a recirculation of money in the community and not new money coming into the community. To ensure that local spectators were not included in the sample, respondents who identified a resident zip code within the host community were excluded from the final analysis. This would ensure that only non-resident visitor spending (i.e. money coming from outside the region) was included in the economic impact calculation. An estimate of visitor numbers was derived from the questionnaire asking for the total number of people in the immediate group (see Table 23.1). Race 5 had the highest total number of visitors with 440 individuals. Race 1 had the lowest number of total visitors with 228 individuals.

The researchers assumed that the proportion of out-of-town visitors who were surveyed represented a similar proportion of the total number of visitors. Therefore, when estimating the total spending by all visitors, these were discounted by the proportion of all visitors estimated to be from the host community. Additionally, some visitors may not have made an extra trip to the region for the event. These visitors, while not from the host community, would have planned to visit the area regardless of whether the event happened. Spending from these individuals is not counted in the total spending because they probably would have spent a similar amount in the host community without the event, so there is assumed to be no additional spending from these individuals. Once an accurate visitor count was established, the next step involved determining the length of stay.

Step 2: Identify the length of stay

Because the ski racing events were multi-day competitions it was important to determine how many days visitors stayed in the area. Across the six races, visitors averaged between 1.82 days and 3.97 days spent in the

Table 23.1. Ski racing event tourists. Authors' own table.

Race	Tourists (no.)	Average group size	No. of race event days	Average no. days spent in the area for the event	Proportion of non-locals making an extra trip[a]
Race 1	228	5.46	4	3.23	0.73
Race 2	248	9.08	2	2.54	0.80
Race 3	358	5.58	3	2.82	0.78
Race 4	234	3.91	3	2.36	0.70
Race 5	440	3.77	2	1.82	0.68
Race 6	391	4.21	3	3.97	0.93

[a]Proportion of non-local visitors who made the trip to the area specifically to attend a ski racing event.

community (see Table 23.1). Once the length of stay was established, the research team needed to calculate the average daily spending of individuals. This will provide the basis for determining total visitor spending.

Step 3: Calculate visitor spending

Knowledge of how visitors spend money in a community provides a direct link to the value that an event may have on local businesses. In this case, the survey asked respondents to identify expenditure type (i.e. admissions, recreation, restaurants, groceries, retail shopping, lodging, transportation) and the amount spent in each category within the community. Table 23.2 provides a summary of average daily visitor spending by each expenditure category across all of the racing events. As shown in Table 23.2, lodging was the highest expenditure item, followed by recreation. In contrast, retail shopping and groceries had the lowest expenditures. Knowing where visitors spend money is useful to determine which sectors of the economy experience direct economic impact from the event.

Although having an estimate for average daily spending is valuable, it is more insightful to see how these figures are impacted by the total number of visitors at each race and across the season. Therefore, daily expenditures are multiplied by how many days the visitor stayed in the area and the number of visitors for each event. To do this, first, calculate daily expenditures by summing the daily average spending in each category for each race. For example, during race 1, visitors spent $7.60 (admissions) + $18.23 (recreation) + $16.00 (restaurants) + $5.95 (groceries) + $0.16 (retail) + $46.57 (lodging) + $6.82 (transportation) = $101.33 (average daily spending, per person). Next, calculate total event spending for race 1 by multiplying the average daily spending by the average length of stay, then multiplied by total visitors ($101.33 × 3.23 × 228) = $74,623.47. Lastly, complete this step for each race to calculate total spending (learners are asked to do this as part of discussion question 6, see Table 23.3 in the 'Analysis' section).

Step 4: Correct for local spending and those not making an extra trip

Once the total spending for each race is calculated, use the proportion of non-locals who are making an extra trip for the event to only account for the additional spending that would have occurred because the event took place. For example, race 1 generated $74,623.47 total spending. From Table 23.1, 73% of the spending from race 1 was additional expenditure induced by the event. The additional spending is then calculated as $74,623.47 × 0.73 = $54,475.13.

Step 5: Apply multipliers

Although having an estimate of total additional expenditure is valuable, it does not consider the ripple effect of new funds circulating in the local economy. Each expenditure moves money throughout a local economy and can generate new job creation, personal income, and sales that flow throughout the supply chain. Multipliers provide a coefficient estimate that accounts for the recirculation of money to determine the local economic activity generated by an event. However, the multiplier might overestimate the impact because not all local spending is

Table 23.2. Average daily visitor spending by category. Authors' own table.

Race	Admission	Recreation	Restaurants	Groceries	Retail	Lodging	Transport (car)
1	$7.60	$18.23	$16.00	$5.95	$0.16	$46.57	$6.82
2	$21.61	$23.32	$13.69	$7.16	$12.10	$33.28	$10.68
3	$20.56	$19.25	$19.61	$3.75	$2.82	$23.58	$8.15
4	$14.78	$11.95	$13.66	$4.05	$1.70	$19.82	$5.34
5	$12.66	$31.50	$13.47	$3.97	$3.14	$7.19	$9.36
6	$23.85	$20.95	$20.65	$5.46	$6.74	$39.04	$7.77

fully captured in a local region's economy. For example, if an individual purchases an imported ski helmet from a local retailer, only the retail margins enter into the local economy. Multiplying the capture rate by the multiplier results in a more accurate economic impact estimate.

Economic multipliers are often determined using input-output models which provide a relationship between every industry to estimate how spending in one particular sector may result in additional spending in every other sector. There are complex software programs to estimate local multipliers from these models. A more practical method for a small to medium non-profit organization is to use results from a statistical analysis of these models to predict local multipliers, which only requires two population measures, the natural log of the population (measured in millions) and the population density (measured in thousands of people per square mile) (Chang, 2001). The estimated relationship is as follows:

$$\text{Sales multiplier} = 1.6 + 0.5 \times \ln(\text{population}) - 0.01 \times (\text{population density}) \quad \text{(Eqn 23.1)}$$

The ski races occurred in a metropolitan area of approximately 700,000 people (0.7 million) at a population density of 300 people/square mile (0.3 thousand/square mile). Plugging these into Eqn 23.1 yields a multiplier of 1.42.

$$\text{Sales multiplier} = 1.6 + 0.5 \times \ln(0.7) - 0.01 \times (0.3) = 1.42$$

A fraction of this is captured by the local economy, estimated between 60% and 70% of the multiplier (Stynes, 1997). Correcting for the range of this capture rate, the sales multiplier is estimated to be between 0.85 and 0.99 as shown below:

$$\text{Lower bound}: 0.6 \times 1.42 = 0.85$$

$$\text{Upper bound}: 0.7 \times 1.42 = 0.99$$

The research team uses the total additional spending they already calculated and estimates how those expenditures stimulate economic activity in the local community. To do this, the team must multiply total additional expenditures by the range of the multiplier above. The expenditures were calculated as $444,325.08, using the multipliers with the capture rate yields the following range:

$$\text{Lower bound}: 0.85 \times \$444{,}325.08 = \$377{,}676.32$$

$$\text{Upper bound}: 0.99 \times \$444{,}325.08 = \$439{,}881.83$$

The result is that the ski races yielded a total local economic impact estimated between $377,676.32 and $439,881.83.

While economic impact studies are often viewed as a complex analysis, organizations can estimate an economic impact for smaller-scale events relatively easily and with low time and financial cost. Surveying event attendees and calculating their expenditures associated with their visits using the technique outlined above yields a range for the economic impact of an event. The economic impact can provide evidence to garner support for these events by justifying the hosting costs with the associated economic benefits. To prepare their final report, the team must now write an executive summary of their findings to present to the Board of Directors.

Discussion Questions

1. Across the entire racing season, how much did tourists spend in the area? Which race produced the highest visitor spending?

2. Which race produced the greatest economic impact? What was the total economic impact?

3. Which type of businesses benefitted the most from the ski racing events? Which type of businesses benefitted the least from the ski racing events? Did spending patterns differ across racing events? Speculate what might be driving these differences.

4. How should the board use the economic impact study to position the economic value of the club to local businesses to try to secure donations or sponsorship?

5. What are some of the strengths and limitations of using economic impact studies as a way to secure financial support (e.g. donations, sponsorship)?

6. Determine total visitor spending for each race, total visitor spending for the entire season, and average spending in each category (use templates Table 23.7 and 23.8 in the Appendix, and the data provided in Tables 23.1 and 23.2). Also calculate the data on the proportion of non-locals making an additional trip, determine the total additional spending for each race and economic impact for each race and for all of the racing events. Economic impact should be reported with a lower and upper range (use templates Table 23.9 and 23.10 in the Appendix).

Teaching Notes

Leveraged as a means for economic development, sporting events and outdoor recreation continue to be positioned as a powerful way to drive local, regional and national economies. Economic impact studies are widely used in sports and event tourism as a tool to analyse visitor spending in an area. These reports seek to determine the economic stimulus from outside dollars that enter a local community. This is typically calculated by ascertaining the number of visitors, the average amount of money spent during the trip, and the use of appropriate multipliers (Crompton, 2010). For example, when athletes, coaches, parents, sporting officials and other spectators attend events, they spend money on entrance fees, lodging, food, travel, recreation and memorabilia. Collectively, these expenses reflect 'new' money entering into a community and thus generate positive economic impact for local businesses and residents. One purpose of an economic impact analysis is to provide residents and elected officials a way to measure a return on investment (Crompton, 2010). When a community hosts an event, it will incur the costs associated with putting on the event; calculating the economic value of the event is critical to ensure the benefits are also counted. Recreation events are often 'high-cost' budgetary items that many community members may not see as valuable; however, understanding the economic return on investment provides stakeholders with the information to make informed decisions on the value of hosting an event (Crompton, 2010, p. 16). The reported economic value that sports and outdoor recreation bring to a community is often used to justify hosting events and to build the needed infrastructure such as stadiums, trails, pathways, lodging, or to seek commercial sponsorship (Crompton, 2010). Although large-scale sporting events can have a significant economic impact within a community, the economic benefit of smaller-scale sporting events may provide a more sustainable model of sport tourism that communities should consider (Gibson *et al.*, 2012). For small service providers, knowing how much a community benefits from hosting an event can provide critical information for marketing, fundraising, sponsorships and communications with various stakeholder groups.

Below is a summary of instruction that provides an outline of how to sequence content and learning activities, the time required to complete each session, and suggested teaching strategies. In addition, recommended assessment activities and student spreadsheet templates are offered.

Target audience

The target audience is undergraduate or graduate students studying tourism, hospitality, recreation, events or business management. The activities are best suited for small group work (two to three students) but could be completed individually.

Teaching methods and equipment

This case study is designed to take place over a few class sessions or one session that spans several hours. The first session will require students to engage in discussion and mapping concepts using a whiteboard or flipchart paper. Additionally, instructors may wish to utilize a slideshow for the presentation of basic concepts. To calculate and analyse the economic impact, it is recommended that students have access to Microsoft Excel or a comparable spreadsheet platform (e.g. Google Sheets). To complete the final learning activity (executive summary), we suggest students have access to a word processing program (e.g. Microsoft Word, Google Docs). In the following teaching instruction plan we provide a general class outline and recommended sequencing of instruction.

Teaching instruction plan.

Suggested sequence of teaching points	Time required (min)	Teaching strategies
Session 1: Introduction to economic impact analyses for events	10	Utilizing a think>pair>share strategy, have the students identify and write down the last event they attended (e.g. concert, sports, festival). Once the students have identified the event, have them list all of the expenditures that went into going to that event (e.g. registration, travel, fuel, food, memorabilia, lodging). Next, have the students form a pair to discuss and compare their events and the associated expenditures. Have students identify any similarities/differences that they notice in their expenses and what factors are likely to have an impact on any similarities/differences. For instance, are there certain categories of expenses (e.g. transportation, lodging, shopping, fuel, food/beverages, restaurants) that were more/less expensive than others? Were there categories where there was no expenditure? Why? What type(s) of service providers and businesses most benefitted from your trip to the event? Have small groups share with the larger group their findings. Discuss any patterns observed.
Session 1: Lecture on foundational concepts related to economic impact studies in tourism	30	The purpose of this lecture is to familiarize students with the foundational concepts used in economic impact studies. Key concepts that should be defined include: (i) economic impact analysis; (ii) visitor versus resident spending; (iii) multipliers; and (iv) capture rate.
Session 1: Assessment	10	In the last 10 min of class, have students create a concept map that links visitor spending to a destination's economic impact. Key concepts that should be included in the map are resident versus visitor spending, multipliers, capture rate and economic impact.
Session 2: Review of economic impact case study analysis	10	Briefly review the purpose of economic impact analyses and why Always Winter Ski Club is conducting an economic impact study. Highlight how economic impact studies can be used to illustrate to local businesses and elected officials the value of hosting events that bring visitors into a community.
Session 2: Lab exercise	60	To calculate economic analysis, students should have access to some type of spreadsheet (e.g. Excel, Google Sheets) to run the calculations. Students should consult sample templates (Tables 23.7–23.10, see Appendix) to create a spreadsheet to calculate total expenditures across each spending category and race. Once total expenditures are calculated, students should apply the appropriate multiplier to determine economic impact relative to sales and personal income. These findings should then be reported in a final economic impact report (see assessment). To help guide the process, students should be guided through each of the steps and outlined tasks. Tasks 1–3 must be completed in the order provided. Below are some checkpoints that instructors may want to use to ensure that students are accurately interpreting the data presented in Tables 23.1 and 23.2.

Continued

Teaching instruction plan Continued.

Suggested sequence of teaching points	Time required (min)	Teaching strategies
		Following Step 1, use Table 23.1 to assess the following: (i) race(s) that produced the highest/lowest number of user days; (ii) race(s) that brought visitors who travelled in larger groups; and (iii) race(s) that had the highest/lowest proportion of non-locals making the extra trip. Upon completion of Task 1, identify: (i) the amount of money tourists spent in the region; (ii) the races that resulted in the greatest/least amount of visitor spending; and (iii) hypothesize about any factors that may be driving the results. Upon completion of Task 2, determine: (i) the race that had the highest proportion of spending from non-locals making an extra trip; (ii) the race that had the lowest proportion of spending from non-locals making an extra trip; and (iii) hypothesize about any factors that may be driving these results. Upon completion of Task 3, assess: (i) the estimated economic impact range for each race; and (ii) the estimated economic impact range for all of the racing events combined.
Session 2: Assessment (take-home assignment or additional lab exercise)	60–90	To assess the accuracy of students' calculations and their understanding of the implications of their results, it may be helpful to engage in a discussion (see questions below) before writing the executive summary. **Discussion Questions** **1.** Across the entire racing season, how much did tourists spend in the area? Which race produced the highest visitor spending? **2.** Which race produced the greatest economic impact? What was the total economic impact? **3.** Which type of businesses benefitted the most from the ski racing events? Which type of businesses benefitted the least from the ski racing events? Did spending patterns differ across racing events? Speculate what might be driving these differences. **4.** How should the board use the economic impact study to position the economic value of the club to local businesses to try to secure donations or sponsorship? **5.** What are some of the strengths and limitations of using economic impact studies as a way to secure financial support (e.g. donations, sponsorship)? **6.** Determine total visitor spending for each race, total visitor spending for the entire season, and average spending in each category (use templates Tables 23.7 and 23.8 in the Appendix, and the data provided in Tables 23.1 and 23.2). Also calculate the data on the proportion of non-locals making an additional trip, determine the total additional spending for each race and economic impact for each race and for all of the racing events. Economic impact should be reported with a lower and upper range (use Tables 23.9 and 23.10 in the Appendix). Have students write a brief executive summary (approximately three pages). The summary should be accompanied by visitor spending in a table and/or chart format. Total spending should be defined by category and race event with a final column that includes the estimated total economic impact.

Duration

About 3–3½ hours (180–210 min)

Analysis

1. Across the entire racing season, how much did tourists spend in the area? Which race produced the highest visitor spending?

Answers: $547,694.39; Race 6.

2. Which race produced the greatest economic impact? What was the total economic impact?

Answers: Race 6 ($152,721.06–$177,875.12); The economic impact of all of the racing events is estimated to be between $377,676.32 and $439,881.83.

3. Which type of businesses benefitted the most from the ski racing events? Which type of businesses benefitted the least from the ski racing events? Did spending patterns differ across racing events? Speculate what might be driving these differences.

With the exception of one race (race 5), lodging was the greatest source of visitor spending. The next two greatest areas of spending were admission tickets and recreation (at the ski resort). Simply based on visitor expenditures, lodging and the ski resort saw the greatest percentage of visitor spending. Retail shopping and grocery stores experienced the smallest share of visitor spending.

A number of factors probably contributed to the variance observed across races in visitor spending such as:

- length of the race event (e.g. visitors stayed in the area longest during race 6, while visitors stayed the least amount of time in the area during races 4 and 5);
- type of race (e.g. championship versus regular season race);
- where other teams were coming from (e.g. out of state versus in state);
- group size (race 2 had the largest group size, race 5 had the smallest group size); and
- age of athlete (e.g. younger athletes may have travelled with their family instead of as a team, and families may be more budget conscious resulting in spending behaviour such as purchasing groceries instead of eating at a restaurant).

4. How should the board use the economic impact study to position the economic value of the club to local businesses to try to secure donations or sponsorship?

Each visitor spent an average of $288.41 per race. (The average visitor spending per race can be calculated by dividing the average total spending of all races by the average number of visitors at all races. Refer to Tables 23.1 and 23.4 to obtain the data to complete these calculations.) These are unique visitors who came specifically for the racing events and otherwise would not have come to the area. Now that the board has an estimate of visitor spending, it can approach local businesses who are most likely to benefit from the races (e.g. ski resort, hotels, restaurants) to seek sponsorship or other financial support (e.g. donations). Further, the board can use these data to make future projections about hosting races in future seasons.

5. What are some of the strengths and limitations of using economic impact studies as a way to secure financial support (e.g. donations, sponsorship)?

The strengths are as follows. Economic impact reports provide a tool that organizations can utilize to assess the cost–benefit of hosting events. They can also help organizations make strategic decisions around which type of events to host (or put a bid in for) by looking at which races produced the greatest return on investment. Organizations can also use economic impact data to illustrate the economic benefits of hosting events to local stakeholders (local government, business owners, residents) as a method to seek support. Some evidence exists that economic impact data results in favourable attitudes towards sports sponsorship among tourism-related businesses, such as lodging, restaurants and transportation (e.g. Keshock, 2004).

For tourism organizations that seek to solicit community support by sharing the economic value of hosting smaller events, there is no guarantee that a local business can or will provide support. While some research shows that businesses have positive attitudes to sponsoring sporting events when presented with economic impact data, it may be difficult to causally link sponsorship into greater sales for that business. Thus, it is important for tourism providers to have realistic expectations on how others may interpret and act on the results, regardless of how much specific businesses and individuals within the community may benefit.

6. Determine total visitor spending for each race, total visitor spending for the entire season, and average spending in each category (use templates Tables 23.7 and 23.8 in the Appendix, and the data provided in Tables 23.1 and 23.2). Also calculate the data on the proportion of non-locals making an additional trip, determine the total additional spending for each race and economic impact for each race and for all of the racing events. Economic impact should be reported with a lower and upper range (use templates Tables 23.9 and 23.10 in the Appendix).

Instructors should refer to Tables 23.3–23.6 to evaluate the accuracy of student answers.

Appendix: Templates and instructions for learners' assignment

Students will complete a three-page executive summary that consists of the following: (i) a one-page introduction and summary of key findings; and (ii) two or three tables or charts that show daily visitor spending, total spending by category and race event, and economic impact. Table templates are provided in Tables 23.7–23.10.

Table 23.3. Instructor's solution for spending per person, per day, by category. Authors' own table.

Race	Admission	Recreation	Restaurants	Groceries	Retail	Lodging	Transport. (car)	Total daily spending
1	$7.60	$18.23	$16.00	$5.95	$0.16	$46.57	$6.82	$101.33
2	$21.61	$23.32	$13.69	$7.16	$12.10	$33.28	$10.68	$121.84
3	$20.56	$19.25	$19.61	$3.75	$2.82	$23.58	$8.15	$97.72
4	$14.78	$11.95	$13.66	$4.05	$1.70	$19.82	$5.34	$71.30
5	$12.66	$31.50	$13.47	$3.97	$3.14	$7.19	$9.36	$81.29
6	$23.85	$20.95	$20.65	$5.46	$6.74	$39.04	$7.77	$124.46
Average	$16.84	$20.87	$16.18	$5.06	$4.44	$28.25	$8.02	$99.66

Table 23.4. Instructor's solution for total visitor spending by category and race event.[a] Authors' own table.

Race	Admission	Recreation	Restaurants	Groceries	Retail	Lodging	Transport (car)	Total event spending	Total additional spending
1	$5,596.94	$13,425.30	$11,783.04	$4,381.82	$117.83	$34,296.01	$5,022.52	$74,623.47	$54,475.13
2	$13,612.57	$14,689.73	$8,623.60	$4,510.23	$7,622.03	$20,963.74	$6,727.55	$76,749.45	$61,399.56
3	$20,756.55	$19,434.03	$19,797.47	$3,785.85	$2,846.96	$23,805.42	$8,227.91	$98,654.20	$76,950.28
4	$8,162.11	$6,599.27	$7,543.60	$2,236.57	$938.81	$10,945.40	$2,948.96	$39,374.71	$27,562.30
5	$10,138.13	$25,225.20	$10,786.78	$3,179.18	$2,514.51	$5,757.75	$7,495.49	$65,097.03	$44,265.98
6	$37,021.64	$32,520.06	$32,054.38	$8,475.39	$10,462.30	$60,600.62	$12,061.14	$193,195.52	$179,671.83
Total	$95,287.94	$111,893.59	$90,588.87	$26,569.04	24,502.44	$156,368.94	$42,483.57	$547,694.39	$444,325.08

[a]To calculate total visitor spending for each category, multiply the category of spending by the number of tourists by the average number of days at the race. For example, for race 1, category admission: $7.60 (admission, race 1) × 228 (number of tourists at race 1) × 3.23 (average days at the race 1) = $5,596.94.

Table 23.5. Instructor's solution for sales multiplier calculator. Authors' own table.

Population (millions)	Population density (thousands per square mile)	Multiplier 1.6 + 0.5*ln(pop) - 0.01*(popdensity)	Lower bound: capture rate 60%	Upper bound: capture rate 70%
0.700	0.300	1.42	0.85	0.99

Table 23.6. Instructor's solution for calculations on economic impact. Authors' own table.

Race	No. of visitors	No. of days	Total additional expenditure	Lower bound estimate (multiplier = 0.85)	Upper bound estimate (multiplier = 0.99)
Race 1	228	3.23	$54,475.13	$46,303.86	$53,930.38
Race 2	248	2.54	$61,399.56	$52,189.63	$60,785.56
Race 3	358	2.82	$76,950.28	$65,407.73	$76,180.77
Race 4	234	2.36	$27,562.30	$23,427.95	$27,286.67
Race 5	440	1.82	$44,265.98	$37,626.08	$43,823.32
Race 6	391	3.97	$179,671.83	$152,721.06	$177,875.12
Total	1899	–	$444,325.08	$377,676.32	$439,881.83

Table 23.7. Sample template to calculate average daily visitor spending by category.

Race	Admission	Recreation	Restaurants	Groceries	Retail	Lodging	Transport. (car)	Total daily spending
1								
2								
3								
4								
5								
6								
Average								

Table 23.8. Sample template to calculate total spending by category and race event.

Race	Admission	Recreation	Restaurants	Groceries	Retail	Lodging	Transport (car)	Total event spending	Total additional spending
1									
2									
3									
4									
5									
6									
Total									

Table 23.9. Template for sales multiplier calculation.

Area population (millions)	Population density (thousands per square mile)	Multiplier 1.6 + 0.5*ln(pop) – 0.01*(popdensity)	Lower bound: capture rate 60%	Upper bound: capture rate 70%

Table 23.10. Template for executive summary table – economic impact.

Race	No. of visitors	No. of days	Total additional expenditures	Lower bound estimate	Upper bound estimate
Race 1					
Race 2					
Race 3					
Race 4					
Race 5					
Race 6					
Total					

Note: This case study was created from a class research project on sustainable tourism. The data presented in this case are factual, but the names of individuals and organizations have been changed to protect the confidentiality of those involved.

References

Chang, W.-H. (2001) Variation in multipliers and related economic ratios for recreation and tourism impact analysis. PhD thesis, Michigan State University, East Lansing, Michigan.

Crompton, J.L. (2010) *Measuring the Economic Impact of Park and Recreation Services*. National Parks and Recreation Association, Ashburn, Virginia, pp. 3–68.

Gibson, H., Kaplanidou, K. and Kang, S. (2012) Small-scale event sport tourism: a case study in sustainable tourism. *Sport Management Review* 15, 160–170.

Keshock, C. (2004) *The Effects of Economic Impact Information on the Attitudes of Potential Sports Sponsors Operating in Mid-size and Small College Communities*. Florida State University, Tallahassee, Florida.

Stynes, D.J. (1997) *Economic Impacts of Tourism: a Handbook for Tourism Professionals*. University of Illinois, Urbana, Illinois.

US Ski & Snowboard Association (2019) US Ski & Snowboard 2019 Annual Report. Available at: http://usskiandsnowboard.org/sites/default/files/files-resources/files/2021/USS_19AnnualReport_FINAL_lowres.pdf (accessed 20 May 2022).

24 Operations Research in Tourism: Developing Tourist Circuits

Senthilkumaran Piramanayagam*
Manipal Academy of Higher Education, Manipal, India

Abstract
Operations research (OR) is a science that supports rational decision making. It is an analytical method for decision making and problem solving that is helpful in managing organizations. The OR process consists of: (i) identifying a problem that needs to be solved; (ii) developing a model around the problem resembling a real organization and variables; and (iii) providing a comprehensive and complex solution aimed to minimize the cost, maximize the revenue or optimize the use of resources. OR is an interdisciplinary branch of mathematics that integrates statistics, algorithms and mathematical modelling to arrive at an optimal decision. In tourism and hospitality, OR helps demand forecasting, market segmentation, pricing and revenue management, productivity analysis, crowd management, network and value chain analysis, event scheduling and tourism circuit development. The current case study aims to familiarize the tourism and hospitality students with an OR tool known as the 'travelling salesman problem' (TSP) in creating tourist circuits.

Subject: operations research (OR), tourism management, tourism circuit development

Learning Objectives

At the end of the case study, the learner will be able to:

1. Understand the application of the 'travelling salesman problem' (TSP) technique in tourism to create tourism circuits.
2. Identify the destinations and create the shortest travel paths to create a tourism circuit.
3. Develop a tourist circuit with the shortest distance possible with a minimum cost of travel.

Case

The rich natural, cultural, historical and religious resources provide a considerable opportunity to develop tourism in India. In 2020, tourism in India contributed about US$121.9 billion to the gross domestic product (GDP) and was expected to reach about US$512 billion by 2028 (IBEF, 2022). In the same year, tourism accounted for 39 million jobs which are about 8% of the total employment generated in the country. It is expected to contribute about 53 million jobs in 2029

*senthil.kumaranp@manipal.edu

DOI: 10.1079/9781800621022.0024

(IBEF, 2022). According to India tourism statistics, in 2020 the country witnessed about 10.93 million foreign tourists, an increase of 0.37 million compared with the previous year. The annual growth rate was estimated to be 5.20% in 2020 (IBEF, 2022). In terms of international tourist arrivals, India ranked 23rd among countries. However, in terms of global tourism receipts, India ranked in 12th position (Ministry of Tourism, 2022).

Why is tourism important?

Tourism is considered a vital economic activity that contributes to the success of economies around the globe. Tourism renders many benefits to the tourism destination, community and the country. The main reasons attributed for the motivation for tourism development by destinations are job creation, infrastructure development, contribution to GDP and foreign exchange earnings (FEEs), cultural exchange between residents and the visitors, improved standard and quality of life. In terms of job creation by tourism, it is classified into direct, indirect and total jobs. Tourism not only creates jobs in the tourism sector but also employment opportunities in other economic sectors such as agriculture, communication, education and health sectors. A visit by a tourist extends the job opportunity for hotels, shopping centres, transport service providers, farmers, and other businesses operated to fulfil the needs and wants of the tourists. The potential benefits of tourism are presented in Fig. 24.1.

For a government, the tourism-related business contributes to the revenue through tax incomes. Development of highways, airports, leisure facilities, parks, schools, hospitals and improved public services benefits the tourist and the community. More than its economic benefits, tourism is valued for its other benefits, which cannot be measured in terms of money. It helps community members of the destination in terms of capacity building as they have to deliver quality services to tourists to retain and sustain the benefits. The infrastructure and other support services aimed at tourists also facilitate the health, leisure and well-being of the community. Tourism helps build community pride and social cohesion as each community member needs to work together to portray, preserve and share the benefits. In terms of cultural benefits, tourism helps in preserving local culture, traditions, craft and culinary specialities, which tourists most value and they act as a significant tourist attraction in the region. In terms of environmental benefits, the income generated through tourism helps protect and preserve

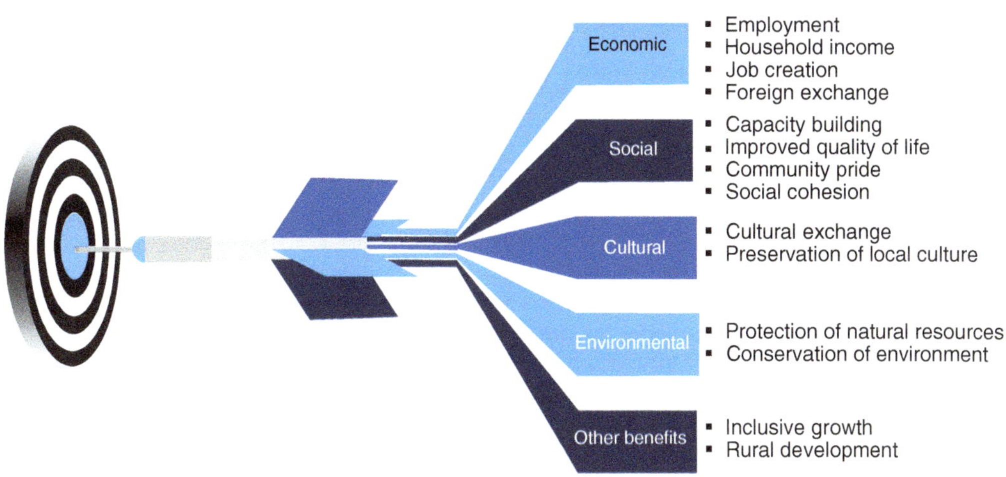

Fig. 24.1. The potential benefits of tourism. Author's own elaboration.

natural resources and the environment. Countries aim to ensure inclusive growth, rural development and pro-poor support: tourism acts as a tool for this.

Share of foreign tourists

Despite the continued growth and opportunity for tourism, the share of foreign tourist arrivals to India during 2019 was about 1.23% and accounted for 4.97% of international tourist arrivals in the Asia Pacific region (Ministry of Tourism, 2022). Domestic tourism is the most significant contributor to the local economy as it has been growing at a compound annual growth rate of 20% almost for the last 5 years (Ministry of Tourism, 2022). However, the dispersion of domestic tourism in the country indicates that domestic tourism flow is not equal across the states and regions. India is being currently valued as a most important tourist-generating region, as more citizens are opting for a holiday abroad. The outbound tourist numbers are almost double the foreign tourist arrivals to the country. In 2019, the number of outbound tourists reached 26.91 million, while it stood at 21.87 million in 2016 (Ministry of Tourism, 2022). Experts of tourism in the country opined that the government needs to develop strategies to attract more foreign tourists by supporting domestic tourism equally in all states and regions and encouraging the growing numbers of outbound tourists to explore the unmatched hidden tourism attractions and destinations in the country. After a prolonged discussion, the government introduced a flagship programme called 'Swadesh Darshan', meaning 'Experience your country', to create theme-based tourism circuits. A tourist circuit is a travel route or path where at least three or more major tourism destinations or attractions are located adjacent to others, however, not in the same city or a town. In its introduction, the government stated that the scheme aimed to identify and develop many theme-based tourism circuits based on the rich historical, cultural, religious and natural heritage that will support job creation and tourism development.

Swadesh Darshan

The main objectives of the scheme are:

- to emphasize tourism as a major source of job creation and economic growth;
- to create tourism circuits that have great potential to attract tourists;
- to promote heritage and cultural values of the nation to generate employment;
- to develop world-class infrastructure;
- to develop community-based tourism and a pro-poor approach to tourism by creating awareness of the significance of tourism among local communities in terms of scope for income, improved standard of living and overall development (i.e. harnessing the benefits of tourism); and
- to make full use of infrastructure and national cultural characteristics as a strong point for every region by introducing theme-based circuits.

Some tourism circuits developed under Swadesh Darshan are presented in Table 24.1.

XTour – an emerging tour operator

Ms Harshita, the CEO of a small travel start-up firm XTour in Karnataka, a principal state in India, decided to develop unique travel circuits in the state, which is well known for its religious and pilgrimage centres. Karnataka is blessed with a pleasant climate and natural, cultural and heritage resources. The tourism tagline of the state is 'one state, many worlds'. The state is proud to have three United Nations, Scientific and Cultural Organization's (UNESCO) World Heritage sites. Karnataka is the sixth largest state in the country with abundant hill stations, information technology (IT) hubs, historical monuments, sculptures, modern cities, coastal regions, hill ranges, forests, paddy fields, spice plantations and a wide variety of fauna and flora. XTour, the travel start-up, developed its distinctiveness among competitors by offering niche tourism packages where experience is the core of travel. Harshita, a graduate in IT and a travel enthusiast, founded XTour, which aimed to offer customized tour packages for honeymooners, senior travellers, babymoon trips, adventure and wildlife

Table 24.1. Tourist circuits developed under the Swadesh Darshan programme. Author's own table.

Theme	Tourism circuits – some examples
Buddhist	Madhya Pradesh Buddhist Circuit
Coastal	Maharashtra Sindhudurg Coastal Circuit
Desert	Rajasthan Development of Desert Circuit
Eco	Kerala Eco circuit
	Jharkhand Eco Circuit
Heritage	Puducherry Heritage Circuit
Himalayan	Jammu and Kashmir Himalayan Circuit
Krishna	Haryana Tourism Infrastructures in Kurukshetra
North East	Tripura North East Circuit 2
Ramayana	Uttar Pradesh Ayodhya under Ramayana Circuit
Rural	Rural Circuit Malanad Malabar Cruise Tourism
Spiritual	Bihar Mandar Hill and Ang Pradesh
Tirthankar	Parkala – Moodbidri Circuit
Tribal	Chhattisgarh Tribal Circuit
Wildlife	Assam Wildlife Circuit

photographers. As the start-up is more concerned about sustainability and preserving natural resources, the firm offers tour packages to individuals, couples and groups with a maximum number of seven travellers. Harshita used to promote the travel packages through social network sites as she wanted XTour to focus on experience-seeking and -sharing individuals on social media. Her well-wishers warned her that it is not a good idea to serve a small group of 'niche' tourists with unique travel motives. However, XTour is successful and positioned as a most-valued tour operating company where the major clients are from multinational companies and young travellers who value experience more than just travel. XTour has been recently joined with a popular television channel that broadcasts wildlife stories around the globe, in which XTour provides all the travel-related services for the wildlife photographers.

Opportunity from a new market – senior travellers

Harshita observed that there are a lot of enquiries from senior citizens. At her recent meeting with the marketing team, they also expressed the strong demand from senior travellers. Most of the enquiries indicate that these people want to visit all the famous religious destinations within the state but without much discomfort and unnecessary travel. Though they have a strong intention to see all the sacred places in the state, the major constraint is their age, which restricts extensive travel in a single day. It necessitates a travel package that covers all the destinations with the shortest possible route and with adequate rest during travel. Considering this as an emerging opportunity for XTour, Harshita decided to create niche travel products for senior citizens. As the temples and religious sites are spread around the state, she would like to develop unique tourist circuits specifically for senior citizens in order to fulfil their need for travelling the shortest distance in comfort. As a preliminary process, she called various cab drivers associated with XTour to develop a possible tourism circuit based on the needs of senior citizens. However, the process became futile as there is a disagreement between the cab drivers on proposed circuits as each has their perspective on distance and major attractions to be covered during the tour programme.

She decided to approach a professor in tourism in an institution well known for its tourism and hospitality business management programmes. Dr Pillai, the Professor of Tourism, suggested using the 'travelling salesman problem' (TSP), a combinatorial optimization tool used in operations research (OR) to find the shortest possible route to visit each destination exactly once and return to the origin of the tour. He explained to Harshita that the various steps and processes to develop a tourism circuit were:

- Step 1: Identify the destinations to be covered in the tour package;

- Step 2: Develop a distance matrix between each destination;
- Step 3: Apply computer programs/apps to generate travel packages using the TSP algorithm;
- Step 4: Evaluate the tourism circuits and make necessary modifications; and
- Step 5: Create an innovative tour package, with costings and marketing.

As suggested by Dr Pillai, Harshita followed the process. To avoid using the names of each destination, she allocated each one with a code to make the distance matrix simple.

STEP 1: IDENTIFY THE DESTINATIONS TO BE COVERED IN THE TOUR PACKAGE. About 11 religious tourism destinations spread throughout the state were selected to be promoted as a Hindu tourism tourist package. The 11 selected destinations and their codes are shown in Table 24.2.

Table 24.2. Selected destinations in the tour package along with codes (from the state capital Bengaluru). Author's own table.

Name of the destination	Code
Bengaluru	1
Udupi	2
Murdeswar	3
Kollur	4
Kalaburgi	5
Gokarna	6
Vijayapura	7
Bidar	8
Dharmasthala	9
Shravanabelagola	10
Karkala	11

STEP 2: DEVELOP A DISTANCE MATRIX BETWEEN EACH DESTINATION. The distance by road between each destination was taken from Google Maps. The estimated distance matrix is presented in Table 24.3.

STEP 3: APPLY COMPUTER PROGRAMS/APPS TO GENERATE TRAVEL PACKAGES USING THE TSP ALGORITHM. Harshita adopted the software package 'TSP Solver and Generator: V 0.1.4 (Beta 2)' to generate the best travel sequence with the minimum travel distance between all the tourism destinations.

STEP 4: EVALUATE THE TOURISM CIRCUITS AND MAKE NECESSARY MODIFICATIONS. The software generated the following shortest travel path: Bengaluru → Shravanabelagola → Dharmasthala → Karkala → Udupi → Kollur → Murdeshwar → Gokarna → Bidar → Kalaburgi → Vijayapura → Bengaluru.

The travel path from Bengaluru to all the other religious destinations covering the entire Karnataka state is 2050 km as per the TSP model using the branch and bound technique. The longest distance is between Gokarna and Bidar (617 km) in the resulting travel path, followed by Vijayapura to Bengaluru (525 km). Since the mode of travel is to be by road, this is cumbersome for senior citizens to travel. Based on this limitation, an alternative travel path was considered with Mangaluru as the start and end destination. Mangaluru is coded as 'City 1' replacing Bengaluru, with the other ten destinations coded as shown in Table 24.1. The distance matrix from Mangaluru to all the ten religious destinations is presented in Table 24.4.

Table 24.3. Travel distance (km) matrix 1 (Bengaluru as a source and end destination). Author's own table.

	City 1	City 2	City 3	City 4	City 5	City 6	City 7	City 8	City 9	City 10	City 11
City 1	0	402	489	429	636	483	525	650	297	142	361
City 2	402	0	102	75	661	177	500	740	106	274	41
City 3	489	102	0	60	561	78	401	674	206	360	137
City 4	429	75	60	0	575	136	415	712	160	300	102
City 5	636	661	561	575	0	503	166	115	663	596	620
City 6	483	177	78	136	503	0	343	617	282	379	212
City 7	525	500	401	415	166	343	0	280	567	509	523
City 8	650	740	674	712	115	617	280	0	771	819	733
City 9	297	106	206	160	663	282	567	771	0	168	63
City 10	142	274	360	300	596	379	509	819	168	0	230
City 11	361	41	137	102	620	212	523	733	63	230	0

Table 24.4. Travel distance (km) matrix 2 (Mangaluru as a source and end destination). Author's own table.

	City 1	City 2	City 3	City 4	City 5	City 6	City 7	City 8	City 9	City 10	City 11
City 1	0	56	156	129	679	232	555	793	75	222	53
City 2	56	0	102	75	661	177	500	740	106	274	41
City 3	156	102	0	60	561	78	401	674	206	360	137
City 4	129	75	60	0	575	136	415	712	160	300	102
City 5	679	661	561	575	0	503	166	115	663	596	620
City 6	232	177	78	136	503	0	343	617	282	379	212
City 7	555	500	401	415	166	343	0	280	567	509	523
City 8	793	740	674	712	115	617	280	0	771	819	733
City 9	75	106	206	160	663	282	567	771	0	168	63
City 10	222	274	360	300	596	379	509	819	168	0	230
City 11	53	41	137	102	620	212	523	733	63	230	0

The travel path from Mangaluru to all the other religious destinations covering the entire Karnataka state is 1993 km. The resulting path is Mangaluru → Dharmasthala → Shravenabelagola → Bidar → Kalaburgi → Vijayapura → Gokarna → Murdeshwar → Kollur → Udupi → Karkala → Mangaluru. In the new circuit, as suggested by TSP, the longest distance between Shravenabelagola and Bidar (819 km) is too challenging. To reduce the travel distance between destinations and maximize the tourism benefits for the ten religious tourism sites, Harshitha divided the tourism circuit into two different travel packages. The proposal did not consider a circuit consisting of Bidar, Kalaburgi and Vijayapura due to the non-availability of air connectivity to any of these destinations and the limited rail network with major cities of the state. The two religious tourism circuits that were proposed had Mangaluru and Bengaluru as the source destination, as these destinations had air connectivity with other national and international destinations. She started the process one more time from the beginning.

Circuit 1: Bengaluru as a source and end destination

The Bengaluru circuit consists of Mysuru, Tumkuru, Shravenabelagola and Srirangapatna. Srirangapatna is an important Hindu religious centre of pilgrimage and has a collection of other tourist attractions nearby, which is why it has been included in the circuit. Mysuru has the popular Chamundeswary temple and Someswara temple, and Tumkuru has the religious destinations of Aralaguppe, Kaidala and Tiruverkere. The destination coding for this circuit is given in Table 24.5.

The distance matrix from the new proposed tourism circuit from Bengaluru as a source and end destination is shown in Table 24.6.

The shortest travel distance covering the new tourism circuit based on Bengaluru is 421 km, and the resulting travel path is: Bengaluru → Tumkuru → Shravenagbelagola → Mysuru → Srirangapatna → Bengaluru. The travel from Srirangapatna to Bengaluru is the longest travel stretch with 125 km of travel. Considering the road network in Karnataka, this distance may be covered within a couple of hours, which is not too long a time for a senior citizen to travel.

Circuit 2: Mangaluru as a source and end destination

The second tourism circuit proposed in this research had the source and end destination as Mangaluru, along with Udupi, Karkala, Kollur, Murdeswar, Gokarna and Dharmasthala. The destination codes allotted for each destination are presented in Table 24.7.

The distance matrix for circuit 2 is shown in Table 24.8.

Solving the TSP problem using the branch and bound technique resulted in the following travel path: Mangaluru → Dharmasthala → Karkala → Kollur → Murdeshwar → Gokarna → Udupi → Mangaluru. The total distance to be covered in this circuit is

612 km. The longest travelling distance is 177 km, between Gokarna and Udupi. This distance may be covered by the tourist within a couple of hours considering the National Highway network in this region.

Table 24.5. Destination codes of proposed circuit 1 (from the state capital Bengaluru). Author's own table.

Name of the destination	Code
Bengaluru	1
Mysuru	2
Shravanabelagola	3
Srirangapatna	4
Tumkuru	5

Table 24.6. Travel distance (km) matrix of proposed circuit 1 (Bengaluru as a source and end destination). Author's own table.

	City 1	City 2	City 3	City 4	City 5
City 1	0	146	143	125	71
City 2	146	0	81	34	158
City 3	143	81	0	63	110
City 4	125	34	63	0	136
City 5	71	158	110	136	0

Table 24.7. Destination codes of proposed circuit 2 (from Mangaluru). Author's own table.

Name of the destination	Code
Mangaluru	1
Udupi	2
Karkala	3
Kollur	4
Murdeswar	5
Gokarna	6
Dharmasthala	7

Conclusion

The preliminary results of the TSP show that if Bengaluru is considered the start and end destination for the proposed chain of destinations combining all religious destinations in the state, the minimum distance to be covered is 2050 km. Considering the long distance and the time the tourist will have to spend on the road, an alternative path to religious destinations was suggested assuming Mangaluru as the source and end destination. The result yielded the total length to travel is 1993 km. The longest distance to be travelled is 819 km, which is exhausting for the tourist to travel by road. As a result the tour of religious destinations was divided into two small circuits with shorter travel distances, one circuit based on Bengaluru and the other on Mangaluru. The circuit based on Bengaluru that was proposed has a total travel distance of 421 km, the longest travel distance in the circuit being 125 km (Srirangapatna to Bengaluru). The tourism circuit based on Mangaluru is 612 km in which the longest distance to be travelled is 177 km (Gokarna to Udupi).

After identifying the right combination of tourist circuits, Harshita instructed her team to develop these as a tourist package, including tourist attractions to be covered, possible choice of accommodation and travel arrangements. After going through the whole process of applying TSP in her travel

Table 24.8. Travel distance (km) matrix of proposed circuit 2 (Mangaluru as a source and end destination). Author's own table.

	City 1	City 2	City 3	City 4	City 5	City 6	City 7
City 1	0	56	53	129	156	232	75
City 2	56	0	41	75	102	177	106
City 3	53	41	0	102	136	212	64
City 4	129	75	102	0	60	136	160
City 5	156	102	136	60	0	78	203
City 6	232	177	212	136	78	0	282
City 7	75	106	64	160	203	282	0

business, Harshita is so excited about the use of OR in tourism and wants to explore other tools.

Discussion Questions

1. Using the distance matrix provided, calculate the cost for each tourism circuit evolved through the application of the software. Consider the cost as US$2.5/km.
2. Explore other possible operations research (OR) techniques which may help in improving efficiency in tourism and hospitality-related business operations.

Teaching Notes

The intention of the current case is to familiarize the learner with the 'travelling salesman problem' (TSP), a mathematical model used to find the shortest travelling path that supports a travelling salesperson to reduce the cost and unnecessary travel. The origin of TSP is unknown. W.R. Hamilton, an Irish mathematician and Thomas Kirkman formulated the TSP with mathematical orientation. The same has also been called a messenger problem, which is used to provide the shortest route for a postman to deliver the message. The primary objective is to create awareness about the application of mathematical and OR concepts in solving the problem of running an efficient day-to-day operation of a business that deals with travel. The primary input for the TSP is associated with two sets of data. First is the number of cities/destinations and second is the distance between the cities/destinations. Although there are various approaches to solve the TSP, the branch and bound method is most commonly used to develop the shortest possible tour route where the visitor visits once and returns to the same starting point. As tourism students have been exposed to various tourism packages during their career with travel agencies, learning and understanding the TSP could help them create and manage multiple tourist circuits of different themes and different costs according to the target travel market. The case study also provides scope for interdisciplinary teaching and learning between management science and tourism.

Target audience

This case can be used in postgraduate-level courses on introduction to tourism, tour operation management or travel agency management.

Teaching methods and equipment

The case can be practised in the classroom using the concept of tourism circuits, their need, functions and way to create a tourism circuit using OR tools. The case can also be adopted as a means to interdisciplinary teaching and learning, where teachers from management science or OR may collaborate with teachers from hospitality and tourism. The teachers from management science or OR may provide the foundations on evaluation, scope, various approaches, application of the TSP in other domains of knowledge, basic assumptions and approaches. Later the teachers from hospitality and tourism may explain the concept of the tourism circuit, its importance and existing tourism circuits in the country or regions. Discussion may be extended to the application of the TSP using Bloom's taxonomy of learning (Bloom, 1956), where tutors may teach and evaluate the students based on the learning hierarchy.

The idea is to use the case as a demonstration of how the TSP may be solved, and then get the students to develop their own tourist circuit(s) using the TSP. Prior to the teaching, the tutor needs to identify the destinations to be included in the tour programme. It is most desirable that the tutor needs to teach the application of TSP before using the software and application to familiarize themselves with the calculation part of the TSP and whether the tourism tutor is capable of calculating the TSP using a simple tour of four to five destinations. As the primary aim of the case is to familiarize students with the application of the TSP, it is wise to avoid a very complex level of analysis. A whiteboard, a manual on the link to the software and instructions on how to use it including data entry and interpretation of results need to be provided. Charts and coloured pencils may be provided for students to use in order to help them present the assigned task. The final application of software to solve the TSP may be done using a personal laptop or in a computer lab. A liquid crystal display (LCD) may be used to demonstrate the application of the TSP using the software.

The case is most suitable for group work rather than individual assignments. The group size may vary between three and five individuals. The number of destinations to be included in the case may be increased based on the size of the group. The tutor's role becomes prominent as he/she has to identify various combinations of destinations using different themes and optimal solutions for all the proposed tourism circuits. In the case of collaborative, interdepartmental teaching, tutors need to identify, collaborate and discuss the problems that need to be solved using TSP with teachers from management science or OR.

Teaching instructions

The case can be taught in multiple phases as indicated on the teaching instruction plan below. In the first phase, the concept of tourist circuits may be introduced, followed by the origin and application of TSP in the second session. In the third session, students need to apply the TSP to develop a suitable tour package with a strong rationale either based on the cost or shortest route in the light of the needs and wants of the target travel group. In the first phase of introduction to tourist circuits, the tutor may elaborate on the evolution of tourist circuits, existing tourist circuits in the country, uniqueness and benefits. In the second phase, the tutor needs to introduce the TSP by explaining its origin, the basic formula for calculation, data requirements and real-time application of TSP using a simple travel solution. In the third phase, the tutor may provide a list of destinations along with the themes for creating tourism circuits and the demographic profile of the travel market. Each student group may be provided with the same data, however, each group should produce a travel circuit for a particular target group of travellers such as senior citizens, honeymooners, babymooners, adventurers or groups of women business travellers. Each group/team's performance may be evaluated using a structured rubric that emphasizes the students': (i) understanding of the concept; (ii) proposed travel circuit that appeals to the target group; and (iii) rationale for the tour package considering the demographic profile of the travellers. Once the students are familiar with applying the TSP, the session may be extended with more tourist destinations in the circuit.

Teaching instruction plan

Discussion point	Time required (min)	Teaching strategies
Phase 1		
Introduction of the concept	10	The tutor asks the students what is meant by a tourist circuit. The instructor needs to allow multiple interpretations and explanations.
Lecture on introduction to tourism circuits, origin, definition and advantages	20	The tutor needs to record the keywords on the board. After receiving answers from the students, the tutor needs to provide a short lecture on tourism circuits, their origin, definition and advantages.

Continued

Teaching instruction plan Continued.

Discussion point	Time required (min)	Teaching strategies
Scope and objectives of theme-based tourism circuits	20	Now the tutor randomly asks a few students about their knowledge of theme-based tourism circuits in their country or region and its objectives. If the students' answers reflect good understanding, the tutor may ask the students to look at or find an example of an existing theme-based tourism circuit online (e.g. available at: http://swadeshdarshan.gov.in/index.php?Theme (accessed 30 September 2022)) to explore the destinations included and the distance between each destination and make a rough calculation on the cost of the tour operation.
Conclusion and evaluation of the session	10	To conclude the session, the tutor once again provides a summary of tourism circuits and their objectives. The tutor may post a Google Form link to check the students' understanding using a few multiple-choice questions.
Phase 2		
Introduction to quantitative methods	7	The instructor begins with the introduction, significance and application of quantitative methods (QM) in the tourism and hospitality business. The tutor may ask the students to explain their knowledge and previous usage of QM during their undergraduate programme. The tutor needs to record the critical operational areas where the QM are adopted.
Introduction to operations research (OR) techniques	8	The tutor provides a lecture on the evolution, significance and scope of OR in business. After giving an example of the application in a generic business context, the tutor may ask the students to specify the areas where they can apply the tools of OR in tourism and hospitality.
Evolution, objectives and application of TSP	10	The tutor needs to introduce the TSP by specifying its origin, application, key terms and various approaches, including the branch and bound technique to solve the TSP. The instructor may randomly ask a few students to volunteer themselves to explain the terms used in the TSP.
Solving a simple TSP problem 1	20	Now, the instructor needs to introduce a simple TSP and teach methods of solving the TSP using the branch and bound technique.
Solving a sample TSP problem 2	15	After demonstrating how to solve the simple problem, the tutor now provides another simple TSP for the students to solve. Students should be advised to tackle the problem step by step with the tutor checking each step before they move on to the next one.Once the problem has been solved, the tutor may appreciate the students who have successfully solved the TSP by themselves. The tutor now introduces the software or applications used to solve the problems. After providing a summary of the session, the tutor may randomly ask a few students about their impressions of using the TSP and its calculation.
Phase 3		
Applying the TSP to develop a suitable tour package for a target travel group	10	The tutor divides the learners into groups of three to five individuals and allocates each group a theme for creating a tourist circuit and the distance matrix (in kilometres or miles). The tutor provides the students with a tentative list of destinations. However, the learners need to make a choice of destinations from this list according to the purpose of travel (e.g. the destinations may differ between a honeymoon package and a religious tourism package). After the students have made their choice, the tutor may ask them to measure the distance between each destination using Google Maps (primarily using roads/National Highways).

continued

Teaching instruction plan Continued.

Discussion point	Time required (min)	Teaching strategies
Identification, installation and data entry in a suitable software package to solve the TSP	15	After creating a distance matrix, learners need to enter the distances in the software according to the requirements of the software. The tutor may guide the learners on data entry using an LCD.
Running the program and interpreting the output	10	Now the tutor may instruct the learners to run the software program and find out the shortest route.
Repeat the process once again and answer discussion question 1	10	To ensure that students have learnt the concept, the tutor may instruct each group to change their theme and repeat the process once again. Based on the distance of each circuit, students need to calculate the cost for each circuit (i.e. answer discussion question 1).[a]
Evaluation and conclusion	15	The new travel route developed based on the new theme needs to be evaluated by the tutor. Students may be tested orally or a PowerPoint presentation may be used to evaluate each group's performance. Each group may be evaluated based on: (i) their choice of destination; (ii) its relevance to the given theme; and (iii) the rationale behind the choice. Note: The case may be extended to another one or two sessions by instructing the students to develop an entire tour package itinerary, along with the cost, using the newly developed tourist circuits.

[a]The tutor may ask students to answer discussion question 2 as a take-home assignment (time needed not shown), to check they have understood the lecture 'Introduction to OR techniques' given in Phase 2 and encourage them to find other examples for applying OR techniques in hospitality and tourism.

Duration

180 min

Analysis

1. Using the distance matrix provided, calculate the cost for each tourism circuit evolved through the application of the software. Consider the cost as US$2.5/km.

The cost for each tourist circuit is calculated as per the given cost per kilometre. The cost of the circuits described in the case study are shown below.

Circuit A

Bengaluru → Shravanabelagola → Dharmasthala → Karkala → Udupi → Kollur → Murdeshwar → Gokarna → Bidar → Kalaburgi → Vijayapura → Bengaluru.
Total distance: 2050 km
Cost: 2050 km* US$2.5 = US$5125

Circuit B

Mangaluru → Dharmasthala → Shravenabelagola → Bidar → Kalaburgi → Vijayapura → Gokarna → Murdeshwar → Kollur → Udupi → Karkala → Mangaluru
Total distance: 1993 km
Cost: 1993 km* US$2.5 = US$4982.5

Circuit 1: Bengaluru as a source and end destination

Bengaluru → Tumkuru → Shravenagbelagola → Mysuru → Srirangapatna → Bengaluru.
Total distance: 421 km
Cost: 421 km* US$2.5 = US$1052.5

Circuit 2: Mangaluru as a source and end destination

Mangaluru → Dharmasthala → Karkala → Kollur → Murdeshwar → Gokarna → Udupi → Mangaluru.
Total distance: 612 km
Cost: 612 km* US$2.5 = US$1530

2. Explore other possible operations research (OR) techniques which may help in improving efficiency in tourism and hospitality-related business operations.

There is a broad scope for applying all OR tools in hospitality and tourism. Various OR techniques and their possible areas of application are as follows:

- *Linear programming* can be used for revenue management, cost minimization, allocation of resources, maximizing the profit, optimal mix of the menu, the optimal combination of variables, investment decisions, production planning and inventory control.
- *Assignment and transportation models* can be used for movement of assets, logistics, purchase and delivery of raw materials, sales territory allocation and working out the shortest route.
- *Inventory models* can be used for managing raw materials and inventory management.
- *Network models* can be used for event planning, hotel project management and scheduling of events.
- *Decision theory* can be used for capital investment, investment under uncertainty and decisions under risk.
- *Game theory* can be used for pricing strategy and pricing, booking limits for reservations and decisions under risk.
- *Queuing and simulation* can be used for capacity and resource planning and decisions under uncertainty, modelling of demand, average waiting time in a queue, average waiting time for service, utilization of service facilities, package handling in airports and employee scheduling.

References

Bloom, B.S. (1956) *Taxonomy of Educational Objectives: the Classification of Educational Goals*, Vol. 1. Longmans, Green & Co., London.

IBEF (India Brand Equity Foundation) (2022) Tourism & Hospitality. Available at: https://www.ibef.org/download/1658817568_Tourism-and-Hospitality-June-2022.pdf (accessed 30 September 2022).

Ministry of Tourism (2022) India Tourism Statistics 2021. Ministry of Tourism, Government of India. Available at: https://tourism.gov.in/sites/default/files/2022-04/India%20Tourism%20Statistics%202021.pdf (accessed 30 September 2022).

Further Reading

Garfinkel, R.S. (1973) Technical Note – On partitioning the feasible set in a branch-and-bound algorithm for the asymmetric traveling-salesman problem. *Operations Research* 21(1), 340–343. https://doi.org/10.1287/opre.21.1.340

Hamdy, A.T. (2017) *Operations Research: an Introduction*, 10th edn. Pearson, London.

OSDN (Open Source Development Network) (2022) TSP Solver and Generator. Available at: https://osdn.net/projects/sfnet_tspsg/releases/ (accessed 12 January 2022).

25 Sustainability of Artists and Craft Persons During COVID-19: a Case Study of Kalakriti Creations

Kamakshi Maheshwari* **and Chandradeep Singh**
Indian Institute of Tourism and Travel Management, Gwalior, India

Abstract
India is a land of unique culture bonded with unmatched art and crafts in every nook and corner. Apart from being one of the most important components of tourism, the handloom and handicraft sector makes a major contribution to employment generation. This case study helps the learner understand the application of a SWOT (strengths, weaknesses, opportunities and threats) analysis on the handloom and handicraft industry and the efforts made by artisans and craft persons. It will also provide an opportunity for: (i) critical thinking regarding the opportunities and threats to the handicraft industry; and (ii) analytical thinking on promoting handicrafts as a product and how artisans survived during the coronavirus disease 2019 (COVID-19) pandemic with the help of their art and crafts.

Subjects: Indian culture and heritage, tourism marketing, business management, entrepreneurship development, strategic management

Learning Objectives

At the end of the case study, the learner will be able to:

1. Understand how to apply a SWOT (strengths, weaknesses, opportunities and threats) analysis to a business model.
2. Identify the various steps involved in the process of developing business and employment opportunities.
3. Understand the application of the glocal concept (i.e. reflecting both local and global considerations) in development of art and craft.

Case

The handloom and handicraft industry is the backbone of India. Indian tourism is highly dependent on authentic culture, handicrafts, handloom products, antiques and antiquities to attract tourists from all over the globe, not only to enjoy the nature, beauty and heritage of the land but to take back memories in the form of souvenirs. On the outbreak of coronavirus disease 2019 (COVID-19), every sector suffered considerable losses in the form of resources. The tourism and handloom and handicraft

*kamakshi.maheshwari@iittm.ac.in

 Teaching Cases in Tourism, Hospitality and Events (eds. S.K. Dixit and S. Piramanayagam)

DOI: 10.1079/9781800621022.0025

industries are very fragile and also interdependent on each other. According to analysis by the United Nations Educational, Scientific and Cultural Organization (UNESCO), more than 30 million people globally are engaged in culture and handicraft and of these most are women who are dependent on their trade for social and economic security (UNESCO cited in EY, 2015), and these people have been badly affected by the pandemic.

On the morning of 14 August 2020, Mr Vikas Singh, founder of Kalakriti Creations (a cluster of artists and craft persons in India), received a call from Mithila, Bihar, from Mr Adarsh, a professional goldsmith and ancestral Madhubani artist, describing the desperate condition of his family since his shop had been shut down due to the COVID-19 pandemic. He also depicted the conditions of other families in different regions of Bihar due to the outbreak of COVID-19. He asked Mr Vikas Singh to help these families to overcome the hand-to-mouth conditions they were trying to cope with while simultaneously reducing the negative impact of COVID-19. As Mr Vikas Singh belongs to Bihar, he felt a responsibility towards his motherland. He immediately invited his core managing team to design a sustainability plan for the craft persons and artists of Bihar which later on may be used in the other parts of the country too.

This was a turning point; 2020 had been a tough year for everyone around the globe due to the COVID-19 pandemic. Many had lost their jobs and businesses. Understanding the situation narrated by Mr Adarsh, Mr Singh and his team decided to form a cluster of such artists, craft persons and men and women doing handmade work who had lost opportunities of making a living due to COVID-19 and were struggling financially. This initiative had the following objectives:

- to build an e-commerce platform for such artist/craft persons who could not earn during the pandemic;
- to revive the endangered arts and crafts and transform their authenticity by making products that people can use today but that are made with authentic materials and skill; and
- to bring these products to market in order to create awareness about the handmade and handicraft products.

As a result, the team decided to dedicate a fully formed e-commerce platform named ‘Kalakriti Creations’ to support local artists, craft persons, potters, etc., with unique hand skills. To begin with, the team decided to focus on the region of Mithila and the city of Bhagalpur in the state of Bihar. A strategy to understand the condition of the artists and craft persons here and their willingness to continue their art work and Madhubani painting in particular was evolved. Mr Adarsh was chosen as the link to undertake a survey and as a result approximately 30–40% families from Mithila came forward to take up hand painting as their occupation. Furthermore, in the survey conducted in Bhagalpur it was found that more than 70% of families were ready to take up handloom weaving as their occupation, making Geecha silk, Bhagalpuri bedsheets and other handloom woven products, provided they had regular sales.

The way ahead

On understanding this situation and critically examining the conditions, team Kalakriti Creations (team KC) decided to support these art saviours, as explained above. However, there were many challenges for the team on the way ahead such as:

1. Due to the pandemic, many state borders were not open so it was difficult to send the raw materials to artists in order for them to work, and also to receive the finished products.
2. Investment was needed in order to provide enough work for every family.
3. It was not clear how the team could make the artists and crafts persons understand the blends of designs and colour scheme required for the new utility products.
4. Many families had left the art forms in lieu of their jobs so refresher training was required for them.
5. Most significantly, it was not clear how these products would be marketed.

6. Another major challenge for the team was pricing. Technology has made everything that used to be made by hand available in replica form but made by machines, so for example the power loom and printing machine have led to competitively priced or low-priced machine-made products compared with hand crafted, handmade and handloom products.

Taking all these challenges into consideration, team KC prepared a plan or work cycle to facilitate all or many of the problems, as shown in Fig. 25.1.

The cycle followed by team KC began with the raw design of each product, then identifying and procuring the raw material to be used to make the product, followed by approaching artists from various states and understanding their skills, identifying the skill gaps and providing online training for the artists so they could follow the design and colour scheme. After developing their designing and colouring skills, the raw material is sent to the artists for the development of products followed by marketing of the products through their own e-commerce site and sites belonging to other corporations with whom they have 'tie-ups' (agreements) and physical exhibitions by Kalakriti Creations, and the artists will be paid for their work. By following this cycle team KC created a synergy between the crafts person and the market. The aim was to devise products that are in greatest demand according to the national market.

As it was the time of the pandemic, team KC started with making two-layered face masks using hand-painted cotton. This has now become a product range of 50 different face masks with a variety of paintings and crafts.

From Bihar to Assam, Kerala, Gujarat, Rajasthan, West Bengal and Maharashtra,

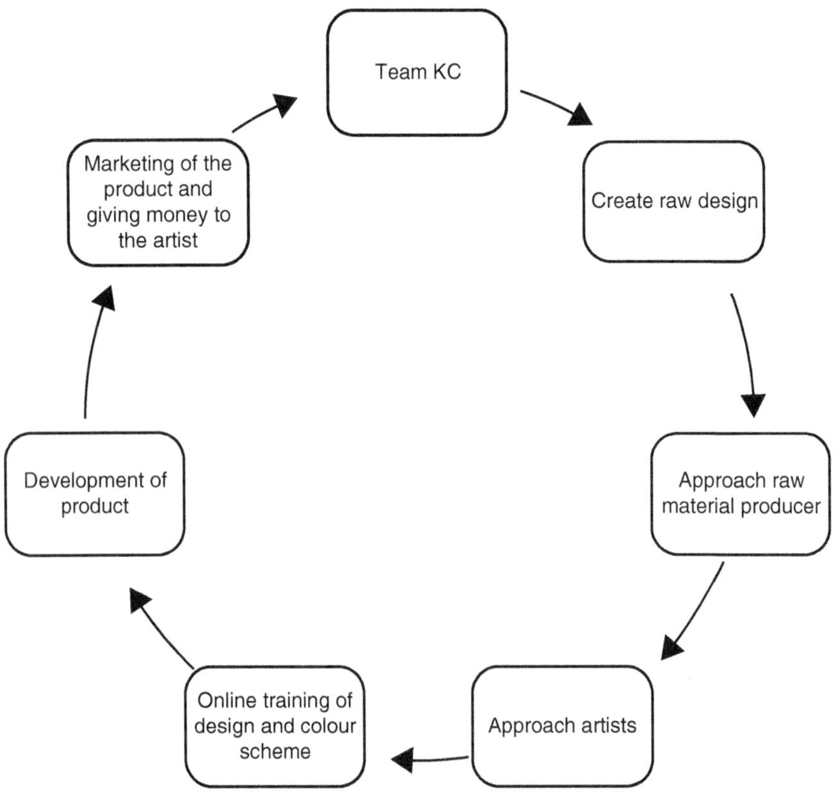

Fig. 25.1. Work cycle of Kalakriti Creations. Authors' own image.

team KC is spreading its wings to cater for every possible art and craft form and to give a reason to the artist for not leaving practising their art and craft.

To overcome the challenge of marketing and branding, team KC developed a website www.kalakriticreations.in (accessed 26 September 2022) to take orders. Apart from this, products are available for sale through social media such as Instagram, Facebook and WhatsApp. Another method taken up was to display the products in exhibitions.

Today, Kalakriti Creations is rightly fulfilling its tagline '*Mitti ke rang Kalakriti ke sang*' (meaning 'experience the essence and colours of soil with Kalakriti') with a customer base of 2000 customers from all over India and 40+ artisan families on board creating more than 50 product lines.

SWOT analysis

From the initial phase Kalakriti Creations wanted to maximize the opportunities for the artists and to capitalize the handloom and handicrafts sector. Over time, team KC went through several ups and downs which can be represented in the form of a SWOT (strengths, weaknesses, opportunities and threats) analysis (Table 25.1).

Strengths

This section of the SWOT analysis describes the strengths possessed by the organization. One strength is that Kalakriti Creations has a team of professionals who were working in the heritage art and craft sector. They had prerequisite experience and knowledge about art and crafts. This knowledge and experience helped the team to identify the importance of a particular art and try to devise strategies to bring it into mainstream products and product lines. Another strength is the need for work opportunities as has already been explained in the case. Also it had been known from various sources that the pandemic wiped out new work opportunities and badly affected existing businesses and people's work. So there is the need for new work opportunities that Kalakriti Creations can generate.

Lastly, there is a large and diverse potential market, as it is already known around the globe that India is a country with a rich and varied culture which gives birth to many diverse forms of art and craft. This gives the team the objective to mingle the old art and crafts with products with modern uses and assists them in expanding their product lines.

Weaknesses

As Kalakriti Creations started working in the tough times of the COVID-19 pandemic the organization faced a scarcity of funds and human resources to work things out smoothly. In the pandemic the team observed that potential investors were hesitating about investing in any form of business. Starting with a few artists in the Bihar region, Kalakriti Creations faced difficulties in locating other arts and craft persons who were willing to take up the work. The majority of artists had moved to mainstream jobs leaving their artistic occupations behind. Even after finding the artists and craft persons willing to agree to work, it was a task for the team to train them to use their art with the element of modern design and colour scheme. It was particularly difficult during the pandemic as physical movement was restricted and the team had to make the artist and craft person understand the design and colour through online or video conferencing.

Opportunities

Despite the tough task of training the artists and craft persons, it is an opportunity for Kalakriti Creations to create a blend of old and new. Taking the traditional art and craft and blending it with the items used for present-day convenience has many advantages, for example: (i) it helps in making the art form more saleable; (ii) it helps to preserve the endangered and dying art and craft forms; and (iii) it creates employment opportunities for the artists and craft persons. Regardless of getting the products ready to meet the demands of present-day convenience, the process remains incomplete if there

Table 25.1. SWOT analysis of Kalakriti Creations. Authors' own table.

Aspects of the business	Description
Strengths	Knowledge and experience of heritage, art and craft
	Need for work opportunities
	Large, diverse and potential market
Weaknesses	Lack of funds
	Lack of human resources
	Lack of network
	Lack of trained artists for the mainstream designs
Opportunities	Development and fusion of ancient art and crafts with modern trends
	Use of e-commerce as a means of distribution channel
Threats	Competition with the big brands that are already working in this area
	Chance that artists and crafts persons may return to their original work places post-pandemic and no longer wish to work in the handloom and handicraft industry
	Technological advancements such as the power loom and machine printing can be used to reproduce the designs in bulk, making the products more cheaply and more quickly, but these products lack the authenticity of the product produced by the artist
	Lockdown restrictions brought about during the pandemic to prevent the spread of the virus caused difficulty for the organization to move goods (e.g. raw materials to the artist)

is not a suitable market place for the products. However, during the pandemic when movement of goods was a struggle, many businesses turned to using technology to sell their goods. So Kalakriti Creations' developed an e-commerce platform to be used as an effective market place. It provides the organization with an expanded reach to customers and at the same time it helps in creating the brand value of Kalakriti Creations.

Threats

Kalakriti Creations started working at a time when there were already big players working in the industry. It turns out that it is difficult to propose a working strategy to cope with the situation and the prevailing competition. Another threat is that although Kalakriti Creations motivated artists to work with the organization, it was during the pandemic when there was a lack of opportunities, so people started moving towards any sort of opportunity available. However, post-pandemic the organization is facing issues related to reverse migration, in that some people started moving back to their original work places as soon as the restrictions brought about by the pandemic were lifting.

As every coin has two sides, although technology can be useful it also has another side which enables the artist to produce their art using technology in bulk, more quickly and cheaply, using power looms, machine printing, automatic stone cutting and grinding machines, etc. The use of technology, no doubt, creates huge volumes of product but at the same time it is depleting the authenticity and ethnicity of the art. Art and craft always has an element of the human touch but the use of the above-mentioned technologies strips that uniqueness and human element from the art and craft.

Due to movement restrictions during the lockdown, imposed by the government in the pandemic to prevent the spread of the virus, it became difficult for the team to move goods. For example to move the raw materials from the supplier to the artists and also the finished product back to the team for them to sell.

Business model

The business model adopted by Kalakriti Creations is shown in Fig. 25.2.

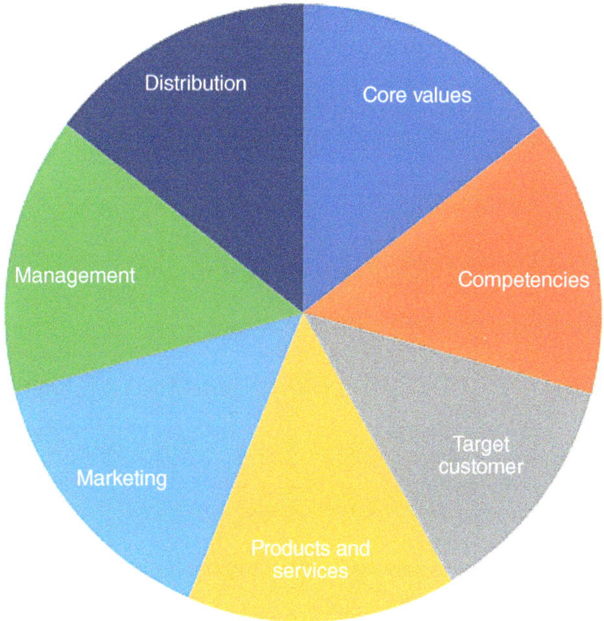

Fig. 25.2. Business model adopted by Kalakriti Creations. Authors' own image.

Core values

The organization works with the core values of preserving the art and protecting the artists. It works to provide the artists and crafts persons with the opportunities to work and earn a living and the chance to save the art forms.

Competencies

The organization works with artists who are already capable in their respective type of art and crafts. Most of the artists did this work as an ancestral profession but somehow moved to some other profession in search of better economic opportunities. Kalakriti Creations works to utilize the competencies of these artists and craft persons and to provide them with the same opportunities they were getting in some other professions before the pandemic.

Target customer

The organization began with the idea of making more people aware of these endangered and dying arts and crafts. For the same purpose Kalakriti Creations brought forward the idea of using the art forms in making modern products that people can use today such as products for home decor or clothing. The target customer of the organization is household clients. Kalakriti Creations' products are marketed through their e-commerce platform, social media and print media to make them visible to the target customer.

Products and services

As mentioned above, the company has developed its product lines in and around the home decor and clothing segment. Since the initial days the organization has developed various and diverse product lines as shown in Table 25.2.

Marketing

The organization is primarily using technology for the purpose of marketing. The e-commerce platform has provided an expanded reach to customers and the organization is receiving orders from across India. Other than that network marketing is

Table 25.2. Product lines of Kalakriti Creations. Authors' own table.

Product line	Description of items		
Clothing	Saree	Shawl	Male scarf
	Duppata[a]	T-shirts	Palazzo pants
	Stole	Kurta[b]	
Home decor	Trays	Tissue paper holder	Photo frames
	Tea coasters	Tissue box	Key chains
	Wall clocks	Luggage tag	Glass jars
	Toran[c]	Cushion covers	Tic-tac toe game
	Doorside latkan[d]	Fridge magnets	
	Key holders	Car flags[e]	
Bamboo items	Lamps	Pen stand	Water bottle
	Tea cups	Tray	Mobile phone stand
River grass items	Bottle bag	Shopping bag	Dining table mats
	Mat/chatai	Aasan/pooja mat	
Bags	Jute hand clutch	Jute bottle bags	Cloth shopping bag
	Mobile phone pouch/ purse	Jute shopping bags	
	Backpacks	Jute potli bag	
Footwear	Canvas shoes	Mojri[f]	Cloth belly shoes
	Female chappals[g]	Kids' shoes	
Paintings	Madhubani, Bihar	Patchitra, Odisha	Saura , Odisha
	Warli, Maharashtra	Phad, Rajasthan	
	Gond, Madhya Pradesh	Manjusha, Bihar	
Kids' wear	Boy	Girl	
Office essentials	Book shelf	Office organizer/desk organizer	Diaries
	Book marks	Wooden pen	Whiteboard
	Pen stand		
Clay pottery	Cookware Handi	Agarbatti stand	Water bottles (painted and unpainted)
	Cookware Tawa	Dhopbatti stand	Planters
	Cookware Kadai	Laxmi-Ganesh figurine	Diya/lamps
	Curd setter Handi	Glass/tumbler (painted and unpainted)	Piggy bank
	Bowl		

[a]A duppata is a type of shawl worn by women.
[b]A kurta is a loose collarless shirt usually worn by men.
[c]A toran is a decorative piece used across doorways and archways.
[d]A latkan is a hanging decorative piece made of coloured glass, stones, cloth and painting.
[e]Car flags are a set of paintings on small wooden sheets attached with a string which can be tied to the rear windscreen of a car.
[f]Mojri are a type of decorative shoe or slipper.
[g]A chappal is a type of leather sandal.

used and for the local visibility, exhibitions have been used as marketing tools.

Management

The management of the organization work includes product portfolio management, human resource management and operations management.

Product portfolio management works with all the aspects of the products and the product line. This begins with identifying the need for the product, and is followed by modifying the products with respect to the art and craft, procuring the raw material, sending the raw material to the artist, conveying the concept to the artist, and receiving the final product from the artist (Fig. 25.3). As artists and crafts persons working with Kalakriti Creations come from states all over India (Table 25.3) sending them the raw materials and receiving the finished product from them is a challenging operation.

Human resource management covers all the aspects of dealing with the artisans and craft persons. This includes identifying the particular type of art and craft required, locating the artisans and craft persons, and finalizing the agreement between those artisans and craft persons and the organization.

Operations management deals with all the work delegated among the team, including corporate tie-ups, strategy formulation, stock management and revenue management.

Distribution

Once the organization receives the final product the distribution process starts. For this, Kalakriti Creations had corporate tie-ups with e-commerce websites (as well as their own website), courier companies and local delivery agents.

The future post-pandemic

As the government lifts many of the restrictions imposed during the COVID-19 pandemic, will Kalakriti Creations operating mainly in the online market be able to compete with large companies such Fab India and Tribes India that are operating in all the major cities in the country? Team KC need to consider what strategies should be taken up by Kalakriti Creations in order to hold on to the substantial share of the market they have gained and compete successfully with these giants.

Discussion Questions

1. Identify the SWOT elements for the handicraft industry.

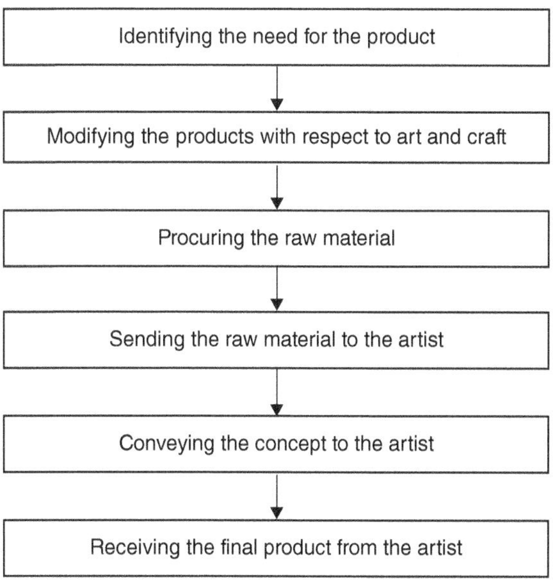

Fig. 25.3. How product portfolio management works at Kalakriti Creations. Authors' own image.

Table 25.3. List of states and art forms Kalakriti Creations is working with. Authors' own table.

State	Art
Assam	Weaving
Bihar	Madhubani
	Manjusha
	Handloom
Madhya Pradesh	Gond
	Chitera
	Batto Bai
	Chanderi
	Mandana
	Mandala
	Chanderi loom
	Maheshwar loom
	Bagh block print
Maharashtra	Warli
Gujarat	Ajrakh loom
	Kutch embroidery
Odisha	Pattachitra painting
Rajasthan	Phad painting
	Kathputli work
	Inlay work
	Meenakari
Kerala	Murals
West Bengal	River grass
	Bamboo
North East	Bamboo work
	Jute work

2. Identify two art/craft forms for development as sellable commodities.
3. Determine various marketing and advertising strategies which can be applied in the global handicraft industry to position handicrafts as a commercial commodity.
4. In your opinion, what could be done by an organization working on humanitarian grounds to sustain and grow during tough times when the organization is competing with massive companies who already have a large market share?

Teaching Notes

The case study is intended to familiarize learners with the concept of strategic planning, entrepreneurship development, awareness of social work and developing the planning process for business management. The objective is to equip postgraduate students and research scholars of tourism and management to identify and appraise the various stages in building up a business cluster or how to promote a business while dealing with all the possible challenges. Analysing the trend in the market for handloom and handicraft products, employment generation through this sector, the role of authentic handicrafts in promoting Indian exports and the tourism industry, increasing foreign exchange earnings, and considering entrepreneurship opportunities would result in students thinking beyond the classroom discussions and developing a real-time approach to understanding the significance of traditional art and handicraft works globally.

However, students are expected to devise a new and better strategy for developing this cluster and creating employment opportunities for the artists who are already on board. It is recommended that the facilitator or tutor discusses the various possibilities to support these artist groups, and tries to help students to devise a new business model and provide a strong base for research areas in this sector.

Target audience

This case is targeted at undergraduate and postgraduate students of tourism and travel management.

Teaching methods and equipment

This case can be activated through scenario analysis and group discussion. The decision and answers to each discussion question should be based on a simple discussion, situation analysis and critical thinking of the case and the conditions described. The instructors may ask students to read related articles and research studies in order to justify their answers to case discussion questions. There is no need to provide other equipment, although the use of a computer, projector and a flipchart may be used to discuss the case.

If students are not familiar with marketing and advertising strategies, to help students answer discussion question 3 the instructor could provide examples of a well-planned marketing mix and branding and advertising approaches.

Teaching instructions

This case can be taught for just over an hour in the classroom.

To start the case the instructor may ask the class to share their experience of any artists or artforms from their area and try to draw a link about how the students think they could bring them on to a platform such as Kalakriti Creations. The discussion will help the tutor to ensure a common understanding of what the case is about. The case may be discussed and analysed to understand the situation under which Kalakriti Creations has taken shape, to identify different stages from development of a product to reaching the customer, and the strategies used by the organization to sustain the artists during the difficult time of the COVID-19 pandemic and for their long-term survival post-pandemic.

Teaching instruction plan

Discussion point	Time required (min)	Teaching strategies
Introduction	10	The instructor begins the case analysis by asking the students questions about any artist/craft persons' family who they have come across. The instructor randomly asks a few students to volunteer to explain how these artists were affected by the COVID-19 pandemic. This can lead to the instructor making students aware (if they are not already) of how artists and crafts persons were so badly affected economically by the lockdowns. The instructor can then ask if the students had been in such a position, what would they have done to sustain these artists and crafts persons? Then the instructor will describe various factors related to the building of Kalakriti Creations. The instructor may note down the main points on the board.
Briefing on the steps involved in developing a product into a sellable commodity	5	The instructor may brief the students on the steps involved in developing a product using an intangible resource (e.g. art and craft) into a sellable commodity and reaching out to the customer. They may conceptualize various possibilities that can be evolved.
Summary of the case	10	The instructor briefly discusses the case of Kalakriti Creations in terms of the background information, the scenario, the dilemma, the decisions made and the challenges faced. The instructor may outline what is required in general for a SWOT analysis for better understanding. Finally, the instructor asks the students to get into groups or teams of about two to four individuals (depending on the size of the class) in order to answer the discussion questions.
Assignment question 1	10	**Identify the SWOT elements for the handicraft industry.** The instructor asks each student group/team to think about the various strengths, weaknesses, opportunities and threats for a business such as Kalakriti Creations. Then the instructor asks each group/team to justify the rationale behind their answer.

Continued

Teaching instruction plan Continued.

Discussion point	Time required (min)	Teaching strategies
Assignment question 2	10	**Identify two art/craft forms for development as sellable commodities.** The instructor asks each student group/team to identify two art/craft forms for development as saleable products (or both may be used in a single saleable product). Ideally products should be something that everybody uses daily so there is a ready market for the product. The instructor now allows each team to discuss their choices. Each team must defend their choice(s) using facts and 'on the ground' reality.
Assignment question 3	10	**Determine various marketing and advertising strategies which can be applied in the global handicraft industry to position handicrafts as a commercial commodity.** The instructor asks the student groups/teams to list various promotional and marketing strategies which could be used to make the cluster of artists/crafts persons more active and accessible. If students are not familiar with marketing and advertising strategies the instructor could provide examples of a well-planned marketing mix and branding and advertising approaches.
Assignment question 4	10	**In your opinion, was the decision taken by Mr Singh and the KC team right?** Yes No When the students have answered the question, the instructor may then ask the groups/teams to give their opinions on what could have been done and to justify their answers.
Closing remarks and discussion	10	The instructor concludes the case analysis session with his/her analysis and answers to the discussions questions, including how to apply a SWOT analysis to a business model.

Duration

1 hour 15 min

Analysis

1. Identify the SWOT elements for the handicraft industry.

In the class session student groups are asked to think about the strengths, weaknesses, opportunities and threats for a business such as Kalakriti Creations, so their answers and reasons should be similar to those shown in Table 25.1 of the case. However, this question can be expanded into an additional assignment following the class session, by asking the students to conduct their own research into the wider scope of the handicraft industry. Each student could select a particular art or craft form from a list of various region-based art and handicrafts, and make a study of it in order to understand the domain-specific characteristics of the particular art or craft. A SWOT analysis can be applied taking the domain and the region as the basis of the study.

2. Identify two art/craft forms for development as sellable commodities.

Experience speaks more than words. This applies in the handicraft industry as well. Art and crafts applied to products that people use daily can result in wider adaptability and acceptability in the commodity market. Research and data based on the export industry can help the learner understand the size of the market for intangible resources (e.g. art and craft) that are transformed into commodities.

3. Determine various marketing and advertising strategies which can be applied in global handicraft industry to position handicrafts as a commercial commodity.

Study of a well-planned marketing mix and branding and advertising approaches may lead the learners to devise the best possible marketing-mix model for the handicraft industry. Study of general marketing principles will also be of great help to the learners.

4. In your opinion, what could be done by an organization working on humanitarian grounds to sustain and grow during tough times when the organization is competing with massive companies who already have a large market share?

This question can have multiple answers. This question is added in order to analyse the perspective of the students and their knowledge about the competitive business environment. As is clear from the case, Kalakriti Creations has to compete with the giants of the industry such as Fab India and Tribes India. Knowledge of the competing brands and of the varied opportunities and threats the organization has to face while dealing with such competitors has to be gathered and considered before formulating strategies to sustain and grow an organization that is aimed at supporting artists and crafts persons from a humanitarian aspect as well as preserving their traditional skills.

Reference

EY (2015) Cultural times: the first global map of cultural and creative industries. Available at: https://en.unesco.org/creativity/sites/creativity/files/cultural_times._the_first_global_map_of_cultural_and_creative_industries.pdf (accessed 3 October 2022).

26 Sage Bliss: the Question of Employee Equity

Senthilkumaran Piramanayagam[1]* and Saurabh Kumar Dixit[2]
[1]*Manipal Academy of Higher Education, Manipal, India;* [2]*North-Eastern Hill University, Shillong, India*

Abstract
Workplace equity is an organizational process that insists on treating every employee equally. A workplace does not rely solely on one employee for its success, and employees need to be treated equally. Although the words equity and equality look similar, both are distinct concepts. Equality seeks to provide equal access to resources and opportunities, while equity refers to the relationship between intrinsic workplace behaviour and extrinsic reward for employee behaviour. Equity is distinct as each employee may not be given equal access to all the organization's resources. Stacy J. Adams (1965) emphasized that people are motivated to act in a given situation based on their perceived equity and fairness. She claimed that when employees perceive inequity in an organization, it has significant implications for organizational success and business outcomes. This case study aims to understand the significance of employee equity and apply Adams' equity theory of motivation through a hypothetical hospitality organization called Sage Bliss. The case will help learners comprehend the problems related to perceived employee inequity in an organization and their consequences, and differentiate between the symptoms, surface causes and underlying causes in order to develop strategies to manage inequity effectively.

Subject: motivation, organizational behaviour, employee equity

Learning Objectives

At the end of the case study, the learner will be able to:

1. Identify and diagnose the issues related to perceived employee inequity in the hospitality business.
2. Apply the concepts of Adam's equity theory of motivation and develop strategies and tactics to manage inequity effectively.
3. Differentiate between symptoms, surface causes and underlying causes for employee behaviour and organizational issues and challenges that affect its survival.

Case

Dr Kapoor is a renowned cardiologist most sought after by people from the corporate

*senthil.kumaranp@manipal.edu

 Teaching Cases in Tourism, Hospitality and Events (eds. S.K. Dixit and S. Piramanayagam)

DOI: 10.1079/9781800621022.0026

and political world in the country. Dr Kapoor graduated from a top medical college in the capital and later pursued his fellowship in a well-known university hospital in the UK. He then moved to the USA where he headed several research projects on behalf of the federal government and international institutions before returning to his home country. Dr Kapoor recognized that globalization had brought many changes in his home country. It has changed the way people work, organizations operate, family structure, lifestyles, diversity in the workplace, and the role of women in both economic activities and traditional homebound activities and their aspirations. These changes have led to a rise in stress levels among citizens and concerns related to health, rejuvenation and well-being so people are seeking opportunities to relax and get away from the day-to-day stresses of everyday life. Dr Kapoor created Plum Villa Resorts and Hotels Inc. to establish the first destination spa and resort in the country. Sage Bliss, the destination spa, is the brainchild of Dr Kapoor, the founder and the current chief operating officer (COO).

His dream of creating a world-class spa and resort spurred him to set up Sage Bliss spa in a hill region in the north-east part of the country. The hill region is also known as the 'Gateway to Heaven'. The city overlooks a sacred river, considered to be godmother by people living in the nearby community. The river is also a lifeline to millions of people who live downstream. The environment is serene and pollution free and the location is well known for its traditional health system and spiritual practices, including numerous yoga and meditation centres. A week-long International Mediation and Yoga Festival held in the location attracts thousands of domestic and international visitors. Leading spiritual leaders used to conduct comprehensive yoga and meditation classes and boot camps here throughout the year. The city enjoys excellent connectivity with other regions of the country through an extensive network of airlines, trains and coaches. The distance between the airport and the spa is about 12 km and the closest railway station is just 5 km away.

Current market structure

Spas and wellness centres are currently classified into various categories, for example: (i) salon spa; (ii) hotel and resort spa; (iii) medical spa; (iv) destination spa; (v) mineral spa; and (vi) thermal spring spa. The spa industry is an emerging global phenomenon due to the convergence of industries, therapeutic practice and traditional therapies. Spa therapies have been familiar around the globe since ancient times and practised in different forms. Now, globalization, increased consciousness of personal wellness, exposure to Western media and changing lifestyles have enabled the rediscovery of these practices and branded them to create a new industry. The global spa market was valued at US$95 billion in 2021 and is expected to reach US$185.5 billion by 2030 (Businesswire, 2022). The current market structure and list of spa and wellness service providers in the context of this case study are presented in Table 26.1.

Considering the scope and growth of the spa industry, Dr Kapoor believed that Sage Bliss has a great opportunity to become a dominant player in the market with new and innovative spa services. As most of the spas in the current location focus only on Ayurveda, Dr Kapoor wants to distinguish itself from others by integrating all traditional healthcare practices with modern medicine. Being his first venture into the spa business, he believed that being unique is vital to differentiate Sage Bliss.

Management and organizational structure of Plum Villa's destination spa

The destination spa has multiple buildings that are spread over 24 acres of land. It consists of villa-type guest cottages, leisure facilities, a medium-sized banquet hall on one side, an amphitheatre, an air-conditioned mediation hall and beautiful spa clinics. As a small property with 54 cottages built in the foothills, Dr Kapoor decided to be the head of the organization and manage the entire operation with the support of a team of physicians and employees. As per the new

Table 26.1. Major spa service providers in India, revenue and market share. Authors' own table.

Name of the service provider	Market share (%)	Revenue in US$ (million)		
	2021	2021	2020	2019
Nirvana Spa	26.9	197.4	84.7	213.2
Vedha Spa	22.9	168.2	64.2	201.4
Sacret Spa	15.5	112.2	42.2	183.2
Manas Spa	10.5	76.4	21.3	98.4
Myths and Angels	6.1	44.6	16.44	69.4
Other brands	18.1	132.7	34.8	178.6

trend, he declared himself the COO of the business. He has recruited two-thirds of his workforce from local people to look after maintenance, cleaning and other support services. He felt that it is his moral responsibility to provide job opportunities to the local community. However, he has recruited other employees from other regions based on their experience and expertise, including: (i) resident physicians with different specialties; (ii) consultants; (iii) masseurs; (iv) yoga, Ayurveda, beauty, skincare and energy therapists; and (v) lifestyle coaches. The organization structure of Sage Bliss is shown in Fig. 26.1.

Laundry, linen and security services are outsourced to support the operation and control the costs related to human resources. Six doctors who specialized in cosmetology, naturopathy, Unani, Siddha, Ayurveda and homeopathy were appointed as consultants to provide services to the guests. Based on guests' requests, doctors from other specialties are invited as guest consultants. Currently there are 37 employees at Sage Bliss. The number of employees has been kept low intentionally to provide a salary that is considered to be the best in the destination. It is essential to keep the salary high compared with other spas and wellness centres to retain the employees, as there is always high employee attrition due to the freezing cold winter. Dr Kapoor ensured that at least one physician was available as a resident consultant throughout the year. After a lengthy discussion with the doctors and consultants, he decided to provide the guests with extensive treatments and wellness services.

A good beginning

The spa was inaugurated successfully with an average occupancy of 85% almost for 6 months. For Sage Bliss, it is an overnight success. The success is mainly attributed to the ambience, excellent service, Dr Kapoor's approach and good customer relationships. Now, Sage Bliss has become another landmark and reason for a visit to the destination. Dr Kapoor's approach of combining allopathic medicinal practices with complementary, alternative and traditional medicine systems is now slowly spreading to other spas. Sage Bliss has been highly rated on all social network sites and travel platforms compared with other top-performing spa services. Client reviews on social media have played a significant role in attracting new customers. As the business started expanding, he recruited his relatives to fill major job roles in order to facilitate and manage the day-to-day operations of the spa. Dr Kapoor spent a lot of time with guests and allowed doctors and staff to manage their day-to-day operations on their own.

During his interaction with guests who have experienced spa treatments worldwide, Dr Kapoor understood that visitors were looking for an integrative spa experience to be amalgamated with Swiss spa therapy, which could be a significant attraction for international travellers. As Dr Kapoor wants to ensure the best quality and right kind of mix of treatments and wellness services, he decided to visit Greece, Ireland and other countries that are well known for their world-class spa services. He invited Ms Monesa, a relative

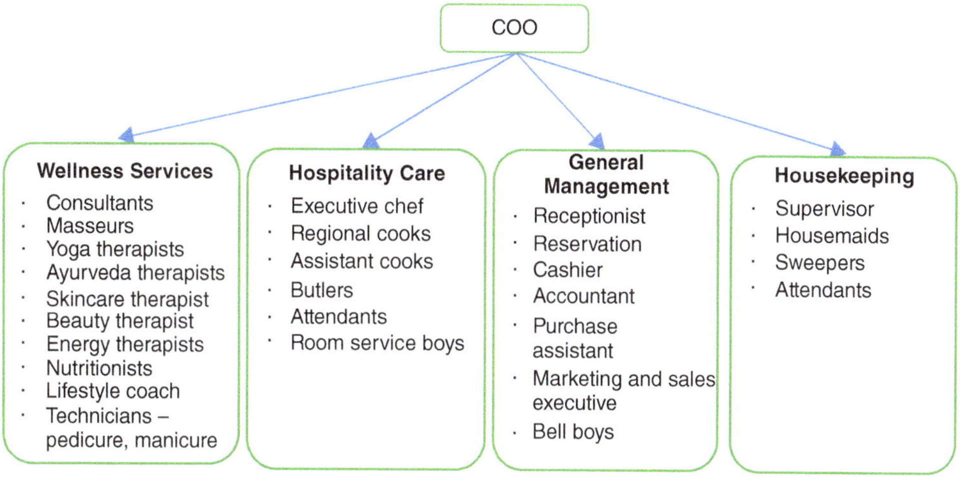

Fig. 26.1. Organization structure of Sage Bliss spa. COO, chief operating officer. Authors' own elaboration.

who has recently graduated in hotel management from a reputed hospitality school in the country, to manage operations in his short absence. Monesa is enthusiastic about her assignment as a walk-in manager for a destination spa; she believed it was an excellent opportunity to learn the operations of a destination spa. As it is her first job assignment after graduation, she wants to do her best in the current role.

Dr Kapoor instructed Monesa to use his office for a while as her temporary office, and he promised that she would be given a well-furnished and air-conditioned office if she is willing to continue the assigned task. Dr Kapoor called all the employees and associates together and introduced Monesa to the team, instructing them to cooperate with her to support the smooth running of day-to-day operations. When some physicians and the executive chef requested an appointment with Dr Kapoor, as they wanted to discuss something, he insisted that they should all speak to Monesa. He informed them that he was making an urgent trip to the capital city to process his visa and make other arrangements for his visit to destination spas. As Monesa has to set up her office and accommodation, she assured them she would meet them next week.

A challenging first job

For Monesa, all started well. She requested Dr Kapoor's secretary to provide details about Sage Bliss as she wants to understand the nature of the business, the organization's structure, the current financial position, the services offered, the profile of the existing customers and the business model. For 2 weeks, she spent her time reading files and analysing financial reports and ignored the request made by employees to meet her. After 2 weeks, while she was walking towards the office she noticed that more than half of the employees were standing in front of her office. She also observed that many employees were not in their designated uniforms and were whispering among themselves. When she greeted everyone, very few responded. Before entering her office, she informed them that she wanted to meet each employee alone and requested others to return to their respective clinics. At first, a senior physician practising Ayurveda entered her office and submitted his resignation letter, explaining that his family was located far away in the south and they were struggling to manage alone without him. To her surprise, he asked Monesa to settle his accounts immediately. He hinted that he had been disturbed by the behaviour of some employees and therapists who are relatives of Dr Kapoor. He narrated that a few therapists demanded extra money from guests and came on duty after consuming alcohol. He specified that he has no control over these employees as he fears that it may lead to unfavourable consequences. Understanding his plight, Monesa requested him to wait for a while until the return of Dr Kapoor or for a formal approval by mail to settle the accounts. At the time, she wanted to meet the next employee. However, Dr Kapoor's secretary entered the room and informed her that everyone standing outside her office had the same request and letter of resignation. Listening to the secretary's words, Monesa was shocked and asked the secretary to arrange a meeting with all employees after the lunch break, as she thought this was suitable time due to the few appointments for treatments.

Before meeting all the employees, Monesa decided to visit the facilities of Sage Bliss as she had not done so before due to the heavy flow of both in-house and walk-in guests during the last few weeks. Now, along with Dr Kapoor's secretary, Ms Monesa visited the guest cottages, followed by the leisure centre, the mediation hall and spa clinics. After seeing the guest villas, she felt awkward that although inside the villas were clean, outside areas were not and the windows were covered with cobwebs. There was a heap of plastic bottles close to the leisure centre. She was shocked to see a housekeeping assistant sleeping peacefully in the meditation hall. Four out of the five imported luxury full-body zero-gravity massage chairs kept in the leisure centre were under repair, and the external leather covers were torn. Half of the bulbs in the chandelier in the banquet hall had fused and had not been replaced due to a fault in an imported mechanized ladder. She found imported Swiss-made cheese and fresh vegetables thrown in the garbage bin in the kitchen. In the food and beverage service pantry, she found a pile of broken dinner plates, saucers, soup tureens and various items of glassware. When she asked who was in-charge of this, everyone turned to point towards the executive chef. She found a massage therapist rudely asking for extra money from a walk-in guest in the spa clinic.

Noting all the miseries that were spoiling the image of Sage Bliss, Monesa walked into the meeting hall. The secretary welcomed Monesa and all the other employees to the meeting and asked Monesa to address the employees. After briefly telling them about her background, Monesa started describing her observations on the current state of Sage Bliss and instructed section heads to explain the reason for the poor maintenance, damaged equipment, waste of ingredients and inappropriate employee behaviour. The executive chef informed her that he was spending most of his time training his junior staff against the careless use of expensive raw materials, preventing damage to expensive equipment and solving petty issues among the cooks, room service boys and butlers. He also indicated the low number of cooks and employees available in hospitality care.

On the other side, housemaids voiced their anger that they have been treated poorly by the other employees from the front office, kitchen and food and beverage service. Further, they expressed their concern about the long working hours and low and disproportionate service charges paid to them. The employee who slept in the meditation hall sadly narrated that he had been at Sage Bliss for the last 3 days working continuously in break shifts and sleeping in the meditation hall due to lack of rest and a designated place for rest. He stated that he had not been able to go back to his home, which was in the town, for the last 3 days. In tears, he asked Monesa to accept his resignation.

The secretary narrated that she was spending most of the time recruiting and providing some basic training for recruits as there is high employee turnover at Sage Bliss, especially in room service and the spa clinics. She said that Dr Kapoor was aware of the issues, and he wanted to induct more people from his family to solve them. The physicians from homeopathy and Siddha medicine, who had worked in various clinics and spas, expressed their dissatisfaction that they have not been treated well by the allopathy doctors. Listening to the complaints, a group of physicians practising allopathy stood up together and stated loudly that all of them wanted to resign now because irrespective of their experience, efficiency and performance, they have all been paid the same wages. The Ayurveda doctor, who had handed in his resignation letter, cited that despite providing consultation and treatment to more than 70% of the guests, he has not been paid highly irrespective of his workload, which led to his decision to resign.

A lonely marketing and sales executive cautioned that the problem had already taken its toll on the business. After referring to the poor social media reviews and ratings, a corporate group had cancelled its 2-week wellness and rejuvenation programme meant for their top management team. The marketing and sales executive listed the major complaints that frequently appear in online customer reviews: (i) employees' rude behaviour; (ii) delay in service; (iii) the poor quality of the food; (iv) poor maintenance of public areas; (v) long waits for a check-in and check-out; (vi) missed appointments; (vii) long waits for treatment; (viii) poor maintenance of the pool; (ix) malfunction of the jacuzzi; (x) careless treatment by therapists; and (xi) confusion in getting consultations with doctors from different specializations. After listening to all the complaints and grievances of the doctors and employees, she has understood that Sage Bliss now suffers from poor employee morale, motivation, inequity and interdepartmental cooperation. After appeasing the employees, Monesa asked the employees to provide their support to address all the issues highlighted at the meeting. On her return to the office, Monesa started writing an urgent email to Dr Kapoor asking for full authority to manage the issues.

Discussion Questions

1. Do you think Stacy J. Adams' (1965) equity theory of motivation is relevant and sufficient to explain the employees' behaviour at Sage Bliss? Explain why and why not.
2. Distinguish between the symptoms, surface causes and underlying causes that impact employee morale and the behaviours described in this case study.
3. If you were appointed as a consultant to provide solutions to the problem at Sage Bliss, what would be your response?

Teaching Notes

Dr Kapoor, a well-known cardiac surgeon with long-standing experience with both national and international-level organizations, established his own destination spa named Sage Bliss. Integrating modern medicine practices with traditional healthcare systems such as Ayurveda, Siddha, Unani and homeopathy is Sage Bliss's USP (unique selling proposition). A young hospitality management graduate Ms Monesa has been deputed as a manager to temporarily look after the spa in the absence of Dr Kapoor. During her stint as manager, she faced some unique challenges and hurdles that reflect the lack of formal organization structure, poor employee morale, job satisfaction and perceived inequity among the employees. Monesa needs to develop plans and strategies to overcome the challenges that seek to affect the success of this newly developed destination spa. The main aim of the case study is to familiarize learners with the outcomes of workplace inequity. The case has been designed to reflect Stacy J. Adam's (1965) equity theory of motivation.

The case can be adopted to teach undergraduate and postgraduate students of hospitality and tourism management. The case can also be discussed from the perspective of the challenges that entrepreneurs face while creating a new venture: (i) employee motivation; and (ii) organizational culture.

Target audience

The case study is most suitable for bachelor's degree and master's degree students studying hospitality and tourism management. The learners' prior knowledge of various concepts and theories related to employee motivation in business will be an added advantage. A tutor may provide a short lecture and reading materials on Adam's equity theory of motivation to familiarize students before the case discussion session.

Teaching methods and materials

The case can be used for both individual and group learning. The case can be allotted to a group of four to five students to address the discussion questions. Individuals can be instructed to answer each question with relevant analysis supported by theories related to motivation. Each individual or group's response should be evaluated using a structured rubric considering the approach, application, quantifiable objective and measurable outcomes.

The equipment required is a whiteboard or flipchart for the tutor to write down the main points and perhaps an LCD (liquid crystal display) projector to display students' presentations of their answers to the discussion questions.

The case scenario can be replicated through role play where the tutor needs to select two or three students from the class of learners to be the team of consultants brought in to solve the issues. Students need to prepare a managerial note for a meeting to resolve the issues and talk to the employees waiting to submit their resignations. In the role play, the tutor may take the role of Dr Kapoor's secretary and the remaining students (i.e. those not in the team of consultants) as the employees willing to quit their jobs. To enhance the realism, the meeting may be recorded, and students acting as employees may come as a group of doctors, therapists and cooks with different uniforms specific to their group. The role play of the case is expected to facilitate collaborative participation, learning and building empathy as each member of a particular group of employees (e.g. doctors, therapists or cooks) attempts to support its members and argue against other groups of employees. At the end of the case study, the tutor may take the role of Monesa to address the issues where the students will play the role of employees.

Teaching instructions

The case can be taught in a 2-hour class using the following teaching instruction plan. This plan only covers the class session when students present their answers to the discussion questions and does not include the role play described above, but instructors could incorporate a role-play session of the scenario before the session outlined here. It is assumed that students have been previously introduced to the basic concepts relating to employee motivation and motivational theories. Also the students are required to read the case prior to the class session (and prepare some basic ideas about fairness and equity in the workplace) so they are familiar with it; the 'summary of the case' is an overview to highlight the main points to be considered.

Teaching instruction plan

Discussion point	Time required (min)	Teaching strategies
Opening session of the case	10	The tutor may ask the students a few questions to verify their understanding of the basic concepts of employee motivation and motivational theories. The tutor may write down the main points that are considered essential to discuss the case on the board.
Brief lecture on employee motivation	20	The tutor gives a short lecture on Adam's equity theory of motivation and its business implications. Then the tutor may circulate a Google Form with multiple-choice questions to evaluate the students' understanding.

Continued

Teaching instruction plan Continued.

Discussion point	Time required (min)	Teaching strategies
Summary of the case	10	The tutor may ask volunteers to raise their hand to summarize the case and list the case facts (students having read the case prior to the class session). A list of points that aim to provide an overview of the case should be written on the board. Now with input, the tutor may ask the students to present their analysis and answer the discussion questions. If students are working in groups/teams (with four or five individuals in each one), when each team of students presents, the tutor needs to encourage other teams to raise questions on the rationale of their analysis and answers.
Assignment question 1	20	**Do you think Stacy J. Adams' (1965) equity theory of motivation is relevant and sufficient to explain the employees' behaviour at Sage Bliss? Explain why and why not.** The tutor needs to ask each team to present their views, and later teams need to debate their views along with their rationale.
Assignment question 2	20	**Distinguish between the symptoms, surface causes and underlying causes that impact employee morale and the behaviours described in this case study.** The tutor may randomly ask a few students to volunteer to narrate the difference between symptoms, surface and underlying causes. A short vignette may be used that reflects the symptoms, surface and underlying causes before enabling the team to answer the questions.
Assignment question 3	30	**If you were appointed as a consultant to provide solutions to the problem at Sage Bliss, what would be your response?** To avoid overlapping answers, the tutor must ask the teams to submit their presentations before starting the discussion. The tutor now allows the teams to present their suggestions with the strong rationale for each suggestion. The tutor must encourage multiple suggestions to each underlying cause for the issues at Sage Bliss.
Closing remarks and discussion	10	The tutor needs to weigh the pros and cons of solutions provided by the teams at the end of all the presentations. The tutor must emphasize that there is no single 'correct' answer to the discussion questions. Finally, the tutor concludes the case analysis session by briefly summarizing the case facts and outcomes, and outlining his/her analysis and answers for the discussion questions using employee motivation theories.

Duration

2 hours

Analysis

1. Do you think Stacy J. Adams' (1965) equity theory of motivation is relevant and sufficient to explain the employees' behaviour at Sage Bliss? Explain why and why not.

Yes. Adam's equity theory of motivation is relevant in explaining the employees' behaviour as several employees complain that fellow employees are not treating them equally. Similarly, there is a grievance on the inequity in payment and performance. The problems among employees of Sage Bliss indicate that employees' motivation is affected due to employee perceived inequity on their benefit/reward for their effort.

No. Adam's equity theory of motivation is not sufficient to explain the employees' behaviour at Sage Bliss. It is capable of addressing the outcome of perceived inequity on reward and effort. It can even be evaluated with other motivation theories such as Porter and Lawler's (1968) model of motivation (which is based on the assumption that rewards bring about satisfaction and that sometimes performance produces reward in the form of job satisfaction). However, beyond the motivation-related issues, some symptoms at Sage Bliss need to be analysed with theories and concepts related to job design, job specification, person–job fit, training and development and the span of control.

2. Distinguish between the symptoms, surface causes and underlying causes that impact employee morale and the behaviours described in this case study.

These concepts are highly distinctive from each other and being able to identify and differentiate between symptoms, surface causes and underlying causes will help the managers solve the problems or crises efficiently. These concepts are clearly explained in the literature of root cause analysis (e.g. see Preuss, 2013). Symptoms are conditions that are outcomes of the problems. Symptoms are observable outcomes such as a decline in sales, high employee turnover and customer dissatisfaction. However, the causes of the symptoms are divided into two levels: (i) surface causes; and (ii) underlying causes. Surface causes are easily identifiable problems that lead to many symptoms. Moving from the question of why that symptom arises leads to the answer of surface problems. Underlying or root causes are invisible, causing surface problems and observable symptoms. In reality, often we try to answer or solve the symptoms instead of tackling the root causes. Table 26.2 explains the above elements in the case of Sage Bliss.

3. If you were appointed as a consultant to provide solutions to the problem in Sage Bliss, what would be your response?

The case needs to be dealt with considering multiple concepts and principles of management. Considering the symptoms, Sage Bliss has numerous issues that need to be addressed effectively to sustain its success. There is limited data availability; however, it can be stated that the process needs to be strengthened in the following ways (the list is suggestive but not exhaustive):

Table 26.2. Components of root cause analysis: symptoms, surface causes and underlying causes that impacted employee morale and behaviour described in the case of Sage Bliss. Authors' own table.

Symptoms	Surface causes[a]	Underlying causes
Damage of equipment	Careless usage	Inadequate training and lack of motivation
Misbehaviour of employees with guest(s)	Poor training and attitude	Lack of motivation
Quarrel among the employees	Poor employee relationships and lack of professionalism	Feeling of inequity Lack of training
Sleeping in the workplace	Lack of supervision and poor job description	Lack of motivation
Poor maintenance of public areas	SOP(s) is/are not followed Poor training Lack of supervision	Lack of motivation Poor training
Employee attrition	Employee dissatisfaction	Lack of motivation
Broken dinner plates, saucers, soup tureens and a variety of glassware	Careless usage	Lack of motivation Lack of training

[a]SOP(s), standard operating procedure(s).

- *Performance-based incentives* – These should be used to reduce the inequity.
- *Training and development* – There should be training for handling of equipment and interaction with guests.
- *Standard operating procedure (SOP)* – This would include guidelines to operate the machines and equipment that employees should follow.
- *Preventive maintenance* – This should be adopted to ensure the availability of essential machines and equipment at all times in order to facilitate the service.
- *Motivation* – The right motivators, both monetary and non-monetary, should be identified and aligned with employee performance.
- *Decentralization* – The right kind of people to manage each process should be identified according to their specialization (qualifications and/or experience).
- *Autonomy* – Each individual head of department should be given autonomy to control their employees.
- *Employee welfare* – The welfare of each employee should adhere to the necessary condition(s) mandated in the Factories Act 1948 (Ministry of Labour and Employment, nd).
- *Communication* – Regular communication with employees is expected workplace behaviour, sharing positive customer reviews and passing on appreciation for good work. Also communication is needed to emphasize transparency in dealing with employees irrespective of their relationship with the top management.
- *Development of organizational policy* – Organizational policy should guide the actions, strategies and expected behaviour of all employees.
- *Work scheduling* – Work should be scheduled based on the statutory norms.
- *Empathy* – The importance of mutual respect and empathy among employees should be insisted upon, irrespective of the nature of the work.
- *Working environment* – The working environment should be clean and it should be an aesthetically pleasing workspace.
- *Work–life balance* – Work schedules should provide for sufficient work–life balance.

References

Adams, J.S. (1965) Inequity in social exchange. *Advances in Experimental Social Psychology* 2, 267–299. https://doi.org/10.1016/S0065-2601(08)60108-2

Businesswire (2022) The Worldwide Spa Industry is Expected to Reach $185.5 Billion by 2030. Available at: https://www.businesswire.com/news/home/20220307005832/en/The-Worldwide-Spa-Industry-is-Expected-to-Reach-185.5-Billion-by-2030---ResearchAndMarkets.com (accessed 30 September 2022).

Ministry of Labour and Employment (nd) The Factories Act, 1948. Available at: https://labour.gov.in/sites/default/files/Factories_Act_1948.pdf (accessed 30 September 30 2022).

Porter, L.W. and Lawler, E.E. (1968) *Managerial Attitude and Performance*. Richard D. Irwin, Homewood, Illinois, p. 195.

Preuss, P. (2013) *School Leader's Guide to Root Cause Analysis. Routledge, New York.* https://doi.org/10.4324/9781315854960

27 Hotel Dream – Reinventing the Segmentation, Targeting and Positioning for Resilience

M. Selvalakshmi,[1*] R.M. Kabila and V.J. Sivakumar[2]

[1]*Thiagarajar School of Management, Madurai, India;* [2]*National Institute of Technology, Tiruchirappalli, India*

Abstract

Hospitality is a hugely influential industry in the world whose revenue generation is mainly driven by service quality and the customer experience. After the coronavirus disease 2019 (COVID-19) pandemic, which shattered many sectors, the hospitality industry is resuming its business but it faces a major hurdle in servicing its customers whose current concern is safety and a contactless experience. This case presents the challenges faced by a budget hotel, the 'Hotel Dream', located in the heart of Madurai city which is known for its tourism in the state of Tamil Nadu, India. The hotel offers a wide range of amenities and caters primarily for transient customers. However, to match the demand post-pandemic the managing director of the hotel must choose the most appropriate segmentation, targeting and positioning (STP) strategy in order to improve the prospects of the hotel in terms of its service product, target customers and positioning in the market.

Subjects: general management, marketing management and hotel management

Learning Objectives

At the end of the case study, the learner will be able to:

1. Understand the hotel industry and its key challenges.
2. Analyse the basis of segmentation used in the hotel industry and to distinguish between determinant and significant attributes critical for hotel segmentation and particular groups of customers.
3. Apply the concept of the 'flower of service' in the hotel industry. According to this concept, value is built on: (i) core products (the centre of the flower); and (ii) supplementary products (represented by the petals that surround the centre or core).
4. Discuss the prospects of using 'revenge tourism' as an opportunity for segmentation, targeting and positioning (STP).

Case

Mr Prabahar's family name is synonymous with the automobile industry in the Madurai district in Tamil Nadu, running businesses established by his father which deal

*selvalakshmi@tsm.ac.in

 Teaching Cases in Tourism, Hospitality and Events (eds. S.K. Dixit and S. Piramanayagam)
DOI: 10.1079/9781800621022.0027

with car tyres and auto electrical accessories/spare parts since 1968. Their shops are predominantly located in the city centre and took care of the largely untapped business sector of after care in the automotive industry in the Madurai district. When his father acquired real estate in a prime location of the city, it motivated Mr Prabahar to explore opportunities in new prospective business domains instead of sticking to the same line of business. The family supported Mr Prabahar to venture into the hotel business and he launched Hotel Dream in 1993. Other members of the family are in different businesses including the decorating business (selling paints, rollers, brushes, etc.) and running a car distribution dealership.

Hospitality industry

The hospitality sector comprises accommodation, food and beverage services, event planning, theme parlours and travel and tourism, and includes hotels, tourism bureaus, pubs and restaurants. Hospitality businesses that provide a favourable client experience have a better customer retention rate than those that provide a less pleasant encounter. Other businesses benefit from hospitality since it fosters positive client reviews. In India, the 'organized' or contemporary industry consists of around 1423 hotels and resorts with an inventory of 105,292 rooms (Ministry of Tourism, 2021). The number of approved hotels and the available inventory of hotel rooms in the country is presented in Table 27.1.

The majority of these hotels are located in major cities or tourist or business sites. India features a number of international hotel chains. Aside from that, a number of international chains have established themselves in India by franchising or forming partnerships with Indian businesses.

The hospitality business in Madurai, Tamil Nadu, has shown considerable growth in recent years and there is great potential for it to grow even more in the future. The city, which is rich in culture and diversity, has attracted a large number of tourists from all around the world. It has also been renowned as a spiritual tourism destination for both international and domestic visitors.

The hospitality industry confronted a great deal of business-critical circumstances during the coronavirus disease 2019 (COVID-19) pandemic. The entire travel industry suffered and the majority of hotel workers lost their jobs and many moved to different types of employment in various industries. Due to travel restrictions established to restrict the spread of COVID-19, less than 3 million international visitors travelled to India in 2020, a 75% decrease from the previous year (*Hindustan Times*, 2021). India has progressively improved its position in the World Economic Forum's travel and tourism competitiveness ranking, rising from 65th place in 2013 to 34th place in 2019 (World Economic Forum, 2019). According to statistics released by the National Statistical Office (NSO) on 31 August 2020, India's hotel industry was one of the worst-affected sectors of the economy owing to the COVID-19 outbreak, declining by 47% in April–June 2020

Table 27.1. Number of approved hotels and hotel rooms in India as of 28 June 2021. Authors' own table using data from Ministry of Tourism, 2021.

Category of hotels	No. of hotels	No. of rooms
One star	10	346
Two star	27	914
Three star	533	18,193
Four star	419	21,351
Five star	226	28,833
Five star deluxe	149	33,877
Heritage hotel	59	1,778
Total	1,423	105,292

(cited in Ramnani, 2020). The hotel business saw a 77% dip in occupancy in May 2020 compared with the same month the previous year. Domestic hospitality was severely damaged by COVID-19-related interruptions, dropping by 65% in 2021 (Research and Markets, 2022).

Even after the COVID-19 pandemic, hotels could not open according to their desires because of government laws. The cost of all goods was rising continually and to manage their purchasing power, hotel staff requested an increase in pay of around 25%. Yet, the hotel administration was not in the frame of mind to grant this, resulting in a huge attrition rate of employees and hotels being understaffed. Hotel businesses struggled to find staff to provide regular service to customers. However, additional work was required because as a result of the COVID-19 pandemic, customers were very concerned about safety and hygiene, so additional cleaning was required and it is critical to keep up with the current hospitality trends in response to customers' changing behaviour. Further, hotels were rapidly adopting technology in order to meet the industry's evolving demands for a completely contactless experience.

The Hotel Dream

The hotel is situated in the heart of Madurai city. It is a central location, which is a major advantage to customers, as it is easy to reach all the desirable places in the city, including the eminent Meenakshi Amman temple which is only around 1 km from the facility. Moreover, the railway station and bus stand are at a walkable distance from the hotel and the air terminal is around 12 km. The hotel was constructed with state-of-the-art infrastructure and features including luxurious rooms, conference halls with good facilities, exquisite dining places and ornamental banquet halls with a blend of contemporary design and conventional facilities.

The hotel offers a total of 56 air-conditioned and non-air-conditioned guest rooms basically falling into four types of rooms spread across four floors:

- *Deluxe rooms* are air conditioned, aesthetically designed and well furnished, providing a luxurious experience for customers. These rooms are located on the first floor with a room size of 25.5 m^2 (275 ft^2) and king-size beds. The rooms are charged in Indian rupees (INR) at INR1500 for single occupancy and INR2000 for double occupancy excluding goods and services tax (GST).
- *Classic rooms* are air conditioned, polished and sophisticated for a classy experience with a room space of 23.2 m^2 (250 ft^2). They are located on the second and third floors of the hotel. There are 25 of these rooms with king-size beds and they are charged at INR1300 and INR1800 for single and double occupancy, respectively, excluding GST.
- *Family rooms* are air conditioned and support recreation and relaxation for family members who come together for a special occasion such as a wedding or other small function. Five such rooms are located on the fourth floor of the hotel with a room size of 27.9 m^2 (300 ft^2) and two king-size beds. These rooms cost INR3000/day excluding GST.
- There are 15 *non-air-conditioned rooms* on the fourth floor with a room size of 23.2 m^2 (250 ft^2). They are charged at a nominal rate of INR900 and INR1150 for single and double occupancy, respectively.

The average monthly sale occupancy of the hotel is about 60–70%. The hotel follows a dynamic pricing model wherein prices vary slightly based on the demand. However, the average room rent stands at INR1650/night. The hotel operates five main halls with various capacities to accommodate birthday celebrations, meetings and family events.

The guests are provided with various supportive facilities such as:

- a doctor on call;
- a travel desk;
- car rental;
- room service;

- car parking;
- a laundry service;
- a conference hall;
- a deposit locker;
- use of the Internet;
- credit card payment facilities; and
- a 24-hour check-in and check-out facility.

The hotel also organizes sightseeing trips to nearby tourist attractions for those customers who are interested. The rooms can be booked on various online platforms such as Goibibo, MakeMyTrip, Travelguru and Agoda.

Food and beverage production and service

Guests at the hotel enjoy delicacies prepared in the multi-cuisine restaurant which serves Chinese and continental cuisine in addition to the Indian dishes. The restaurant is open from 7 a.m. to 11 p.m. and is purely vegetarian. Most of the food prepared is South Indian although North Indian foods, Chinese food, tandoori and continental foods are occasionally prepared.

The hotel offers different meal plans to the guests to promote business, for example:

- European plan (EP) – meals served only in guest rooms;
- continental plan (CP) – room with breakfast;
- modified American plan (MAP) – room with breakfast and either lunch or dinner; and
- American plan (AP) – rooms with food being served three times a day (i.e. breakfast, lunch and dinner).

The restaurant is open for resident guests as well as outsiders although currently outsiders only use the facility occasionally. It takes orders to deliver food that is prepared in the hotel for small parties and meetings held off the premises. The restaurant generates an average sale of INR25,000–30,000/day and of this the food costs account for 25–30%. The restaurant has tied up with online business partners such as Swiggy and Zomato to enhance business opportunities.

The organization chart of the Hotel Dream is presented in Fig. 27.1.

Customers at the hotel

Table 27.2 presents data on the foreign and domestic visitors to Tamil Nadu and Madurai for 2014–2019, which were active years for tourism before the COVID-19 pandemic. Visitor numbers rose over this period.

The outbreak of the pandemic at the very end of 2019 devastated the Indian tourism industry but it is now showing good signs of revival following the massive vaccination drive by both central and state government. As the tourism industry is recovering, 're-venge tourism', whereby people are making

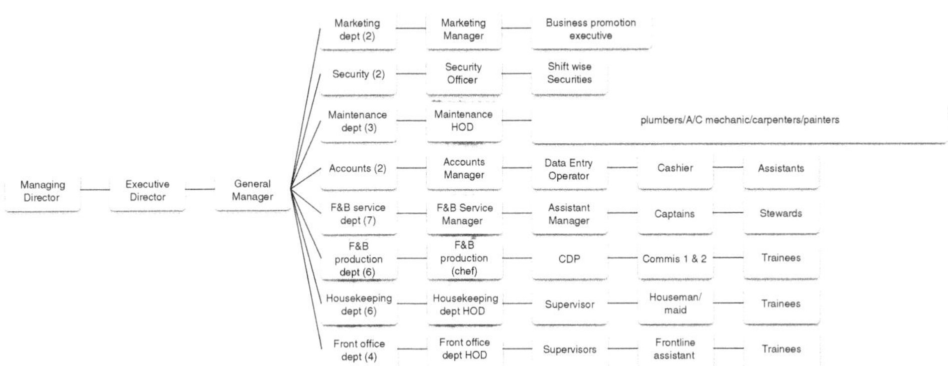

Fig. 27.1. Organization structure at Hotel Deam. The number of staff in each department is indicated in brackets (fourth column from left). A/C, air conditioning; CDP, chef de partie; Commis, comis chef; F&B, food and beverage; HOD, head of department; Authors' own elaboration.

Table 27.2. Domestic and foreign tourist inflow to the state of Tamil Nadu and Madurai city, 2014–2019. Authors' own data compiled from various secondary data sources. Missing data was filled in by interpolation and trend projection.

	Tamil Nadu tourists (hundred thousands)		Madurai tourists (hundred thousands)	
Year	Domestic	Foreign	Domestic	Foreign
2014	3275.55	46.57	116.61	1.21
2015	3334.59	46.85	133.38	1.39
2016	3438.12	47.22	151.28	1.22
2017	3450.61	48.6	164.25	2.17
2018	3859.09	60.74	245.16	2.44
2019	4948.65	68.66	358.68	3.81

up for vacations lost during the pandemic, gives confidence for growth in the number of tourist arrivals for the coming years.

Considering Hotel Dream's customers in particular, there are three main segments: (i) family groups; (ii) business travellers/groups; and (iii) tour visitors. Whenever a family function happens, it is common that the hosts prefer to accommodate their friends and relatives in the local hotels to enable them to attend the function. Around 30% of Hotel Dream's business is generated from such family groups. These customers would not give much importance to the food as they primarily use the rooms only for accommodation. The business group segment of customers visit Madurai for their business development or meetings, and business customers make up around 50% of Hotel Dream's guests. The remaining 20% of customers are the tour visitors. From Table 27.2, it is evident that this tourist segment in Hotel Dream can certainly improve the bottom line if the potential is tapped right. Usually the temple city of Madurai, where Meenakshi Amman temple functions go on throughout the year, draws enormous crowds to attend the holy occasions. At these times rooms offering accommodation across the city are almost totally filled by this segment and such a floating population pay increased rates for their rooms as they are in such high demand.

Service concerns at Hotel Dream

Typically, service-related complaints are associated with the hotel's offerings, resulting in the guests' discontent. When the hotel is near capacity, the front desk receives a higher frequency of service-related complaints on issues such as disturbing neighbours or poor guest experiences due to untidy rooms.

During the night time, guests feel inadequate customer support is provided so that should they want something there is a long response time. As the staff designated for the night service are predominantly outsourced contract labourers, their poor service attitude has often been highlighted in customer reviews. There are occasional grievances about the functioning of equipment such as difficulties with the television, air conditioning or plumbing.

These days, most bookings are generated online. Visitors expect that the image displayed when they are booking ought to match the room they have been allocated. Most online bookings are made through online travel agents (OTAs) on the Web and these OTAs have various designs for displaying pictures of the accommodation on offer in their galleries, so the visitor is often confused about the room they have booked. When the guests arrive at the hotel and look into the room, they figure out that the room designated to them is not the same as the one they saw online when they booked it. This can result in complaints and if the guest's sleep is disturbed while staying in the room, the hotel faces more grumbles from them. Therefore, the hotel attempts to guarantee that the bedding and room is in great shape every day in order to satisfy the guests.

Even with generally excellent accommodation facilities, visitors can complain about the hotel, and often the explanation concerns the food and beverages not being up to their expectations. With their earnest attempts, Hotel Dream has managed to receive complimentary reviews about the extraordinary culinary encounter experienced in the hotel.

The dilemma

Mr Prabahar, the managing director, went through the weekly report forwarded by the general manager, which indicated the increasing demand for services after regular hours. He went through the reviews presented by the customers who visited the hotel recently. They had complimented the hotel for being budget friendly at a prominent location and providing excellent 'value for money'. The food offered was relished but concerns were raised by the staff responsiveness especially at night time.

Night life in Madurai is minimal except for cinema theatres which run night shows until 2 a.m. Essentially, the hotel guests tend to spend most of the night time at the hotel during their stay. The availability of trained staff at the hotel for service during night time is a huge challenge. The support staff at night are predominantly college students who are employed only on a part-time basis and whose turnover tends to be relatively higher than that of regular members of staff. Mr Prabahar wondered whether he should be investing in hiring permanent employees or training the employees to ensure 100% service quality.

At present Hotel Dream is concentrating on the transient segment of customers who come to the hotel as individuals or small groups occupying less than ten rooms per night, and often they are walk-in guests or book at the last minute as they require a short-term stay. Should the hotel cater for other segments too to improve their bottom line?

The food offered at the hotel is primarily focused on supporting the requirements of the guests and three to four functions which happen in the banquet halls on a weekly basis. With the provision and expertise to offer multi-cuisine food, should the hotel focus more on the food industry along with the accommodation? By networking with travel agents, could Hotel Dream present itself as Madurai's special food corner, taking into consideration the traffic, difficulties with parking and inconvenience that may be caused to existing customers?

Mr Prabahar wishes to meet with a family consultant who will guide him to take suitable decisions in terms of designing his service product, the type of customers to target and market positioning.

Discussion Questions

1. What are the challenges faced by Hotel Dream in its positioning strategy?
2. What is the basis of segmentation used in the hotel industry?
3. If Mr Prabahar decides to open up the restaurant to more outsiders, how should he design the new food service compared with the regular hotel services offered to resident guests?
4. Discuss the opportunities created by 'revenge tourism' for reinventing the segmentation, targeting and positioning (STP) of Hotel Dream.

Teaching Notes

Summary

The hospitality industry is a hugely influential industry in the world whose revenue generation is mainly driven by service quality and the customer experience. After the COVID-19 pandemic, which shattered many sectors, the hospitality industry is resuming its business but it faces a major hurdle in servicing its customers whose current concern is safety and a contactless experience. This case presents the challenges faced by a budget hotel, the 'Hotel Dream', located in the heart of Madurai city, which is known for its tourism in the state of Tamil Nadu. The hotel offers a wide range of amenities and caters primarily for transient customers. However, to match the demand post-pandemic the managing director of the hotel is left in a dilemma about whether to improve the prospects of the hotel in terms of its service product, the type of customers targeted and positioning in the market, and to help him make the decision he is analysing its segmentation, targeting and positioning (STP).

Target audience

This case is intended for use in postgraduate courses in general management, marketing management and hotel management. Also it is appropriate for hotel managers/tourism partners on executive management programmes.

Teaching methods and equipment

The case can be introduced to the learners through the video 'A Day in the Hotel Industry' available on YouTube using the following link: https://www.youtube.com/watch?v=r_VPsyxYz2k (accessed 30 September 2022).

The case can be taught through a class discussion session led by the facilitator encouraging individual students to contribute their thoughts in answer to the discussion questions (as described in the suggested teaching instruction plan below). An alternative, depending on the class size and time available, is to have students work in groups of four or five in order to answer the discussion questions, preparing their answers as a presentation to the class. Each group could take the role of the family consultant who would guide Mr Prabahar and suggest solutions for the design of the service product, type of customers targeted and positioning in the market. This would enable the learners to relate closely with the scenario. When each group presents, the facilitator should encourage other groups to raise questions on the reasons for the group's analysis/answers, in this way opening up the discussion.

The only equipment needed is a whiteboard or flipchart and, if necessary, facilities to make PowerPoint presentations.

Teaching instructions

In preparation for the class discussion session, before the class students should read: (i) the case, so they are familiar with it; and (ii) the article by Bansal *et al.* (nd) on the basis of segmentation used in the hotel industry. The suggested teaching instruction plan that follows includes an introduction, discussion of the assignment questions and a conclusion summarizing the insights of this case. Ideally a class session lasting 60 min is required.

Teaching instruction plan

Discussion topic	Time (min)	Teaching pedagogy
Introduction – hotel industry	10	The facilitator discusses the factors driving the hotel industry and recent trends.
Assignment question 1	10	**What are the challenges faced by Hotel Dream in its positioning strategy?** Discussion based on the inputs provided in the case (students having read the case prior to the class session).
Assignment question 2	15	**What is the basis of segmentation used in the hotel industry?** Discussion based on the article (Bansal *et al.*, nd) provided for pre-reading by the students. The facilitator should lead the discussion about the basis of segmentation and how to differentiate between determinant and significant attributes for STP.

Continued

Teaching instruction plan Continued.

Discussion topic	Time (min)	Teaching pedagogy
Assignment question 3	15	**If Mr Prabahar decides to open up the restaurant to more outsiders, how should he design the new food service compared with the regular hotel services offered to resident guests?** The facilitator can introduce the concept of the 'flower of service' to the students using the board to draw the flower with its component petals. Further, it can be used to help learners understand the difference in focus between the hotel service and restaurant service.
Assignment question 4	5	**Discuss the opportunities created by 'revenge tourism' for reinventing the STP of Hotel Dream.** To start the discussion the facilitator asks two or three different students for their thoughts on the opportunities for reinventing the STP. The facilitator may write down suggestions on the board to encourage other students to come forward with their suggestions.
Conclusion	5	The facilitator concludes the case discussion session by briefly summarizing the main points of the case and his/her answers to the discussion questions.

Duration

60 min

Analysis

Introduction about the hotel industry

The hotel industry makes a major contribution to the Indian economy as it functions in collaboration with several other industries such as food and beverages, transportation, health and sanitation, and entertainment. It also provides employment opportunities to a large number of people and supports the livelihood of many citizens in the region. It forms a part of the hospitality industry which addresses the relationship between the guest and the host. The term hospitality industry thus comprises businesses which help people to feel welcomed and enables them to relax and enjoy themselves (Westcott and Anderson, 2021). '*Atithidevo Bhava*' the Sanskrit words which highlight the Indian Hindu philosophy of revering and honouring guests as a god, forms the basic value and virtue of this noble industry in India.

The hotel industry in particular is capital intensive and becomes highly competitive especially in seasons when demand far exceeds availability. The Indian hotel industry remains grossly under served with about 105,292 hotel rooms in total. According to industry estimates, the price of hotel rooms per person in India is equivalent to the international standard (Agrawal, 2016).

The success of a hotel is determined predominantly by variables such as:

- *Location* – A hotel located in a less travelled area might not generate as much revenue even though it offers quality food and amenities. Also proximity to, for example, a flyover creating traffic noise or a sewage support system in the area will affect the popularity of the hotel. Further, if there is any construction taking place in the locality, this will affect the parking facilities available and thereby affect the comfort of the visitors to the hotel.
- *Global perspective* – Hotels need to constantly watch the evolution of the industry across the globe and try to scale up their operations to meet global standards. Amenities such as television/air conditioning/Wi-Fi in guest rooms which were once considered as augmented benefits, today have become the basic elements of the service product.

- *Quality standards* – Due to the high prevalence of the people factor in hotel services, establishment of quality standards to meet the changing customer expectations is challenging. The 'moment of truth' created at various touch points during service interactions are critical in customer acceptance and adoption.
- *Collaboration* – The hotel industry has evolved phenomenally in recent years with the practice of using the aggregator model. The hotel aggregator is a business that rents a hotel on lease (either a fixed number of rooms or the entire hotel) and takes over the operational and marketing aspects of the hotel. This type of collaboration is enabled chiefly by technological support. The digital footprint of the hotel created by the stakeholders including guests, hosts and the travel agent are therefore great influencing factors.

Points of discussion

1. What are the challenges faced by Hotel Dream in its positioning strategy?

Hotel Dream has a history of nearly three decades of successful operations in the competitive market. It has clear advantage due to the locational factor and has good information technology (IT)-enabled infrastructure. It has quality measures in alignment with the industry standards and has been open to work with compatible collaborative partners including Swiggy, Goibibo, MakeMyTrip, Travelguru and Agoda.

However, it faces multiple challenges due to the dynamic external environment. The challenges faced by Hotel Dream in its positioning strategy may be identified as follows:

- The perceived image of the hotel is 'value for money' which may hamper attracting high-value customers.
- Upgrading the hotel infrastructure including, for example, installation of digital locks, provision of uninterrupted Wi-Fi and centralized air conditioning, may not match with the current customer profile who are primarily cost conscious.
- Customer segments which the hotel attracts include the family function/festival crowd (30%), business travellers (50%) and tourists (20%). These appear to be a narrow range of customers under which to operate, and with regards to the demand from the family function/festival crowd this is a derived demand (i.e. their stay in the hotel is a consequence of demand for something else – a particular family event or festival, which presumably happens infrequently).
- There is service inconsistency due to contract employees who support the system especially at night time.

2. What is the basis of segmentation used in the hotel industry?

Segmentation essentially is the process of dividing the population or potential customers into groups based on similar traits or attributes.

The basis for market segmentation generally includes:

- geographic – based on the area, language, population, etc.;
- demographic – based on the customer profiles in terms of age, income, gender, nationality, etc.;
- psychographic – based on the customers' values, attitudes, personality, interests, etc.; and
- behavioural factors – based on the usage, occasion, benefits, etc.

Based on the user types, customers may further be classified as:

1. *Transient* – These are guests who travel alone without being a part of a company or a group. They decide, book and stay based on their personal preferences.
2. *Corporate* – These customers visit places and stay in hotels as part of their company assignments and available discounts and other benefits on behalf of the company. As their clientele is crucial for the hotels, these guests enjoy more privileges from the hotel.
3. *Group* – During festival times or special occasions such as family functions, such customers stay as a group. Their preferences are common to the group and they enjoy bulk discounts and other flexible provisions during their hotel

stay such as support for travel arrangements and a free guide for temple visits.

4. *Wholesale* – This comprises the hotel aggregators/travel agents who deal with tour operators and book in bulk for a group. The individuals in the group may not be alike as they may be from different backgrounds and tend to have unique requirements.

This classification based on the user types tends to reflect their behaviour in terms of: (i) the length of their stay in the hotel (transient guests tend to stay for less time than groups); (ii) days of the week when they stay (individuals and corporates tend to stay on weekdays while families tend to stay at weekends); (iii) revenue per room (individual customers tend to pay more than group customers who enjoy discounts for group bookings); (iv) revenue per client and booking lead time (corporate customers enjoy discounts and may need less booking lead time compared with individual users); and (v) cancellation and no-show ratio (individual customers are more likely to cancel or fail to show up than wholesale customers).

Critical for hotel segmentation and targeting is the distinction between significant attributes and determinant attributes:

- **Significant attributes**: These attributes are those considered to be essential in the service offered when considering a list of hotels, such as amenities, facilities and goodwill.
- **Determinant attributes**: These attributes determine the customer's final choice, as determinant attributes serve as significant differentiators among the competitors. This might include the customer's past experience, a word-of-mouth reference or value for money they perceive in the service.

Thus, segmentation for the hotel industry may not be based on a single factor but a mix of factors which would help in customer profiling. The segment serviced by Hotel Dream may be profiled using cluster analysis of the basic data collected from the customers during their stay. This would enable the management to understand the gap between their current customers and the potential customers that they are not reaching at present, and explore opportunities to target the new potential cluster.

The value proposition identified by customers as the key factor in selecting Hotel Dream can be used to further position the hotel in the market.

3. If Mr Prabahar decides to open up the restaurant to more outsiders, how should he design the new food service compared with the regular hotel services offered to resident guests?

Hotel Dream is primarily a lodge with dining facilities. Mr Prabakar wants to offer the restaurant services separately to outsiders and yet is sceptical whether it would disturb the quality of the service being offered to the hotel customers. The real implications of the choice of the decisions are way beyond its impact on the customers. It has effect on the way the hotel should organize and focus its operations. This can be clearly understood using the concept of the 'flower of service', which comprehends the core service product and its components (Lovelock and Wirtz, 2011). According to this concept, value is built on: (i) core products (the centre of the flower); and (ii) supplementary products (represented by the petals that surround the centre or core).

With regard to hotel services the core product is the main component that supplies the desired experience for the customer – comfortable accommodation. The supplementary product or services refer to the variety of other service-related activities that accompanies the core product. They augment the core product by facilitating and enhancing the value provided, for example:

- The facilitating supplementary elements include billing and payment, supplying information, consultation and order taking.
- The enhancing supplementary elements include exceptions (e.g. handling special requests), safekeeping (i.e. providing security for customers and their belongings) and hospitality.

Table 27.3 compares hotel service and restaurant service with regards to their core product and supplementary services.

Table 27.3. Comparison of hotel service and restaurant service in terms of their core product and supplementary services. Authors' own table.

Components of service	Hotel service	Restaurant service
Core product	Comfortable accommodation with good amenities and facilities	Good food in an environment with a pleasant ambience
Facilitating supplementary service		
Information	Guidance and directions for local tourist attractions, confirmation of reservations, service hours, etc.	Service hours, menu card
Billing and payment	Invoices, amount due, easy payment modes, etc.	Easy payment modes
Order taking	For seats/tables/rooms, vehicles or equipment rental, etc.	Table reservations
Enhancing supplementary service		
Safekeeping	Locker provisions, security cameras, parking facilities, baggage handling and storage, childcare	Security cameras, parking facilities, childcare chairs and play area
Exceptions	Special requests, handling of complaints, etc.	Handling of complaints on food quantity/quality, etc.
Hospitality	Greetings, facilities and amenities, food and beverages	Greetings, toilet and wash facilities and amenities

Thus, the scope for enabling customer satisfaction is higher in the case of restaurant service, as the core product is more tangible. The restaurant already offers a margin of 70% and has an exclusive menu and expert chef whose services can be extended. Its tie up with online business partners Swiggy and Zomato can aid in cost economies.

The current customers are more transient in nature and hence might often move out of the hotel to functions/business meetings and have sumptuous food elsewhere. So, the excess capacity can be used to support indigenous restaurant services.

4. Discuss the opportunities due to 'revenge tourism' for reinventing the STP of Hotel Dream.

Madurai district has the unique advantage of being a strategic location surrounded by hill stations, waterfalls and spiritual places, as well as having cultural connections and food and heritage places, all of which makes it a prospective tourism destination. Table 27.2 indicates a steady rise in tourist inflows from 2014 to 2019, and the state of Tamil Nadu was one of the top three states for attracting domestic and foreign tourists according to the Economic Survey 2019–2020 (Ramakrishnan, 2020).

After the devastating effect of the COVID-19 pandemic, the hotel industry has started showing signs of 'revenge tourism' whereby vaccinated people are keen to travel having missed so many vacations and tend to move interstate. 'Working from home' has led to the confidence to 'work from anywhere' and hence the Millennials and Gen Z are on a travel spree.

Madurai Collector Dr Aneesh Sekhar has promised to revive the tourism prospects of the district with well-laid plans and support for their execution through private–public partnership (*The Hindu*, 2021). Mass and niche tourism can be promoted by addressing issues such as health, hygiene, traffic management and eco-friendliness. Even if a fraction of this opportunity is captured then it would help Hotel Dream evolve into a very successful business. Hence analysing the STP of Hotel Dream will enable it to be strategically positioned in the market so it can be aligned to attract these new tourists.

Acknowledgement

We would like to express our gratitude to Mr Tushar, Thiagarajar School of Management, Madurai, India and his father Mr Stalin for giving us an opportunity and helping throughout the process of writing this case study.

References

Agrawal, V. (2016) A review of Indian tourism industry with SWOT analysis. *Journal of Tourism and Hospitality* 5(1), 196–200.

Bansal, S.P., Verma, Y., Bansal, H. and Mann, P.W. (nd) Module 6: Segmentation and targeting of services. In: Paper 10: Services Marketing. Content for Post Graduate Courses (Subject: Management). Ministry of Human Resource Development (MHRD) Government of India, New Deehi. Available at: http://www.igntu.ac.in/eContent/IGNTU-eContent-720564575556-MBA-4-Dr.RahilYusufZai-ServiceMarketing-3.pdf (accessed 17 October 2022).

Hindustan Times (2021) Covid: India opens for international travel – What happens next? *Hindustan Times*, 3 November. Available at: Covid: India opens for international travel - what happens next? I Travel - Hindustan Times (accessed 17 October 2022).

Lovelock, C. and Wirtz, J. (2011) *Services Marketing: People, Technology and Strategy*, 7th edn. Pearson Education, New Delhi.

Ministry of Tourism (2021) India Tourism Statistics At a Glance – 2021. Available at: https://tourism.gov.in/sites/default/files/2021-09/English%20Tourisum%202021.pdf (accessed 28 May 2022).

Ramakrishnan, T. (2020) Flow of tourists to Tamil Nadu. *The Hindu*, 1 February. Available at: https://www.thehindu.com/news/national/tamil-nadu/state-a-top-destination-for-tourists/article30707037.ece (accessed 17 October 2022).

Ramnani, V. (2020) India GDP Q1 data: Hotel sector shrinks 47%; it will be a long road to recovery. *Moneycontrol*, 31 August. Available at: India GDP Q1 data: Hotel sector shrinks 47%; it will be a long road to recovery (moneycontrol.com) (accessed 17 October 2022).

Research and Markets (2022) *Hospitality Industry in India – Growth, Trends, COVID-19 Impact, and Forecasts (2022–2027)*. Research and Markets Report. Available at: Hospitality Industry in India - Growth, Trends, COVID-19 Impact, and Forecasts (2022 - 2027) (researchandmarkets.com) (accessed 17 October 2022).

The Hindu (2021) To bounce back, hospitality industry needs to innovate and collaborate: Collector. *The Hindu*, 28 September. Available at: https://www.thehindu.com/news/cities/Madurai/to-bounce-back-hospitality-industry-needs-to-innovate-and-collaborate-collector/article36697744.ece (accessed 28 May 2022).

Westcott, M. and Anderson, W. (2021) History and overview. In: *Introduction to Tourism and Hospitality in BC*, 2nd edn. BC campus OpenEd. Available at: https://ecampusontario.pressbooks.pub/introtourismbc/chapter/chapter-1-history-and-overview/ (accessed 17 October 2022).

World Economic Forum (2019) Travel and Tourism Competitiveness Report 2019. Available at: https://www.weforum.org/reports/ (accessed 17 October 2022).

Further Reading

Musa, M.I., Haeruddin, M.I.M., Haeruddin, M.I.W. and Burhan, M.I. (2019) Analysis of segmentation, targeting, and positioning in hospitality sector: the case of Paputo Beach Cafe. *African Journal of Hospitality, Tourism and Leisure* 8(4), 1–20.

Tewari, J.R. (2009) *Hotel Front Office: Operations and Management*. Oxford University Press, New Delhi.

Index

Note: Page numbers in **bold** type refer to **figures;** Page numbers in *italic* type refer to *tables*